Renewables-Based Technology

Wiley Series in Renewable Resources

Series Editor

Christian V. Stevens, Department of Organic Chemistry, Ghent University, Belgium

Titles in the Series

Wood Modification: Chemical, Thermal and Other Processes
Callum A. S. Hill

Renewables-Based Technology: Sustainability Assessment
Jo Dewulf and Herman Van Langenhove (Editors)

Forthcoming Titles

Starch Biology, Structure and Functionality
Anton Huber and Werner Praznik (Editors)

Biofuels
Wim Soetaert and Erik Vandamme (Editors)

Renewables-Based Technology

Sustainability Assessment

Edited by

JO DEWULF

and

HERMAN VAN LANGENHOVE

Research Group ENVOC, Ghent University, Belgium

John Wiley & Sons, Ltd

Telephone (+44) 1243 779777

Email (for orders and customer service enquiries): cs-books@wiley.co.uk
Visit our Home Page on www.wiley.com

Reprinted March 2007

Other Wiley Editorial Offices

John Wiley & Sons Inc., 111 River Street, Hoboken, NJ 07030, USA

Jossey-Bass, 989 Market Street, San Francisco, CA 94103-1741, USA

Wiley-VCH Verlag GmbH, Boschstr. 12, D-69469 Weinheim, Germany

John Wiley & Sons Australia Ltd, 42 McDougall Street, Milton, Queensland 4064, Australia

John Wiley & Sons (Asia) Pte Ltd, 2 Clementi Loop #02-01, Jin Xing Distripark, Singapore 129809

John Wiley & Sons Canada Ltd, 22 Worcester Road, Etobicoke, Ontario, Canada M9W 1L1

Wiley also publishes its books in a variety of electronic formats. Some content that appears in print may not be available in electronic books.

Library of Congress Cataloging in Publication Data

Renewables-based technology : sustainability assessment / editors, Jo Dewulf and Herman Van Langenhove.
p. cm. — (Wiley series in renewable resources)
Includes bibliographical references and index.
ISBN-13: 978-0-470-02241-2 (cloth : alk. paper)
ISBN-10: 0-470-02241-8 (cloth : alk. paper)
1. Renewable energy sources. 2. Sustainable development.
I. Dewulf, Jo. II. Langenhove, Herman Van. III. Series.
TJ808.R47 2006
333.79′4—dc22 2005033645

British Library Cataloguing in Publication Data

A catalogue record for this book is available from the British Library

ISBN-13 978-0-470-02241-2 (H/B)

Typeset in 10/12pt Times by Integra Software Services, Pvt. Ltd, Pondicherry, India
Printed and bound in Great Britain by Antony Rowe Ltd, Chippenham, Wiltshire
This book is printed on acid-free paper responsibly manufactured from sustainable forestry in which at least two trees are planted for each one used for paper production.

Contents

Contributors

Hans-Jörg Althaus EMPA, Swiss Federal Laboratories of Materials Testing and Research, Switzerland

Adisa Azapagic CES, School of Engineering, University of Surrey, UK

Göran Berndes Department of Energy and Environment, Chalmers University of Technology, Sweden

Stefan Bringezu Material Flows and Resource Management, Wuppertal Institute, Germany

Alle Bruggink DSM Corporate Technology, The Netherlands

Alan Brunskill Procter & Gamble International Operations, SA, Switzerland

Andreas Detzel IFEU, Heidelberg, Germany

Jo Dewulf Research Group Environmental Organic Chemistry and Technology, Ghent University, Belgium

James A. Duffield Office of the Chief Economist, United States Department of Agriculture, USA

Karl-Heinz Erb Institute of Social Ecology, Austria

André P. C. Faaij Copernicus Institute, Utrecht University, The Netherlands

Rolf Frischknecht ESU Services, Switzerland

Helmut Haberl Institute of Social Ecology, Austria

Monique M. Hoogwijk Copernicus Institute, Utrecht University, The Netherlands

Niels Jungbluth ESU Services, Switzerland

Andreas Kicherer BASF Aktiengesellschaft, Germany

Martina Krüger IFEU, Heidelberg, Germany

Alexei Lapkin Department of Chemical Engineering, University of Bath, UK

Michael Narodoslawsky Institute for Resource-Efficient and Sustainable Systems, Graz University of Technology, Austria

Anneliese Niederl Institute for Resource-Efficient and Sustainable Systems, Graz University of Technology, Austria

Peter Nossin DSM Corporate Technology, The Netherlands

Axel Ostermayer IFEU, Heidelberg, Germany

Martin K. Patel Copernicus Institute, Utrecht, University, The Netherlands

William E. Rees School of Community and Regional Planning, University of British Columbia, Canada

Klaus Richter EMPA, Swiss Federal Laboratories of Materials Testing and Research, Switzerland

Peter Saling BASF Aktiengesellschaft, Germany

Erwan Saouter Procter & Gamble International Operations SA, Switzerland

Hosein Shapouri Office of the Chief Economist, USDA, USA

Mark Stalmans Procter & Gamble International Operations SA, Switzerland

Jan-Olov Sundqvist IVL Swedish Environmental Research Institute, Sweden

Gert Van Hoof Procter & Gamble International Operations SA, Switzerland

Herman Van Langenhove Research Group Environmental Organic Chemistry and Technology, Ghent University, Belgium

Wilfried G. J. H. M. van Sark Copernicus Institute, Utrecht University, The Netherlands

Michael Wang Center for Transportation Research, Argonne National Laboratory, IL, USA

Frank Werner Environment and Development, Switzerland

Foreword

For the past two decades, more and more people have become convinced that our economy is shifting quite rapidly towards a biobased economy. Not only has academic interest in renewable resources been growing, but also governmental institutions and companies are paying attention to the revival of the use of renewable resources. It is indeed a revival, since our economy was completely biobased until the beginning of the nineteenth century. It only changed through the advert of petrochemistry and the availability of cheap petroleum-derived products. Due to the limited reserves of fossil fuels (with prices rising to 70 dollars/barrel at this time), the broad availability of renewable resources is appealing as a source of materials and energy. Even companies that have been exclusively active in petrochemistry are creating teams to consider renewable resources.

The use of renewable resources and the importance of white biotechnology form the basis of a move to an economy which is less dependent on fossil fuels, resulting also in the fact that economies will become less dominated by oil-producing countries. This is very important to still developing economies, since they might be able to skip a part of the route that Western economies have passed through. However, the use of renewable resources is not a synonym for a sustainable process. Processes based on petrochemicals can be more sustainable than similar ones based on renewables. The efficiency of many petrochemical processes is often very high. Therefore, the need to have measuring tools to look at the complete process, and the complete value chain, is essential to be able to judge the sustainability of a process or a transformation in industry.

Of course, the greatest advantage of using renewable resources is the impact of the process on the environment and on carbon dioxide emissions, but the 'cradle to grave' approach certainly needs to be looked at carefully to make an unbiased judgement. This volume in the series on renewable resources is therefore crucial for an understanding of the importance of renewable resources for industrial applications.

I am delighted that two colleagues in my department, Professor J. Dewulf and Professor H. Van Langenhove, took on the challenge to give an excellent overview of the different techniques to assess renewable resources applied in industrial processes. Their expertise and their valuable network of colleagues have made this volume a highly respected work that has a central place in this series on renewable resources.

Christian V. Stevens
Faculty of Bioscience Engineering
Ghent University
Belgium
September 2005

Series Preface

Renewable resources, their use and modification are involved in a multitude of important processes with a major influence on our everyday lives. Applications can be found in the energy sector, chemistry, pharmacy, the textile industry, paints and coatings, to name but a few.

The field of renewable resources crosses several scientific disciplines (agriculture, biochemistry, chemistry, technology, environmental sciences, forestry), which makes it very difficult to have an expert view on the complicated interaction. Therefore, the idea of creating a series of scientific books, focusing on specific topics concerning renewable resources, has been very opportune and can help to clarify some of the underlying connections in this field.

In our fast changing world, trends are not only characteristic of fashion and political standpoints, science also is not free of hypes and buzzwords. The use of renewable resources is again more important nowadays, however, it is not part of a hype or a fashion. As the lively discussions among scientists continue about how many years we will still be able to use fossil fuels, with opinions ranging from 50 to 500 years, they agree that reserves are limited and that it is essential not only to search for new energy carriers but also for new material sources.

In this respect, renewable resources are a crucial area in the search for alternatives to fossil-based raw materials and energy. In the field of energy supply, biomass and renewable-based resources will be part of the solution, alongside other alternatives such as solar energy, wind energy, hydraulic power, hydrogen technology and nuclear energy.

In the field of material sciences, the impact of renewable resources will probably be even greater. Integral utilization of crops and the use of waste streams in certain industries will grow in importance, leading to a more sustainable way of producing materials.

Although our society was much more (almost exclusively) based on renewable resources centuries ago, this disappeared in the Western world in the nineteenth century. Now it is time to focus again on this field of research. However, it should not mean a 'retour à la nature', but it should be a multidisciplinary effort at a highly technological level to perform research aimed at developing new opportunities, new crops and new products from renewable resources. This will be essential to guarantee a level of comfort for the growing numbers of people living on our planet. The major challenge for the coming generations of scientists is to develop more sustainable ways to create prosperity and fight poverty and hunger in the world. A global approach is certainly favoured. This challenge can only be faced if scientists are attracted to this field and are recognized for their efforts in this interdisciplinary arena. It is therefore also essential that consumers recognize the place of renewable resources in a number of products.

Furthermore, scientists do need to communicate with each other and the general public and discuss the relevance of their work. The use and modification of renewable resources should not follow the path of the genetic engineering concept with regard to consumer acceptance in Europe. Related to this aspect, the series will certainly help to increase the visibility of the importance of renewable resources.

Being convinced of the value of the renewables approach for the industrial world, as well as for developing countries, I was delighted to collaborate on this series of books focusing on different aspects of renewable resources. I hope that readers become aware of the complexity, the interaction and interconnections, and the challenges of this field and that they will help the debate on the importance of renewable resources.

I would like to thank the staff at Wiley in Chichester, especially David Hughes, Jenny Cossham and Lyn Roberts, for seeing the need for such a series of books on renewable resources, for initiating and supporting it and for helping to carry the project through to the end.

Last, but not least, I want to thank my family, especially my wife, Hilde, and my children, Paulien and Pieter-Jan, for their patience and for giving me the time to work on the series when other activities seemed to be more inviting.

Christian V. Stevens
Faculty of Bioscience Engineering
Ghent University
Belgium
June 2005

Preface

Today, the concepts 'renewable resources' and 'sustainability' are receiving a great deal of attention in the academic, government and industrial communities. Renewables-based technology is believed to contribute to the sustainability of modern mankind. Indeed, it has turned out, particularly during the past decade, that fossil resource-based economies are vulnerable with respect to sustainability. With the current fossil oil consumption rate, the BP Statistical Review of World Energy of June 2004 indicates that we have reserves left for about 40 years. Global warming induced by fossil resource consumption ranks highly on the international scientific and political agenda. European and American authorities are conscious that their national economies are highly dependent on fossil oil-producing countries.

Renewable resources can indeed be of key importance for the development of a sustainable society; they are believed to provide new economic opportunities, to contribute to a high standard of living, and to realize a reduction of the human impact on the natural ecosystem.

If renewables-based technology becomes a key technology in the development of a sustainable society in the next few decades, it is of utmost importance that government policy-makers and business decision-makers are provided with the correct assessment tools, to enable them to quantify the contribution of renewables-based technology to sustainability. Because of its relevance, it became obvious in the writing of this book that development of this type of assessment tool is taking place in university, government and private research centres.

As editors, we are proud that we can present a book that consists of contributions looking at the topic from different angles, a topic which is extremely important if one takes sustainability seriously. The 20 chapters in the book originate from 18 different contributors located in nine different countries, including the UK, the USA, Canada and continental Europe. Contributions are provided by academics (9), government departments (3), private research centres (5), and international companies (3). The list includes well-known names, such as the IVL Swedish Environmental Research Institute, the US Department of Agriculture, the Centre for Environmental Strategies at the University of Surrey, the Wuppertal Institute, EMPA in Zurich, Procter and Gamble, IFEU Heidelberg and BASF.

Apart from the concluding chapter, the book is structured into three main parts. Part I, Chapters 1, 2 and 3, discusses the role renewables can play as a resource in our industrial society and the sustainability metrics that we can consider in technology assessment. In

Part II – Chapters 4–10 – relevant assessment tools are explained. Part III, i.e. Chapters 11–19, essentially consists of case studies. The role of the case studies is twofold. They show renewables-based technologies in quite different product types – from biofuels or bioplastics to biobased pharmaceuticals. But most importantly, they provide relevant experience of sustainability assessment tools as they are practised by academics, authorities and industry.

We hope that this book will be a main source of information for students and researchers, but also for those people in private companies and government organizations who want to learn more about the relationship between renewables-based technology and sustainability.

Jo Dewulf
Herman Van Langenhove

List of Abbreviations

(A/P)CFB	(atmospheric/pressurized) circulating fluidized bed; boiler designs where the fuel is fed to a furnace where an inert bed material (like sand) is kept in motion by blowing (combustion) air from underneath.
6-APA	6-aminopenicilinic acid, ß-lactam nucleus for penicillins
7-ADCA	7-amino-desacetoxycephalosporanic acid, ß-lactam nucleus for cephalosporins
A	land used
A	affluence
A	average weighting
AC-13	EU Accession Countries (Estonia, Latvia, Lithuania, Poland, the Czech Republic, Slovakia, Slovenia, Hungary, Romania, Bulgaria, Turkey, Cyprus, Malta)
ADP	abiotic resource depletion potential
A_E	partial area for energy provision
AE	alcohol ethoxylate
AE7	alcohol ethoxylate
AES	alcohol ethoxy sulphates
A_I	partial area for installations
AIC	Akaike Information Criterion
AIChE	American Institute of Chemical Engineers
AIM	Asian-Pacific Integrated Model
a_{in}	area statistically available to an inhabitant
AOX	adsorbable organic halogen, a category of water emissions
A_P	partial area for product dissipation
AP	acidification potential
APME	Association of Plastics Manufacturers in Europe
A_R	partial area for raw material production
ARMS	Agricultural Resource Management Survey
A_S	partial area for staff
AS	alcohol sulphates
ASF	Atmospheric Stabilization Framework
Aspen	Advanced System for Processing Engineering
a_{tot}	specific total area (per service unit)
A_{tot}	total area (total ecological footprint)

B100	one hundred per cent biodiesel
B30	fuel blend with 30 per cent biodiesel and 70 per cent diesel
B5	fuel blend with 5 per cent biodiesel and 95 per cent diesel
BFB	bubbling fluidized bed
BF_{ex}	exergy breeding factor
bhp-h	brake horsepower hour
BIG/CC	biomass integrated gasification/combined cycle
B_j	total burden from the system
$b_{j,i}$	burden (or intervention) j from process or subsystem i
BOD	biological oxygen demand
BUWAL	Swiss Agency for the Environment, Forests and Landscape
c	environmental compartment (air, water, soil)
C	carbon
C&S	capture and sequestration
CAA	Clean Air Act Amendments of 1990
cap	capita
Cd	cadmium
C_E	price per kWh energy
CExC	cumulative exergy consumption
CFC	chlorofluorocarbons
CFC-11 eq.	ODP factor expressed relative to the ozone depletion potential of CFC-11
CGV	conventional gasoline vehicle
CH	Switzerland
CH_4	methane
CHP	combined heat and power plant
CIS	Commonwealth of Independent States
Cl	Chlorine
CML	Centrum voor Milieukunde Leiden
C_N	price of a material
CNO	coconut oil
CO	carbon monoxide
CO_2	carbon dioxide
$CO_{2\,eq}$	carbon dioxide equivalent
CO_2-e	carbon dioxide equivalents
COD	chemical oxygen demand
COE	cost of electricity
$conc_i$	concentration of a substance
Cr	chromium
Cu	copper
CWRT	Centre for Waste Reduction Technologies
DALY	disability adjusted life years
DEFRA	Department for Environment, Food and Rural Affairs
DGGS	distiller's dried grains with solubles
DGVM	dynamic global vegetation model
DIP	deinked pulp
DMC	domestic material consumption

DME	dimethyl ether
DMI	direct material input
DMM	dimethoxymethane
DOE	United States Department of Energy
E	energy flow
E10	fuel blend of 10 per cent ethanol and 90 per cent gasoline
E85	fuel blend of 85 per cent ethanol and 15 per cent gasoline
EC	European Commission
ECCP	European Climate Change Program
E_D	energy demand
EEA	European Environment Agency
EF	eco-footprint
EFA	ecological footprint analysis
EJ	exajoule
$e_{k,j}$	characterization factor k for burden B_j
$E_{k\cdot}$	relative contribution to impact
ELU	environmental load unit
EMAS	Eco-Management and Audit Scheme
EMR	energy of material resource
EMS	Environmental Management System
EnergRec	energy recovery
EP	eutrophication potential
EPIA	European Photovoltaic Industry Association
EPS	Environmental Priority Strategy
EREC	European Renewable Energy Council
EROI	energy return on investment
ETP	ecotoxicity potential
EU	European Union
EU-15	European Union before 05/2004 (Austria, Belgium, Denmark, Finland, France, Germany, Greece, Ireland, Italy, Luxembourg, The Netherlands, Portugal, Spain, Sweden, the United Kingdom)
EWEA	European Wind Energy Association
F	annual feed or flow
FAO	Food and Agriculture Organization of the United Nations
FCA	fuel-cycle analysis
FCV	fuel cell vehicle
FFV	flexible fuel vehicle
FSC	Forest Stewardship Council
FT	Fischer-Tropsch
GATT	General Agreement on Trade and Tariffs
GDP	gross domestic product
GEF	global environment facility
GHG	greenhouse gases
GJ	gigajoule
GMO	genetic modified organism
GREET	greenhouse gases, regulated emissions, and energy use in transportation

Gt	gigaton
GWe	gigawatt electrical capacity
GWp	gigawatt peak
GWP	global warming potential
h	hour
H	hierarchist perspective
H_2	hydrogen
H_2O	water
H_2SO_4	sulphuric acid
ha	hectare
HA	hierarchist perspective, average weighting
hal.	halogenated
HANPP	human appropriation of net primary production
HC	hydrocarbons or hydrocarbon (emissions into water)
HCl	hydrochloric acid
HF	hidden flows
Hg	mercury
HGV	heavy goods vehicle
HHV	higher heating value
HM	heavy metals
HTP	human toxicity potential
HTP_{jA}	toxicological classification factor for substances emitted to air
HTP_{jS}	toxicological classification factors for substances emitted to soil
HTP_{jW}	toxicological classification factors for substances emitted to water
HTU	hydro thermal upgrading
I	human impact on the environment
IChemE	Institute of Chemical Engineers
IEA	International Energy Agency
IMAGE	Integrated Model to Assess the Global Environment
IO	input output
IP	increased livestock productivity
IP	integrated production
IPCC	Intergovernmental Panel on Climate Change
IPPC	Integrated Pollution Prevention and Control
ISO	International Organization for Standardization
ITTO	International Tropical Timber Organization
kg	kilogram
KOH	potassium hydroxide
kW_{th}/kW_e	kilowatt, thermal or electrical
LAS	linear alkylbenzene sulphonate
LCA	lifecycle analysis or life cycle assessment
LCC	lifecycle costing
LCI	lifecycle inventory
LCIA	lifecycle impact assessment
LHV	lower heating value
LPG	liquefied petroleum gas

LVL	laminated veneer lumber
m^2	square metre
m^2a	square metre times year
m^3	cubic metre
MAFF	Ministry of Agriculture, Fisheries and Food
MBT	mechanical biological treatment
MCPFE	Ministerial Conference on the Protection of Forests in Europe
MDF	medium density fibreboard
MeOH	methanol
MESSAGE	Model for Energy Supply Strategy Alternatives and their General
MFA	material flow analysis
Mg	megagrams
M_i	input flow into process
Mio	million
MIPS	material input per unit service
MJ	megajoule
MPCI	Montreal Process Working Group
MPF	mixed plastics fraction
MPOA	Malaysian Palm Oil Association
MSP	multifunctional Salix plantations
MSW	municipal solid waste
MSWI	municipal solid waste incineration
MT	megatons
MTBE	methyl tertiary butyl ether
MtC	megaton carbon
MW_{th}/MW_e	megawatt thermal capacity/megawatt electrical capacity
N	nitrogen or capacity of a plant
N.A.	not available
n.m.	not measured
N_2O	nitrous oxide
N2O	di-nitrogen oxide (emissions)
NACE	classification system of economic activities in the European Communities (Nomenclature Générale des Activités Économiques dans les Communautés Européennes)
NAMEA	National Accounting Matrix including Environmental Accounts
NaOH	sodium hydroxide
NAS	net addition to stock
NEB	net energy balance
NEnV	net energy value
NEP	net ecosystem production
NER	net energy ratio
NGO	non-governmental organization
NH_3	ammonia
Ni	nickel
NMVOC	non-methane volatile organic compounds
NO_3^-	nitrate

NO_x	nitrous oxides
NPP	net primary production
NPP_0	NPP of the potential vegetation
NPP_{act}	NPP of the actual vegetation
NPP_h	NPP harvested
NPP_t	NPP remaining in ecosystems after harvest
O&M	operation and maintenance
OD	ozone depletion
ODP	ozone depletion potential
OECD	Organization for Economic Cooperation and Development
OSB	oriented strand board
P	phosphorus
P	population
PAF	potentially affected fraction
PAH	polyaromatic hydrocarbons
Pb	lead
PDF	potentially disappeared fraction
PEFC	Pan-European forest certification
Pen.	penicillin
PET	polyethylene terephthalate
PET-1	polyethylene terephtale study by APME, 2002
PET-2	polyethylene terephtale study by Detzel *et al.*, 2004
Pg	petagram
PI	performance indicator
PIOT	physical input-output table
PJ	petajoule
PKO	palm kernel oil
PLA	polylactic acid
PM	particulate matter
PM_{10}	particulate matter $< 10\,\mu m$
POCP	photochemical ozone creation potential
PP	polypropylene
PPM	parts per million
PPWD	(European) Packaging and Packaging Waste Directive
PTW	pump-to-wheel
PV	photovoltaic
PV-TRAC	Photovoltaic Technology Research Advisory Council
R&D	research and development
R_c	rate of renewal of the environmental compartment
RDD&D	research development, demonstration and deployment
RDF	refuse-derived fuel
RER	region Europe
RES	renewable energy sources
RET	renewable energy technology
RFS	renewable fuels standard
RME	rapeseed methyl ester

RMI	reaction mass intensity
RS	ruminant meat substitution
RSPO	Roundtable for Sustainable Palm Oil
S	stream leaving a process or species richness
Sb	Antimony
SETAC	Society of Environmental Toxicology and Chemistry
SFA	substance flow analysis
SME	soy bean methyl ester
SO_2	sulphur dioxide
SO_x	sulphur oxides
SPI	sustainable process index
SRC	short rotation coppice
SRES	*Special Report on Emission Scenarios*
t	(occupation) time or tonnes
T	technology scalar or transport
TAPPS	total annualized profit per service unit
tbe	ton biomass equivalent
TCDD	tetrachloro dibenzo dioxin
THC	total hydrocarbons
tkm	ton kilometre
TMR	total material requirement
toe	ton oil equivalent
TÜV	Technische Überwachungsverein (Technical Inspection Association)
UCPTE	Union for the Coordination of Production and Transmission of Electricity
UHC	unburned hydrocarbons
UNCED	United Nations Conference on Environment and Development
UNCSD	United Nations Commission on Sustainable Development
UNDP	United Nations Development Programme
UNIDO	United Nations Industrial Development Organization
US	United States
USA	United States of America
USDA	United States Department of Agriculture
USSR	The Union of Soviet Socialist Republics
v/v	volume per volume
VE	more vegetarian food and less food wastage
VOC	volatile organic compound
WBGU	German Advisory Council on Global Change
WEA	World Energy Assessment
WEC	World Energy Council
WHO	World Health Organization
w_k	relative importance of impact E_k
W_p	watt peak
WTA	willingness to accept
WTP	well-to-pump
WTP	willingness to pay
x_i	mass or energy flow associated with a subsystem

y	specific yield
YOLL	years of lost life
Zn	zinc
α	renewability degree indicater
η	efficiency indicator
ρ	re-use indicator
σ	recoverability indicator
τ	(non)toxicity indicator

Part I
Renewables as a Resource and Sustainability Performance Indicators

1

The Contribution of Renewables to Society

Göran Berndes

1.1 Introduction

Stocks and flows of biomass are vital components of the biogeochemical system of the Earth. Biomass builds up the ecosystem, which contains the reservoir of genetic and species diversity and provides environmental services such as water purification, waste assimilation, soil fertility rehabilitation, water runoff regulation and flood control. Biomass is also crucial for human subsistence in other ways as it serves as food, and can be used for energy purposes and for the production of, e.g., sawn wood, paper, and various chemicals. Throughout history, human societies have ultimately depended on the management and harvest of biological (land and water) resources, and their inability to sustain their productivity have led to the end of their civilizations (Ponting, 1992).

Thus, human beings have always influenced their habitats, and still today the conversion of ecosystems to land for biomass production is perhaps the most evident alteration of the Earth. However, emissions to air and water also lead to substantial environmental impacts and a large portion of these emissions come (directly or indirectly) from other than land use activities. The industrialized society of today is unique historically in that access to biomass does not impose the ultimate limit: humans have learned to decouple industrial activities from biological productivity by exploiting fossil resources in the form of petroleum, coal and gas and this ability proved a powerful driver of societal development in the twentieth century. The role of biomass as a source

Renewables-Based Technology Edited by J. Dewulf and H. Van Langenhove

of energy has steadily declined and the global energy system is today dominated by fossil fuel use. The petrochemical industry creates synthetic materials and chemicals that successfully compete with biobased products, and also the food sector has undergone dramatic changes: most of our food still comes from agriculture, but is today produced in an intensive manner that relies on fossil fuels and petroleum-based chemicals, where synthetic nitrogen fertilizers are among the crucial causes behind the past century's transformation of world food production (Smil, 2001).

In the twentieth century, the impacts of human society on nature escalated. At the beginning of the twenty-first century, human societies have put almost half of the world's land surface to their service, and have caused extensive land degradation and loss of biodiversity worldwide (Turner II *et al.*, 1990, Oldeman *et al.*, 1991, Groombridge and Jenkins, 2002). Human activities influence global biogeochemical cycles, bringing about environmental effects such as eutrophication, acidification, stratospheric ozone depletion and climate change (Figure 1.1). It is clear that the substitution of biomass with fossil resources (and the intensification of agriculture) have saved large areas from deforestation and conversion to agricultural land. But at the same time, much of the environmental impacts we see today is caused by the intensified land use and the use of petroleum, coal and fossil gas. For that reason, today there are attempts to reduce our dependence on fossil resources and return to relying more on biomass and other renewable resources for our subsistence. Addressing the concerns about climate change, land degradation and other environmental impacts, while providing food, energy and materials for a growing and wealthier global population, will be a formidable challenge.

This chapter will discuss the potential role of biomass as a renewable resource in a future global industrial society. Some analysts, such as Hoffert *et al.* (2002), dismiss biomass as an important future renewable resource, especially in the context of energy system transformation and climate stabilization. Others take the opposite view and propose biomass as one of the major future renewable resources (see Berndes *et al.*, 2003 for a review of 17 studies of the global bioenergy potential). There is no way to narrowly determine the potential contribution of biomass in a future global industrial society, since it depends on a range of parameters that can vary substantially in the future (Hoogwijk *et al.*, 2003). The aim of this chapter is instead to provide some perspectives and point out a few potentially important issues likely to come into focus in a future of extensive use of biomass for energy and as a renewable feedstock in industry. To begin with, a short review of biomass use in society, including a comparison with other major product and resource flows, followed by an outline of the prospects for non-food crop production and agricultural residue utilization in the future – emphasizing some crucial aspects that so far have received less explicit attention in assessments. After that, the drivers behind increased demand for biomass will be described. The case will be made that the demand for climate-neutral fuels and materials (especially fuels) may lead to a dramatically expanded human biomass use, with implications for biodiversity and nature conservation, and competition for land and other resources. Illustrative outlines of possible consequences are given and discussed. Finally, multifunctional biomass production systems are described. Such systems offer a way to meet the growing biomass demand while at the same time promote environmental protection and sustainable land management, thus providing a possible strategy

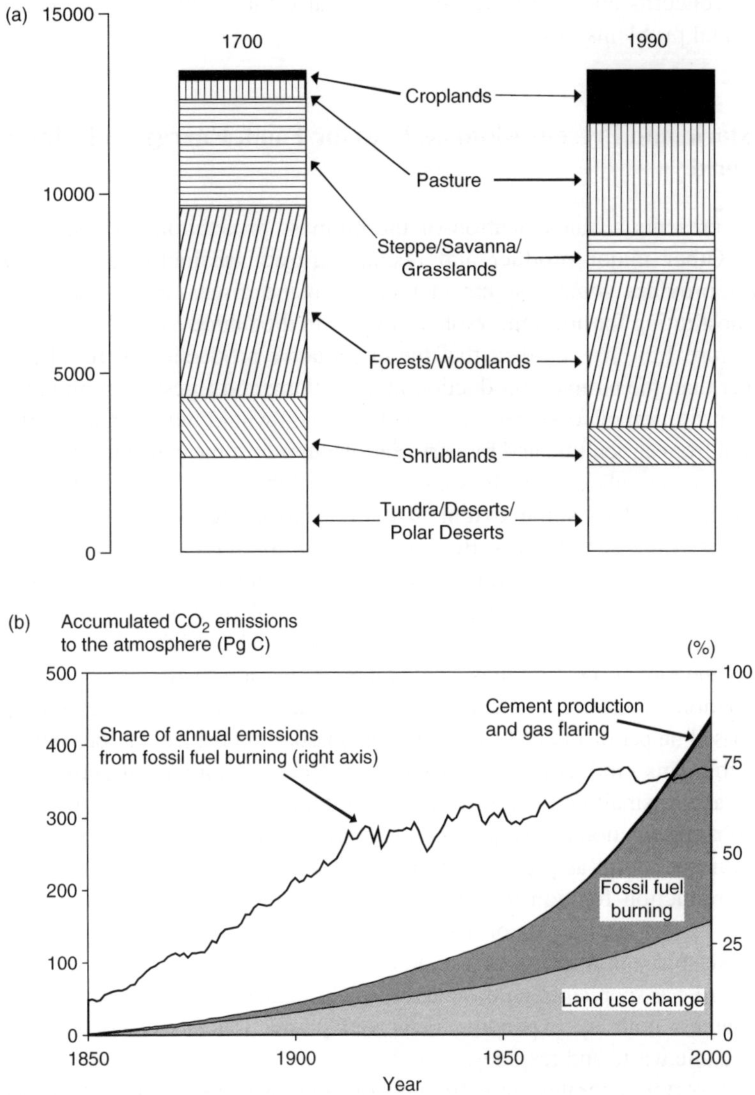

Figure 1.1 *Selected indicators of human influence on the Earth system in the past. Figure 1.1a outlines the land transformation during the past 300 years (units: million hectares). Note that this figure does not capture the far-reaching conversion within each given ecosystem type. For example, the conversion of primary natural forests to secondary production forests has resulted in the elimination of a multitude of critical habitats, leading to negative consequences for the state of biodiversity in forest ecosystems. Figure 1.1b presents the anthropogenic CO_2 emissions to the atmosphere since 1850 (expressed as carbon in CO_2). For many decades fossil fuel burning has been the major source of CO_2 emissions into the atmosphere, presently contributing close to 75% of annual emissions. But more than one-third of the accumulated emissions from 1850 to now have been caused by land use change, primarily conversion of forests to agricultural land. In cement production, fossil CO_2 is released during the calcination process where calcium carbonate is heated to yield lime. CO_2 is also released into the atmosphere when natural gas is 'flared' from petroleum reservoirs*
Source: Based on Klein Goldewijk (2000) and RIVM (2005).

to address concerns about climate change and also many other of the most pressing environmental problems of today.

1.2 Historic and Present Biomass Uses for Food, Energy and Materials in the World

Figure 1.2 presents a quantification of the biomass production for food, energy and materials. Other major product and resource flows are included for comparison. Figure 1.2 provides some insights in relation to the discussion of the prospects for biomass substituting for non-renewable resources in the future.

From Figure 1.2a, it is evident that the quantitative production of fossil resources is much larger than the biomass production in agriculture and forestry, implying that a far-reaching substitution of fossil resources with biomass would require a dramatic increase in the output from agriculture and forestry. Petroleum is to some extent used for the production of plastics and bulk chemicals, some 10–15% of the coal is used in steel production, and fossil gas (and to some extent also other fossil resources) are used for the production of synthetic fertilizers. But it is the use of fossil fuels in the energy sector that is the main source of society's exploitation of fossil resources. Clearly, the decoupling of societal energy use from biological productivity, that took place more than 100 years ago, has now brought us to energy consumption levels that make it difficult to return to a situation where the global society relies solely on biomass for energy.

The situation is different when looking at materials that are presently primarily produced based on petroleum and fossil gas, e.g., plastics, rubber and various bulk chemicals (Figure 1.2b). This production presently uses 5–10% of total annual petroleum and gas production and is small compared to the agricultural output: compare, for instance, the present global production of cereals (the major crop type in agriculture) with the plastics production in the world as presented in Figure 1.2b. It is also evident from Figure 1.2b that crop production for non-food/feed uses presently occupies a very small part of agricultural land use: the major part of society's biomass production for material purposes takes place in forestry. However, as will be shown below, agriculture can play a major future role as supplier of renewable feedstocks to industry, substituting non-renewable fossil resources, both by expanding dedicated production of non-food crops and by utilizing organic waste and residues.

The forest sector generates large amounts of biomass residues, both in the forests and at industrial sites such as sawmills and pulp/paper plants. Over the years, the forest industry has improved the wood utilization efficiency by cascading residue flows to energetic or lower value material uses. But the potential for increased residue utilization in forestry is large: increased wood extraction in connection to thinning operations and final logging may yield substantial increases in biomass output. The prospect for increased stemwood extraction by extending and/or intensifying conventional forestry operations is an issue where standpoints diverge, depending on different views regarding environmental, technical, legal and economic restrictions (Nilsson, 1996). The discussion below will focus on the agricultural sector. But several of the issues treated (e.g., the economics of biomass under an ambitious climate policy regime) are relevant also for the forest sector.

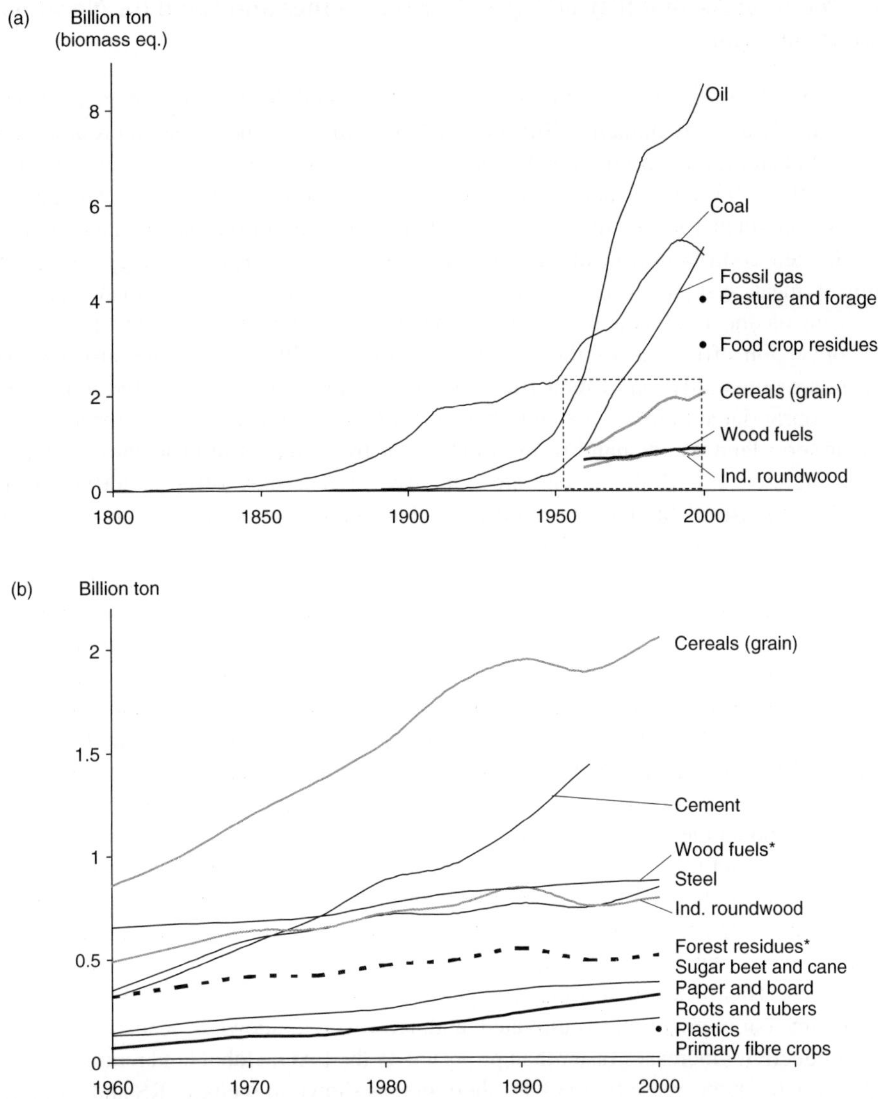

Figure 1.2 *Global annual production of major biomass types in agriculture and forestry, and of selected major products and basic resources. The fossil resources are given on a ton biomass equivalent basis (tbe) in order to facilitate a comparison with the different biomass types (conversion based on 1 ton oil equivalent 42 GJ; 1 tbe = 18 GJ). Figure 1.2b is a scaled-up segment of Figure 1.2a, including additional products. Only data for the year 2000 are presented for the three categories 'Pasture and forage', 'Food crop residues', and 'Plastics' due to lack of time series data*

Note: * 'Pasture and forage' refers to the part eaten by grazing animals. 'Forest residues' is indicative of forest biomass availability linked to industrial roundwood production, based on characteristics given in Johansson et *al.* (1993). 'Wood fuels' (FAO data) does not include all biomass uses for energy. For example, the FAO 'Wood fuels' data for year 2000 corresponds to about 15 EJ, while the global biomass use for energy is estimated at about 35–55 EJ per year (Turkenburg, 2000).

Source: Based on Marland et *al.* (2003), FAOSTAT (2005) and RIVM (2005).

1.3 Potential Availability of Agricultural Residues and Land for Non-Food Crop Production

The total food system appropriation of biological productivity is many times larger than what is finally used by humans. Wirsenius (2003a) estimated the global appropriation of terrestrial plant biomass production by the food system to be some 13 Pg (dry matter) per year in 1992–1994. Of this, about 8% ended up in food commodities eaten. Animal food systems accounted for roughly two-thirds of the total appropriation of plant biomass, whereas their contribution to the human diet was about 13% (gross energy basis). The ruminant meat systems were found to have a far greater influence than any other subsystem on the food system's biomass metabolism, primarily because of the lower feed-conversion efficiency of those systems. Based on this notion, one suspects that: (i) there are potentially major industrial (and energy) feedstocks to be found in the large pool of appropriated biomass not ending up as food; and (ii) there is scope for mitigating the long-term land use demand in the food sector by increases in efficiency (including dietary preferences). Both options are attractive in that they offer opportunities for increasing the use of biomass in industry, and in the energy sector, without imposing further conversion of natural land to agricultural uses.

In order to gain a better understanding of these opportunities, a mass and energy balanced model of the global food system was used to assess how the global biomass potential is influenced by different development paths in the food and agriculture system (Wirsenius, 2003a, 2003b; Wirsenius *et al.*, 2004). The starting point for the analysis was the recent projections of global agriculture up to 2030 made by the Food and Agriculture Organization of the United Nations (FAO) (Bruinsma, 2003). In addition to the 'Reference' scenario, depicting the FAO projection, three explorative scenarios were developed: 'Increased livestock productivity' (IP); 'Ruminant meat substitution' (RS); and 'More vegetarian food and less food wastage' (VE).

The results from the scenarios indicate that if the projections made by the FAO come true, the prospects for non-food crop production will be less favourable. In the scenario depicting the FAO projection, it is estimated that total agricultural land area globally will expand from current 5.1 billion hectares to approximately 5.4 billion hectares in 2030 (Figure 1.3). This means that a major expansion of non-food crops would require even further conversion of natural to cultivated land. However, as shown in scenario IP, if livestock productivity increases faster than projected by the FAO, global land requirement for food may actually decrease to 2030. Furthermore, as shown in scenario RS and VE, if the higher livestock productivity is combined with changes in diets (a 20% substitution of ruminant meat with pig/poultry meat) and reduced food wastage, global agricultural land demand may decrease to 4.2–4.4 billion hectares (Figure 1.3). If the surplus agricultural land was targeted for non-food crop cultivation, a considerable amount of biomass could be produced without claiming land beyond what has already been appropriated.

In the IP, RS and VE scenarios, also the amount of food-system residues and by-products available for non-food purposes will be higher than in the FAO projection (Figure 1.4), mainly due to a lower use of crop residues as feed in those scenarios. In, e.g., the European regions, agricultural land demand decreases also in the Reference scenario, due to decreasing population (-8% from 1998 to 2030) and continuing rises in crop and livestock productivity. This is in contrast with the developing regions, where population growth

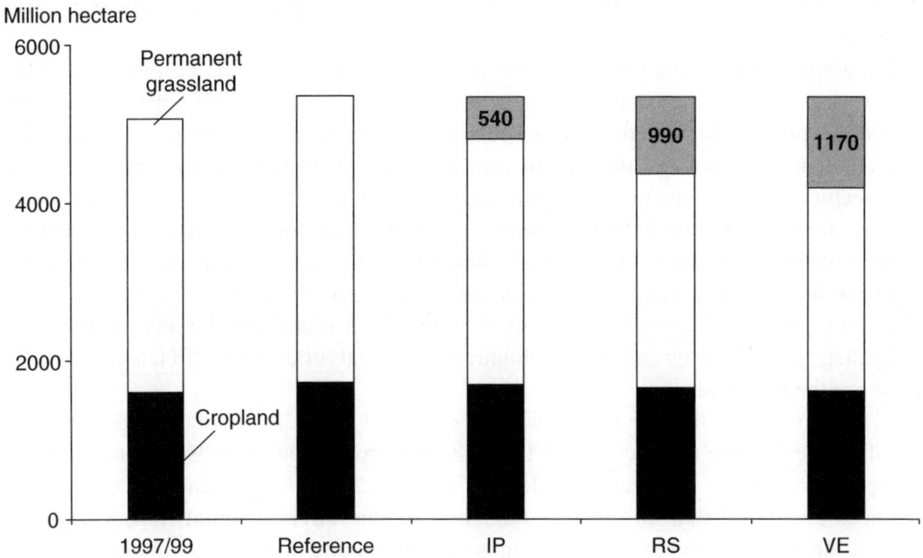

Figure 1.3 *Present and estimated future global extent of agricultural land in the scenarios. The Reference scenario depicts the FAO projection. Alternative scenarios: IP = increased livestock productivity; RS = ruminant meat substitution; and VE = more vegetarian food and less food wastage. The shaded topmost part of each alternative scenario column indicates the difference in land requirements for food production between the Reference scenario and the alternative scenario.*
Source: Based on Wirsenius *et al.* (2004). Reproduced by permission of ETA-Florence/WIP-Munich.

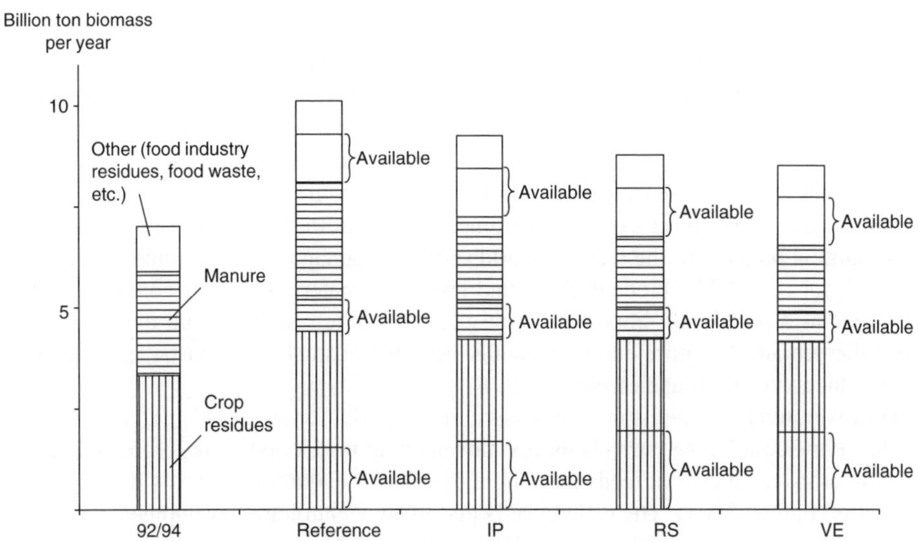

Figure 1.4 *Estimated production of by-products and residues in the global food system. The amounts possibly available for use as feedstock for industry or for energy 2030 in the scenarios are indicated*
Source: Based on Wirsenius *et al.* (2004). Reproduced by permission of ETA-Florence/WIP-Munich.

and increasing food consumption per capita add to rising land demand, as in, e.g., Latin America.

The above scenario exercise indicates that biomass from surplus cropland and from food sector residues may indeed play a large role as a renewable feedstock and help reduce the present dependence on non-renewable energy and materials. The scope for establishment of bioenergy plantations on surplus cropland may be considerable: if the food sector development follows a path similar to that in the RS/VE scenarios, a global biomass supply from plantations of the order of 3–6 billion ton per year does not seem to be impossible with regard to the land requirements of food production. Also the potential supply of biomass residues from the food system is impressive, being of the order of 3–4 billion ton per year. However, it is also clear that food sector development – and especially dietary preferences and the development of animal production efficiency – strongly influence the potential.

It is not axiomatic that the use of biomass resources is environmentally superior to the use of non-renewable resources. Both the dedicated production of feedstock crops and the collection of residues can lead to undesired environmental impacts and it is crucial that practices are found that ensure that reduction of one environmental impact does not increase another. However, if guided in sound directions, a growing biomass demand may be instrumental in promoting sustainable land management. This will be discussed further in a later section, where it will be described how biomass plantations can be located, designed and managed so as to generate environmental benefits in addition to those associated with the substitution of nonrenewable fuels or industrial feedstocks.

1.4 Drivers Behind Increasing Demand for Biomass for Energy and Materials

There are several factors behind today's interest in biomass as industrial feedstock and for the production of fuels and electricity. One early driver was the need to reduce food crop surpluses and find productive use of agricultural land in industrialized countries experiencing overproduction of food. Also, concern about high energy prices connected to the 1970s' oil crisis spurred an interest in the use of domestic energy sources to reduce dependence on foreign oil. At the same time, the insight that the industrial practices and consumption patterns of the western world seriously damage the environment stimulated a search for recyclable, biodegradable and less toxic materials. In this context, biomass was seen as a potentially important domestic, renewable resource, with the potential to meet the demand for more environmentally benign feedstocks in industry as well as for the production of fuels and electricity.

However, today's interest in biomass as a raw material for industry is not without precedents. For example, the farm chemurgy movement in the United States promoted the use of farm crops as industrial feedstocks more than half a century ago, partly due to similar concerns (Finlay, 2004). The difference is that today technology development has put us in a situation where industry can produce biofuels and bioproducts with a quality that satisfy a high consumer demand. Similar to when humans learned to use fossil feedstocks to create advanced synthetic materials with unique properties, the conversion of biomass to fuels and bioproducts continuously develops into increasingly sophisticated processes.

Modern biotechnology, material science, agricultural and process engineering today allow for a number of biobased products such as biodegradable plastics, oleochemicals, biocomposites, bulk chemicals, and biofuels.

The use of bioenergy for the production of heat and electricity has successfully increased in countries like Finland and Sweden. As the utilization has increased, the techniques and technologies to collect, transform and transport biomass have improved so as to reduce costs. Due to such developments the potential bioenergy resources have appeared to increase and there are optimistic scenarios suggesting that biofuels could also be used to replace significant parts of the fossil fuels used for transport. Stimulated by directives and regulations, first-generation liquid biofuels, such as ethanol and biodiesel based on traditional starch, sugar and oil crops, penetrate markets in, e.g., the European Union. Second-generation liquid biofuels, such as Fischer Tropsch fuels, Dimethyl Ether, lignocellulose-based ethanol and hydrogen based on gasification of biomass, are envisioned to become increasingly competitive to their fossil alternatives as technologies develop and allow production based on more abundant and potentially much cheaper lignocellulosic feedstocks.

Thus, technology development in processing biomass to fuels and materials can be expected to make possible a wide range of options for the substitution of non-renewable resources. This includes the continuation, and expansion, of 'traditional' practices such as the use of vegetable oils for lubricants and coatings, and of wood for buildings, and also new uses of established crops, such as natural fibres replacing glass fibre in composites. But it is the substitution of non-renewable (fossil) resources in the energy sector that poses the greatest challenge from the perspective of renewable resources availability (Figure 1.2). Environmental scarcities (i.e., limitations of the capacity of the ecosphere to assimilate societal emissions to air and water) rather than fossil resources scarcity determine the required extent and rate of this substitution, and the concern about human-induced climate change is possibly the most important driver for change.

Radical change of the global energy system is required if we are to reach stringent climate targets. This is a daunting challenge. For example, governments, several scientists, and environmental organizations have argued in favour of an upper limit on the increase in the global annual average surface temperature set at or around 2 °C above pre-industrial temperature levels. Such a target may require that atmospheric CO_2 concentrations are kept below 400 ppm (Azar and Rodhe, 1997), implying that total global CO_2 emissions by the year 2100 would have to drop to around 2 billion ton carbon (C) per year. Assuming a global population of 10 billion people in 2100, global average per capita emissions would then have to drop to about 0.2 ton C per capita and year by 2100. This is below the level that prevails in India today. In fact, even if a less ambitions climate target is chosen, the total global emissions would eventually have to drop to levels below 2 billion ton C per year also for these higher concentration targets (Houghton *et al.*, 2001).

At the same time, global energy consumption is expected to more than double during the twenty-first century. This means that the requirements of CO_2 neutral energy may have to grow to levels several times the present global total fossil fuel use, if we are to avoid venturing into a future with a doubled, tripled or even quadrupled pre-industrial atmospheric CO_2 level.

Surveys of possible future energy sources come up with several candidates capable of supplying large amounts of CO_2-neutral energy, including solar and wind energy, bioenergy, nuclear fission and fusion, and fossil fuels with carbon capture and sequestration (Hoffert *et al.*, 2002). But bioenergy ranks as one of few technological options capable of tackling climate change already today: being a low cost renewable fuel already competitive on some markets, and near penetration into new applications as policies, markets and related technologies develop. Advanced technologies, such as nuclear fusion, may eventually satisfy safety requirements and offer abundant energy supplies, but a prudent strategy for tackling the challenge of climate change cannot rely on those to aid CO_2 stabilization during the twenty-first century. Rather than awaiting the prospective (30–50 years ahead) realization of potential silver bullet solutions, society, people and companies will have to turn to what is available closer in time – regardless of whether the estimated ultimate long-term contribution of these options correspond to 30 or 300% of the present world energy use.

Since the potential biomass supply is low compared to the required levels of CO_2 neutral energy – almost regardless of whether one is optimistic or pessimistic about the potential biomass supply – more costly CO_2-neutral energy sources will have to enter if low CO_2 targets should be reached. When such energy technologies are in place, they will most likely cost substantially more than bioenergy, and therefore bioenergy will remain very competitive even in the scenario where advanced energy technologies have come to dominate the global energy supply. The more costly carbon-free energy sources can be expected to set the energy price at a level that would lead to higher profits for the bioenergy sector. With these higher profits, farmers would get stronger economic incentives to turn to bioenergy unless food prices rise to the point where profits are as large as in the energy sector. Thus, land and food prices are likely to be pushed upwards. The implications of such a development are discussed further in the next section.

1.5 Land Use Competition

The economics of land use will be different in a world where carbon and CO_2 have a price. The value of the carbon flows that are induced by different land use practices may become similar to the value of the primary product output from the very same practices. The cost of inputs such as nitrogen fertilizers and diesel will increase, but – potentially much more important – the food and forestry sectors will also have to face increasing competition from the energy sector. Food and bioenergy interactions and competition for scarce land and biomass resources have been the subject of several studies (Azar and Berndes, 1999; Azar and Larson, 2000; McCarl and Schneider, 2001; Johansson and Azar, 2003; Sands and Leimbach, 2003). Forest sector concerns about increasing energy sector demand for biomass are expressed in, e.g., Dielen *et al.* (2000) and policies stimulating this development are even argued to induce developments towards less eco-efficient use of forest wood (Van Riet, 2003).

Azar (2005) presents detailed modelling as well as some illustrative calculations of the willingness to pay for biomass in a world striving for low emissions. Based on a survey of future energy technology costs, and that the marginal energy price will be set by advanced technologies such as solar hydrogen, it is shown that farmers could sell biomass for energy at a price that is four to five times the estimated production cost. If such a situation were to materialize, it is estimated that land values might increase by an order of magnitude, and food prices might increase by a factor of two to five.

Figure 1.5 illustrates the same prospect, but for the case of electricity generation under a carbon tax/permit price regime. The fossil options are cheapest at low carbon tax rates, but the economic performance changes with increasing tax rates and at about 40 and 110 €/ton C tax rate, biomass electricity becomes cheaper than coal and fossil gas based electricity respectively. Capture and sequestration of CO_2 from coal combustion (carbon C&S) becomes competitive with conventional coal technologies at around 75 €/ton C, and the same happens for fossil gas at about 135 €/ton C.

Figure 1.5 also illustrates the interesting cost development for biomass-based electricity generation combined with carbon C&S: the cost will drop if the plant owner gets paid for capture and sequestration of the carbon. At a 225 €/ton C carbon tax, biomass-based electricity can be produced at no cost since the revenues from the carbon C&S cover the costs of electricity generation. A plant owner running a biomass-fired power plant would obviously be willing to pay much for the biomass in a situation with such carbon taxes/permit prices. For comparison, the Swedish carbon tax on the transportation sector and on household and district heating is presently about 290 €/ton C.

The socio-economic consequences of higher land values and higher food prices, are complex and there are different views about how a large biomass demand would influence development in agriculture. On the one hand, human demand for conquering more bioproductive lands might lead to the conversion of biodiversity rich ecosystems into monocultural biomass plantations, and poor people might be evicted from their lands. On the other hand, higher land values will stimulate increased land conservation efforts on agricultural land and it might generate income for the rural poor.

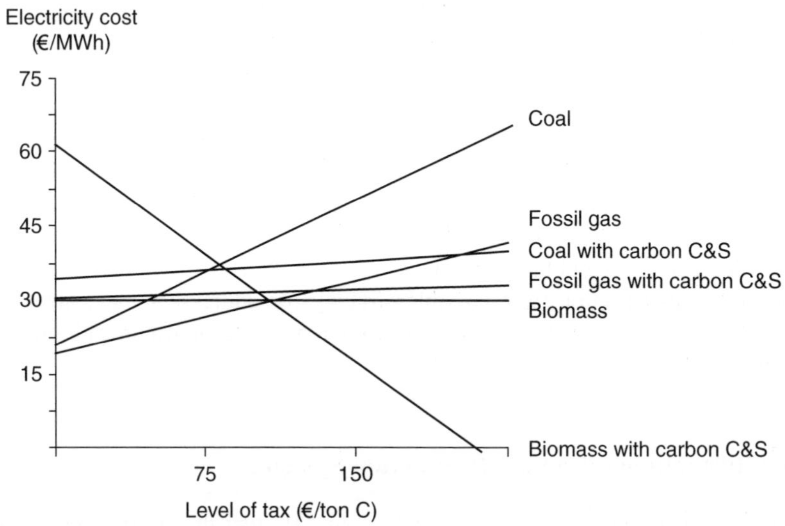

Figure 1.5 *Electricity generation cost for different feedstock and technology options under a carbon tax regime. The term 'carbon C&S' indicates that the CO_2 arising from fuel combustion is captured and sequestered in the ocean or in underground geologic repositories. Addition of carbon C&S leads to increased technology costs but much less sensitivity of electricity cost to carbon taxes. However, the electricity cost increases slowly with rising carbon taxes since not all CO_2 is captured in the C&S operations*

Source: Based on Azar *et al.* (2005), using the exchange rate 1 USD = 0.75 €. Reproduced from *Climatic Change* (in press) 2006, Azar *et al.* with kind permission from Springer Science and Business Media.

Biomass plantations can be established on degraded or otherwise marginal land, where production of food crops is not economically viable. It has been suggested that by targeting such land, farmers could restore soil organic matter and nutrient content, stabilize erosion and improve moisture conditions. In this way an increasing biomass demand could become instrumental in the reclamation of land that has been degraded from earlier over-exploitation and improper management (Hall *et al.*, 1993). However, studies indicate that biomass production on marginal/degraded land may not be the automatic outcome of increasing biomass demand (Johansson and Azar, 2003; Azar and Larson, 2000). If the allocation of land is done by profit-maximizing farmers and forestry companies, prime cropland may be targeted if the higher yields on the better soils outweigh the increased land costs. Biomass plantations may eventually be pushed to marginal/degraded land due to increasing land costs following increased competition for prime cropland, but this competition will likely also be reflected in increasing food commodity prices.

In industrialized countries, this may be less of a problem since food commodity prices only constitute a minor share of retail food prices, and the share of personal consumption expenditure spent on food is moderate. However, in developing countries where food often accounts for a very substantial part of total household consumption, the situation is different. An increase in the prices of staple food crops might cause an increased number of, or a worsened situation for, people chronically hungry and undernourished. Thus, the balance of distributional impacts is difficult to assess. Still, the risk that more people will be affected by hunger must not be disregarded. In a scenario with unequal economic development in the world, a large bioenergy demand with strong paying capacity in industrialized countries may impact food security and food availability in developing countries, creating a moral dilemma in the development of bioenergy strategies.

These potential impacts should not be taken as arguments against policies aimed at reducing CO_2 emissions. Rather, they imply that CO_2 abatement policies cannot be assessed in the absence of distributional considerations and are a clear signal of the importance of global and national efforts to advance development and reduce poverty in the world, especially in developing countries. Synergies and joint action with other multilateral environmental conventions and agreements should be sought, in order to ensure that CO_2 abatement policies do not aggravate the situation in relation to, e.g., food security, water resources and biodiversity.

In the concluding section, one possible strategy for addressing the concerns about climate change and also many other of today's most pressing environmental problems, is briefly presented.

1.6 Multifunctional Biomass Production Systems

Research carried out in Sweden and elsewhere reveals that the environmental benefits from a large-scale expansion of properly located, designed and managed biomass plantations could be substantial, as the negative environmental impacts from current agriculture practices and also municipal waste treatment could be significantly reduced (Berndes *et al.*, 2004; Berndes and Börjesson, 2004; Börjesson and Berndes, 2005; Lewandowski *et al.*, 2004).

Multifunctional biomass production systems can be defined as systems that, besides producing biomass for substitution of fossil resources, also provide additional environmental services. The potential for multifunctional biomass production systems based on Salix (Multifunctional Salix Plantations, MSPs) in Sweden has been assessed and the results are very encouraging: about 15 000 hectares are used for Salix production in Sweden today. An estimated 50 000 hectares could be dedicated to MSP systems providing environmental services having an estimated economic value exceeding the total cost of Salix production. On more than 100 000 hectares, the biomass could be produced in MSPs providing environmental services having an estimated value above, or roughly equal to, half the biomass production cost (Figure 1.6).

Thus, given that additional revenues – corresponding to the economic value of the provided environmental services – can be linked to the MSP systems, the economic performance of such biomass production can improve dramatically. Biomass supply from MSPs could bring substantial improvements in the biomass supply cost and also in other aspects of competitiveness against conventional resources. Establishment and expansion of such plantation systems would also induce development and cost reductions along the whole biomass supply chain. Thus, MSPs could become prime movers and pave the way for an expansion of low cost perennial crop production for the supply of biomass as industrial feedstock and for the production of fuels and electricity.

The production and use of biomass from multifunctional biomass production systems would not only contribute to the development towards more sustainable energy and

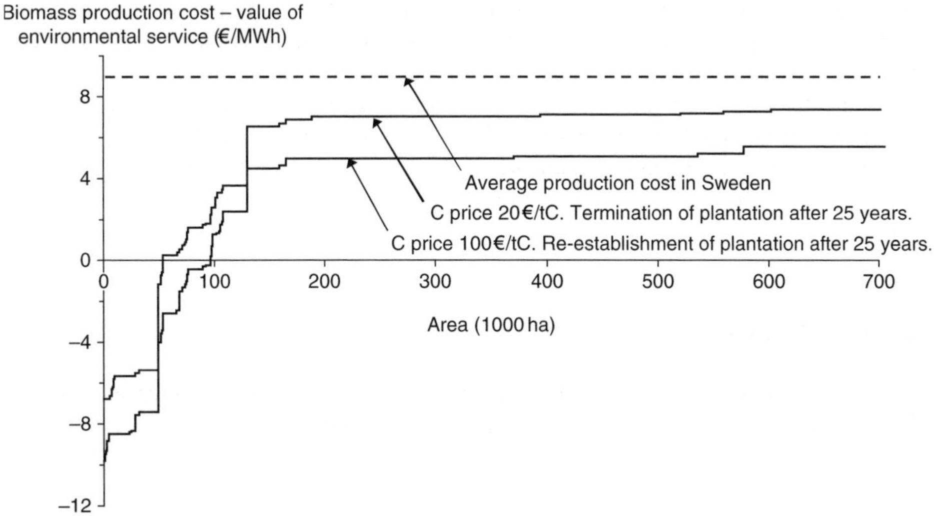

Figure 1.6 *The practical potential for multifunctional bioenergy systems in Sweden, and an illustration of the estimated value of the additional environmental services provided, as they relate to the cost of Salix production. Assessed environmental services include: reduction of nutrient leaching and soil erosion; cadmium removal from agricultural land; increased nutrient recirculation and improved treatment efficiency of nutrient-rich drainage water and pre-treated municipal wastewater and sludge; and provision of habitats and contribution to enhanced biodiversity and game potential*

Source: Based on Berndes and Börjesson (2004). Reproduced by permission of ETA-Florence/WIP-Munich.

industrial production, but also to development towards a more sustainable agriculture and to increased recirculation and efficiency in societal use of essential resources such as phosphorus and other nutrients. This way, multifunctional biomass production systems may become a valuable tool also for meeting additional great challenges such as getting the world's water cleaner and preserving the long-term quality of agricultural soils.

1.7 Summary

The use of biomass as a renewable feedstock in industry, and for the production of fuels, heat and electricity, can help reduce our dependence on non-renewable energy and materials. The question is not whether it is technically possible to supply several billion tons of biomass for energy and industry every year, but rather whether it can be done acceptably from a social and environmental point of view. A key question will be how to make sure that a large-scale expansion of biomass use for energy complies with other urgent environmental objectives, such as reduced erosion and land degradation, reduced eutrophication, good quality groundwater, a rich agricultural landscape, nature conservation and the protection of global biodiversity. These are prerequisites for bioenergy and biomaterials to be regarded as attractive options for the future.

If guided in sound directions, the growing biomass demand can be instrumental in promoting sustainable land management. The case of multifunctional Salix plantations in Sweden points the way for an expanded supply of biomass that should be systematically explored in both industrialized and developing countries. Besides estimating the potential extent and economic value of the environmental services that can be provided, a key issue will be to identify suitable mechanisms to put a premium on the environmental services provided. In some cases, actors can be identified that are willing to pay for a specific environmental service. In other situations, information campaigns and innovative government measures that credit the biomass producer may be required in order to stimulate the establishment of multifunctional biomass production systems. A challenge when implementing such measures lies in the harmonization of the different policies in the energy, environmental and agricultural fields.

Acknowledgements

This chapter builds further on earlier collaboration with Christian Azar and Stefan Wirsenius at Chalmers and Pål Börjesson at Lund University.

References

Azar, C. (2005) Emerging scarcities: Bioenergy-food competition in a carbon constrained world, in Simpson, D., Toman, M. and Ayres, R. (eds) *Scarcity and Growth in the New Millennium*, Resources for the Future Inc./Johns Hopkins University Press, Baltimore, MD.

Azar, C. and Berndes, G. (1999) The implication of CO_2-abatement policies on food prices, in Dragun, A. and Tisdell, C. (eds) *Sustainable Agriculture and Environment: Globalisation and Trade Liberalisation Impacts*, Edward Elgar: Cheltenham.

Azar, C. and Larson, E. D. (2000) Bioenergy and land-use competition in Northeast Brazil, *Energy for Sustainable Development*, **4**, 64–71.

Azar, C., Lindgren, K., Larson, E. D. and Möllersten, K. (2005) Carbon capture and storage from fossil fuels and biomass: Costs and potential role in stabilizing the atmosphere, *Climatic Change*, forthcoming.

Azar, C. and Rodhe, H. (1997) Targets for stabilization of atmospheric CO_2, *Science*, **276**, 1818–1819.

Berndes, G. and Börjesson, P. (2004) Low cost biomass produced in multifunctional plantations – the case of willow production in Sweden, in *2nd World Conference on Biomass for Energy, Industry and Climate Protection*, Rome, Italy, 10–14 May, pp. 289–292.

Berndes, G., Fredriksson, F. and Börjesson, P. (2004) Cadmium accumulation and Salix based phytoextraction on arable land in Sweden, *Agriculture, Ecosystems and Environment*, **103**, 207–223.

Berndes, G., Hoogwijk, M. and van den Broek, R. (2003) The contribution of biomass in the future global energy supply: a review of 17 studies, *Biomass and Bioenergy*, **25**, 1–28.

Börjesson, P. and Berndes, G. (2005) The prospects for willow plantations for wastewater treatment in Sweden, *Biomass and Bioenergy*, in Press.

Bruinsma, J. (ed.) (2003) *World Agriculture: Towards 2015/2030. An FAO Perspective*, Earthscan Publications Ltd, London.

Dielen, L. J. M., Guegan, S., Lacour, P.-A., Mäki, P. K., Stolp, J. A. N. and Rytkönen, A. (2000) *EU Energy Policy Impacts on the Forest-Based Industry, Part 1, Modelling Analysis of the Influence of the EC White Paper on Renewable Energy Sources on the Wood Supply to the European Forest-Based Industries*. Stichting Probos, Wageningen, The Netherlands.

FAOSTAT, *Statistical database of the FAO*, Food and Agriculture Organization of the United Nations, Rome, Italy. Available at: www.faostat.fao.org/Accessed January 2005.

Finlay, M. R. (2004) Old efforts at new uses: a brief history of Chemurgy and the American search for biobased materials, *Journal of Industrial Ecology*, **7**, 33–46.

Groombridge, B. and Jenkins, M. D. (2002) *World Atlas of Biodiversity: Earth's Living Resources in the 21st Century*, University of California Press, Berkeley, CA.

Hall, D. O., Rosillo-Calle, F., Williams, R. H. and Woods, J. (1993) Biomass for energy: supply prospects, in Johansson, T. B., Kelly, H., Reddy, A. K. N. and Williams, R. H. (eds), *Renewable Energy: Sources for Fuels and Electricity*, Island Press, Washington, DC, pp. 593–651.

Hoffert, M. I., Caldeira, K., Benford, G., Criswell, D. R., Green, C., Herzog, H., Jain, A. K., Kheshgi, H. S., Lackner, K. S., Lewis, J. S., Lightfoot, H. D., Manheimer, W., Mankins, J. C., Mauel, M. E., Perkins, L. J., Schlesinger, M. E., Volk, T. and Wigley, T. M. L. (2002) Advanced technology paths to global climate stability: energy for a greenhouse planet, *Science*, **298**, 981–987.

Hoogwijk, M., Faaij, A., van den Broek, R., Berndes, G., Gielen, D. and Turkenburg, W. (2003) Exploration of the ranges of the global potential of biomass for energy, *Biomass and Bioenergy*, **25**, 119–133.

Houghton, J. T., Ding, Y., Griggs, D. J., Noguer, M., van der Linden, P. J. and Xiaosu, D. (eds) (2001) *Climate Change 2001: The Scientific Basis. Contribution of Working Group I to the Third Assessment Report of the Intergovernmental Panel on Climate Change*, Cambridge University Press, Cambridge.

Johansson, D. and Azar, C. (2003) Analysis of land competition between food and bioenergy, *World Resources Review*, **15**, 165–175.

Johansson, T. B., Kelly, H., Reddy, A. K. N., Williams, R. H. and Burnham, L. (eds) (1993) *Renewable Energy: Sources for Fuels and Electricity*, Island Press, Washington, DC.

Klein Goldewijk, K. (2000) Estimating global land use change over the past 300 years: the HYDE database, *Global Biogeochemical Cycles*, **15**, 417–434.

Lewandowski, I., Londo, M., Schmidt, U. and Faaij, A. (2004) Biomass production in multiple land use systems: categorization of feasible land use functions and their quantification by the example of phytoremediation, paper at World Conference and Technology Exhibition on Biomass for Energy, Industry and Climate Protection, Rome, Italy.

Marland, G., Boden, T. A. and Andres, R. J. (2003) Global, regional, and national CO_2 emissions, in *Trends: A Compendium of Data on Global Change*, Carbon Dioxide Information Analysis Center, Oak Ridge National Laboratory, US Department of Energy, Oak Ridge, TN, USA.

McCarl, B. A. and Schneider, U. A. (2001) Greenhouse gas mitigation in US agriculture and forestry, *Science*, **294**, 2481–2482.

Nilsson, S. (1996) *Do We Have Enough Forests*? International Union of Forestry Research Organisations (IUFRO), Vienna.

Oldeman, L. R., Hakkeling, R. T. A. and Sombroek, W. G. (1991) *World Map of the Status of Human-Induced Soil Degradation: An Explanatory Note, October 1991*, 2nd revised edition. International Soil Reference and Information Centre (ISRIC)/United Nations Environment Programme (UNEP), Wageningen, The Netherlands.

Ponting, C. (1992) *A Green History of the World: The Environment and the Collapse of Great Civilizations*, St. Martin's Press, New York.

RIVM, History Database of the Global Environment – HYDE, National Institute of Public Health and the Environment, Bilthoven, the Netherlands. Available at: http://arch.rivm.nl/env/int/hyde/. Accessed January 2005.

Sands, R. D. and Leimbach, M. (2003) Modeling agriculture and land use in an integrated assessment framework, *Climatic Change*, **56**, 185–210.

Smil, V. (2001) *Enriching the Earth: Fritz Haber, Carl Bosch, and the Transformation of World Food Production*, MIT Press, Cambridge, MA.

Turkenburg, W. C. (2000) Renewable energy technologies, in Goldemberg, J. (ed.), *World Energy Assessment* UNDP/UN-DESA/WEC, New York.

Turner II, B. L., Clark, W. C., Kates, R. W., Richards, J. F., Mathews, J. T. and Meyer, W. B. (eds) (1990) *The Earth as Transformed by Human Action: Global and Regional Changes in the Biosphere over the Past 300 Years*, Cambridge University Press, Cambridge.

Van Riet, C. (2003) Impacts of the EU energy policy on the woodworking industries: the sound use of wood from an industries perspective, in *Strategies for the Sound Use of Wood*, Poiana Brasov, Romania.

Wirsenius, S. (2003a) The biomass metabolism of the food system: a model-based survey of the global and regional turnover of food biomass, *Journal of Industrial Ecology*, **7**, 47–80.

Wirsenius, S. (2003b) Efficiencies and biomass appropriation of food commodities on global and regional levels, *Agricultural Systems*, **77**, 219–255.

Wirsenius, S., Azar, C. and Berndes, G. (2004) Global bioenergy potentials: a new approach using a model-based assessment of biomass flows and land demand in the food and agriculture sector 2030, in *2nd World Conference on Biomass for Energy, Industry and Climate Protection*, Rome, Italy, pp. 471–474.

2

The Potential of Renewables as a Feedstock for Chemistry and Energy

Wilfried G. J. H. M. van Sark, Martin K. Patel, André P. C. Faaij and Monique M. Hoogwijk

2.1 Introduction

Since the Kyoto Protocol officially came into force on 16 February 2005, at least part of the world has committed to reduce the amount of greenhouse gas emissions by 2008–2012 and to address in a serious way human-caused climatic change effects. Already now, post-2012 commitments are addressed to prevent the global average temperature from rising more than 2 °C (International Climate Change Taskforce, 2005; Stainforth *et al.*, 2005). As energy and materials demand and supply are of paramount importance for the development of economies and their people, an enormous challenge is posed to mankind in realizing a transition to a fully sustainable energy and materials system. Past examples of transitions to newly developed energy systems, e.g., the change from wood to coal, and the subsequent change from coal to oil and gas, have shown that such transitions require many decades (Nakicenovic *et al.*, 1998). Transitions do not occur by themselves. One needs to follow strategies and formulate visions of the desired end-result. One could adopt the so-called *trias energica* (Lysen, 1996) strategy, i.e., integrating three major energy strategies: (1) reduce the use of energy through efficiency improvements; (2) supply energy via renewable sources; and (3) a cleaner use of remaining fossil fuels.

The use of renewable energy sources (RES) has many advantages with respect to the use of fossil fuels, as reported in the World Energy Assessment (WEA) study (UNDP,

Renewables-Based Technology Edited by J. Dewulf and H. Van Langenhove

2000): (1) diversification of energy sources by increasing the share of a diverse mixture of RES, which leads to an enhanced energy security; (2) RES are more widely available leading to a reduced geopolitical fuel dependence; (3) no (or no net) greenhouse gases (GHG) emissions to the atmosphere; and (4) many renewable energy technologies (RETs) are well suited for small-scale, off-grid applications and hence can contribute to poverty alleviation by improved rural electrification.

The use of scenarios as unfolding images of the possible future is a good means of comparing different views of and strategies towards a sustainable future with renewable energy. In the *Special Report on Emission Scenarios* (SRES) (Nakicenovic, 2000) developed by the Intergovernmental Panel on Climate Change (IPCC), a set of 40 scenarios is discussed that simulate long-term (up to 2100) GHG emissions. These SRES scenarios are based on four storylines describing the world's development over time. Two axes are used to construct the storylines, in short, one denoting economic scale (global, regional) and one denoting societal view (capitalist versus socialist). The so-called A1 and A2 storylines are considered societies with a strong focus towards economic development, while the B1 and B2 storylines are more focused on welfare issues and are ecologically orientated. The A1 and B1 storylines are globally oriented, with a strong focus towards trade and global markets, the A2 and B2 storylines are more regionally oriented. This is reflected in the projected data for population and gross domestic product (GDP) that range (for 2050) from 8.7 billion (A1, B1) to 11.3 billion (A2) and 8.6 10^3 billion US$\$_{95}$/y (A2) to 24.2 10^3 billion US$\$_{95}$/y (A1), respectively, for these four scenarios (Nakicenovic, 2000).

Figure 2.1 shows the total energy demand and the energy mixture in four distinct years of four scenarios. Note that the SRES scenarios are non-policy intervention scenarios.

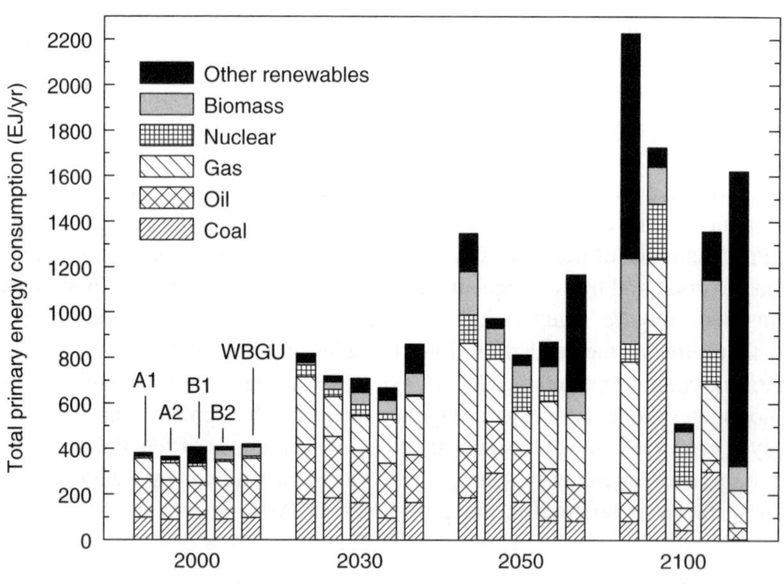

Figure 2.1 *Total primary energy demand and energy mixture for four SRES scenarios of the IPCC (Nakicenovic, 2000) and the WBGU A1T-450 scenario (Graßl et al., 2004). Different energy models are used, i.e., AIM, ASF, IMAGE2.1, and MESSAGE for the scenarios A1, A2, B1, and B2, respectively*

Clearly, the various scenarios show large differences in demand and energy mix, as a result of variations in dynamic driving forces, i.e., population dynamics, and economic and technological development. In all scenarios, biomass, solar and wind energy are expected to contribute at significant levels.

Other scenarios, for example, the A1T-450 from the German Advisory Council on Global Change (WBGU) (Graßl *et al.*, 2004), are more specific in the definition of 'other renewables', totalling 620 and 1400 EJ for 2050 and 2100, respectively, see Figure 2.1. Solar electricity is projected to have the highest share of the RES in 2050 and 2100. In addition, the European Renewable Energy Council (EREC) developed a scenario in which the contribution of renewables to the world energy supply in 2040 is nearly 50% (European Renewable Energy Council, 2004) with the highest share for biomass.

Images of the future as presented in scenarios can be used by policy-makers to develop, e.g., climate or energy strategies. The use of a technology-specific road map may assist this. Imperative here is the assessment of the ultimate potential of the technology. This will be pursued in the following sections.

2.2 Supply of Energy and Materials Using Renewables

Renewable energy is the most abundant resource for energy on our Earth, and most of it directly stems from the sun. The energy content of the *annual* amount of solar irradiation is estimated by the World Energy Council (WEC) at 3 10^6 EJ (World Energy Council, 2004), which is about 10 times the *total* amount of non-renewable earthly resources (oil, gas, coal, uranium). The most important indirect forms of solar energy together constitute only about 1/1000th of the direct solar energy content, i.e., 1260 EJ, 630 EJ, and 90 EJ for biomass, wind, and hydro, respectively (ibid.). The 1998 global primary energy consumption is 402 EJ/y (UNDP, 2000) and has risen to 418 EJ/y in 2001 (UNDP, 2004); the contribution of RES was estimated at 13%, see Table 2.1 (ibid.).

In the following sections we will address photovoltaic (PV), wind, and biomass potentials in a systematic way (Hoogwijk *et al.*, 2003; 2004; Hoogwijk, 2004b). All studies use a

Table 2.1 *Comparison of renewables-based supply in 1998 with the global technical potential in EJ/year*

	Renewables-based supply (1998)	*Global technical potential*	
		WEA	*This study*
Biomass	45	200–500	40–1100 (energy) 100 (materials)
Wind	0.07	70–180	345 (on-shore)
Solar	0.06	1500–50000	1340 (PV)
Hydro	9.3	50	
Geothermal	1.8	5000	
Marine	–	n.e.	
Total	56.2	6820–55730	

Note: Marine energy was not estimated (n.e.)
Source: Data taken from the World Energy Assessment (UNDP, 2000) compared to the present study.

grid cell database for climate and land use data at a $0.5\times0.5°$ grid (longitude, latitude) from the IMAGE model (New *et al.*, 1999; IMAGEteam, 2001); terrestrial area is covered by about 66 000 cells. Four different categories of potential can be identified (Van Wijk and Coelingh, 1993):

1 *theoretical* potential: the theoretical limit of the primary resource;
2 *geographical* potential: the theoretical potential reduced by geographical constraints, i.e., areas which are considered available and/or (partly) suitable;
3 *technical* potential: the geographical potential reduced by losses associated with the conversion from the primary to the secondary resource;
4 *economic* potential: the technical potential possible at cost levels that are competitive with other energy sources;

Of further interest is the *implementation* potential, which is the total amount of technical potential that is implemented in the energy system. This can be influenced by various incentives and societal perception to be higher than the economic potential.

2.2.1 Solar Energy

Solar energy is the most abundant source of energy on Earth, however, the abundance of solar irradiation is not equally distributed over the Earth; the sunniest regions are located around the equator, receiving some 300 W/m^2 on average annually, which translates into about 7 $kWh/m^2/day$ (NASA Atmospheric Sciences Data Center, 2005). Regions further away from the equator still receive enough solar irradiation for applications exploiting solar thermal and photovoltaic conversion techniques.

One generally distinguishes three forms of solar conversion technologies to produce the following: (1) low-temperature heat (solar thermal); (2) high-thermal heat (solar thermal power plants); and (3) electricity (photovoltaics). In this chapter we will focus on photovoltaics.

Photovoltaic (PV) technology involves the direct conversion of (sun) light into electrical energy. It generally exploits semiconductor materials in various device configurations to create and collect charged carriers from light. At present, the most common material used in PV technology is silicon. Solar cells are based on mono- and multicrystalline material (silicon, III–Vs), and amorphous or microcrystalline thin films (silicon, II–VIs). Commercial PV cells have efficiencies from 5 to 15%, while the maximum laboratory efficiencies reach 35% (Green *et al.*, 2004); the Carnot thermodynamic limit is 95% (Martí and Luque, 2004).

Typically, PV systems can be divided in four categories: (1) large centralized, grid-connected systems as analogue to conventional power plants; (2) grid-connected distributed generation systems, mostly in urban areas, to first supply the building with electricity and feed the excess back to the grid; (3) off-grid domestic systems (solar home systems) providing power to local households and villages; and (4) off-grid remote systems to power applications such as telecommunication appliances, water pumping, buoys. Grid-connected systems constitute about 70% of the PV world market, of which over 90% in distributed form (European Photovoltaic Industry Association, 2004). The world annual production capacity of PV manufacturers has surpassed the 1 GWp barrier to reach nearly

1.3 GWp in 2004 (Schmela, 2005), which leads to a cumulative installed PV power of about 3.5 GWp. Market growth has been 30% over the past five years. PV system performance suffers from losses occurring in inverters, cables, shading. The so-called performance ratio is around 0.8 kWh/Wp, which means that annually PV systems deliver about 800 kWh per installed kWp (Bucher, 1997).

The global theoretical potential of PV is calculated using solar irradiation atlases in combination with equations describing seasonal, time of day and geographical variation of solar irradiation (Duffie and Beckman, 1991): one arrives at a value of 633 10^3 EJ/year, nearly a factor five lower than estimated by the WEC.

The geographical potential is the yearly irradiance integrated over the terrestrial area suitable for PV installations, with a distinction between centralized and decentralized grid-connected PV installations. Sørensen (1999) has introduced various suitability factors (how much land can be used for PV) for several land categories, ranging from 0% (urban area) to 5% (desert) for centralized installations. Land use by PV has to compete with other land use options, such as agriculture. Table 2.2 shows calculated regional suitability factors for PV; the total global potential area for PV use is about 1.7% of the total terrestrial area (Hoogwijk, 2004). As calculations are performed on grid cell basis, values are averaged over regions. The global technical potential is taken as the sum over all regions and amounts to 1320 EJ/y or 365 PWh/y (rounded values) (ibid.). High values are found in a wide spread equatorial area. Simply multiplying per region the available area (suitability factor times regional area) with the average irradiation intensity, amount of hours per year, and the PV conversion efficiency (conservatively set at 10.5%) and summing this for all regions, one arrives at 1225 EJ/y or 340 PWh/y (ibid.) This is only indicatively correct, as the variation of irradiation in many regions is large.

Decentralized applications are installed on rooftops and facades in urban areas. Utilization factors for PV amounting to 40% for rooftops and 15% for facades have been reported (Alsema and Van Brummelen, 1993). Assuming a roof area surface per capita of 20 m^2 (average, dependent on GDP and region) and the population, one arrives at a total potential area of about 0.1% of the total terrestrial area. The technical potential is calculated to be 6 PWh/y (22 EJ/y) (Hoogwijk, 2004). Clearly, decentralized PV constitutes less than 2% of the total PV technical potential of 370 PWh/y (1340 EJ/y), which is in sharp contrast to the present situation where this is reversed (European Photovoltaic Industry Association, 2004). As a comparative example, WEC has reported 1575 EJ/y for centralized systems assuming that 1% of unused land is used (World Energy Council, 2004).

Present costs of PV systems are around 5 US$/Wp, which leads to an electricity price of about 0.5 US$/kWh in a wide spread equatorial area. This is far higher than prices for conventional generation. Table 2.2 shows the economic potential for a cut-off price of 0.6 US$/kWh totalling 225 PWh/y (805 EJ/y).

Using assumptions for future perspective PV technologies, the cost of electricity could be reduced to 0.05 US$/kWh (Hoogwijk, 2004). The assumptions include a future efficiency of 25% and system costs of 1 US$/Wp. In the vision report of the Photovoltaic Technology Research Advisory Council (PV-TRAC) these assumptions are considered to be valid around 2030 (Photovoltaic Technology Research Advisory Council, 2004). Present R&D programs are focused on reaching these short-to-medium-term cost and efficiency targets, while long-term projections are aimed at efficiencies approaching 50% and costs around 0.5 US$/Wp thus improving economic potential.

Table 2.2 *Regional and global technical potentials of photovoltaic and wind energy with distribution of terrestrial area, time-averaged irradiance, average wind speed at 10 m height. For PV, suitability factors, technical potential and average kWh-cost for centralized and decentralized grid-connected systems are shown, and technical potential at a cost cut-off of 0.6 US$/kWh. For wind, suitability factors, average wind power density, technical potential and technical potentials at cost cut-off 0.1 US$/kWh*

Region	*Area (Mha)*	*Photovoltaic energy*						*Wind energy*				
		Average irradiance (W/m²)	*Suitability factor (%)*		*Technical potential (PWh/y)*		*Technical potential at cut-off of 0.6 US$/ kWh(PWh/y)*	*Average wind speed (m/s)*	*Suitability factor (%)*	*Average power density (MW/km²)*	*Technical potential (PWh/y)*	*Technical potential at cut-off of 0.1 US$/kWh (PWh/y)*
			Central	*Decentral*	*Central*	*Decentral*	*Total*					
Canada	950	93.6	0.5	0.01	4	0.1	0	4.1	20.9	1.08	19	16
USA	925	127.4	0.92	0.08	12	1	1.2	4.3	26.8	1.02	21	13
Central America	269	175.9	1.38	0.04	6	0.2	1.8	3.3	10.8	0.4	2	1
South America	1761	152.4	0.84	0.02	22	0.4	3.7	3	4.7	0.26	8	6
North Africa	574	203.1	4.5	0.01	49	0.1	45.6	2.9	9.6	0.42	3	0
West Africa	1127	184.1	2.1	0	46	0.1	40.4	1.8	0.4	0.01	0	0
East Africa	583	195.3	2.71	0	30	0	25	2.6	6.5	0.28	3	0
South Africa	676	180.2	2.1	0.01	25	0.1	16.1	2.2	0.4	0.03	0	0
West Europe	372	108.8	0.69	0.26	3	1.1	0.1	4.3	12.6	0.58	4	2
East Europe	116	124.4	0.63	0.08	1	0.1	0	3.1	5.2	0.22	0	0
Former USSR	2183	95.8	0.92	0.01	25	0.2	0	3.4	9.4	0.47	16	7
Middle East	592	198.1	3.32	0.03	37	0.3	27.5	3.1	7.9	0.33	2	0
South Asia	509	193	1.92	0.05	18	0.5	15.6	2.3	2.9	0.12	1	0
East Asia	1108	149.4	2.14	0.06	33	0.9	0	2.4	2.3	0.1	2	0
South. East Asia	442	158.6	0.51	0.05	3	0.3	0.1	2	0	0.01	0	0
Oceania	838	188.5	3.32	0.01	52	0.1	47.2	3.6	23.7	0.91	14	6
Japan	37	126.4	0.23	1.21	0.1	0.5	0	3.3	2.7	0.08	0	0
World	13063	156.2	1.69	0.11	366	6	224.2	3	8.6	0.37	96	53

Source: Data taken from Hoogwijk *et al.* (2004a; 2004b).

2.2.2 Wind Energy

The amount of energy available in wind energy is smaller than available from direct solar energy: only part of the solar energy reaching the atmosphere is converted into wind energy as a result of solar-induced temperature differences on Earth. The rotation of the Earth also contributes to wind speed and direction. The global theoretical wind energy potential has been estimated to be some 2% of the solar energy reaching the atmosphere (Hubbert, 1971). Mankind has used wind power for over 25 centuries. The oil crises in the 1970s prompted large-scale development of wind turbines for the generation of electricity. At that time, typical turbine size was about 30 kW, with a rotor diameter of 10 m. The rapid growth in technology has led to present turbine sizes of 2–5 MW with rotor diameters and hub heights in excess of 100 m. The most common configuration now is the vertical axis, three-bladed rotors turbine, with the rotor in the upwind position. The continuous development of turbines has been going hand in hand with improvements in control and power regulation systems and conversion efficiencies that now are typically around 50%, i.e., 85% of the Betz limit (European Wind Energy Association, 2003). Since the 1990s, offshore wind power has been developed, motivated by the higher and more predictable wind speeds at sea.

The cumulative installed capacity in 2003 was about 40 GW with a growth of 30% annually over the past five years (ibid.). Turbine costs are around 750$/kW leading to electricity prices between 0.03–0.08 US$/kWh depending on the location-dependent wind speed (ibid.). Turbine costs have halved over the past decade (Junginger *et al.*, 2005), and it is expected that this will continue in the coming decade. An increased focus will be on off-shore wind parks of near GW size (European Wind Energy Association, 2003; Junginger *et al.*, 2004).

For the assessment of global and regional potential of on-shore wind energy the IMAGE model is used (IMAGEteam, 2001; Hoogwijk *et al.*, 2004). For the geographical potential a restriction is made to include only on-shore applications. The suitability of certain land areas restricts the geographical potential and is defined in terms of a suitability factor, which depends on the amount of urban and bioreserves area and other land use, altitude and on the wind regime. Areas with average wind speed lower than 4 m/s at 10 m height are not considered suitable. The wind regime is the most severe restriction, and results in a reduction of the suitability factor to zero in some regions, see Table 2.2. The global average value is about 9% of on-shore area.

For determination of the technical potential, average monthly wind speeds are used (New *et al.*, 1999; IMAGEteam, 2001). The technical potential is the product of the amount of full-load hours, i.e., the number of hours a wind turbine operates at its rated power, and the wind turbine power density (MW/km^2) in suitable areas. For suitable areas a value of 4 MW/km^2 is taken, somewhat lower than current practice in wind farms (Hoogwijk *et al.*, 2004). Results based on an availability factor of 95% (maintenance, etc.) and a wind park array efficiency of 90% (Grubb and Meyer, 1993) are presented in Table 2.2. The total global technical potential is 95 PWh/y (345 EJ/y). High values are found in Canada, the USA, the former USSR, and Oceania, totalling more than 70% of the global potential. As an example, calculation of the potential for Western Europe can be performed by multiplication of its area (372 Mha) and density (0.58 MW/km2), which yields 2150 GW installed wind power capacity. Based on the wind speed of 4.3 m/s at

10 m height and hub height of 70 m, the amount of full-load hours is 2200 (capacity factor is 25%). The technical potential thus is 4 PWh/y. The global economic potential at cut-off price 0.10US$/kWh is 53 PWh/y, showing that wind energy currently is economically viable.

The results of other studies vary between 19 and 53 PWh/y, see for a discussion Hoogwijk *et al.* (2004). A sensitivity study shows that the potential depends nonlinearly on wind resource data: a variation of 25% downward and 25% upward leads to a variation from 19 to 210 PWh/y, respectively. In addition, the total wind potential would be about 40% higher if one were to include the off-shore potential, which was estimated at 37 PWh at a depth of 50 m (Leutz *et al.*, 2001).

Future R&D is aimed at further cost reductions and minimization of environmental impacts, in combination with enabling technologies for the increased assimilation of wind power into the electricity grid.

2.2.3 Biomass-Based Energy and Materials Supply

Biomass Resource Categories

To calculate the potential availability of biomass resources for energy and materials, a description of the various resource categories is needed: residues from forestry and agriculture, various organic wastestreams and, most important, the possibilities for active biomass production on various land categories (e.g., for wood plantations or energy crops as sugar cane). The many options available to convert biomass to energy are discussed in detail in Chapter 13; the possibilities for using biomass for the production of (renewable) materials are covered in Section 2.3.2.

Biomass residues potential availability may be divided into:

- *Primary residues*: residues generated before and at harvest of main product, e.g., tops and leaves of sugar cane.
- *Secondary residues*: residues generated in processing to make products, e.g., bagasse, rice husks, black liquor.
- *Tertiary residues*: residues generated during and post-end use (and non-used products), e.g., demolition wood, municipal solid waste (MSW).

In general, biomass residues (and wastes) are involved with a complex of markets. Many residues have useful applications such as fodder, fertilizer and soil conditioner, raw material for, e.g., particle board, medium density fibreboard (MDF) and recycled paper, etc. Net availability as well as (market) prices of biomass residues and wastes therefore generally depend on market demand, local as well as international markets for various raw material and on the type of waste treatment technology deployed for remaining material. The latter is particularly relevant when tipping fees are deployed, giving some organic wastestreams a (theoretical) negative value. Typically, the net availability of organic wastes and residues can fluctuate and is influenced by market developments, but also on climate (high and low production years in agriculture) and other factors.

Energy crops are crops planted on agricultural or other land for energy and material purposes. In practice, there exists competition between land used for energy crops, animal grazing, forestry and food. On the other hand, synergy exists between these sources; the growth in food and forestry products directly implies increased amounts of crop residues with potential use for energy purposes.

Potential of Biomass Residues and Organic Wastes

Residues from agriculture: potential depends on yield/product ratios and the total agricultural land area as well as type of production system. Less intensive management systems require re-use of residues for maintaining soil fertility. Intensively managed systems allow for higher utilization rates of residues but also usually deploy crops with lower crop to residue ratios. Estimates vary between some 15 up to 70 EJ per year. The latter figure is based on the regional *production* of food (in 2003) multiplied by harvesting or processing factors and the assumed recoverability factors (Smeets and Faaij, 2004). These figures do not subtract the potential alternative use for agricultural residues. As indicated by Junginger *et al.* (2001) competing applications can reduce the net availability of agricultural residues for energy or materials significantly.

Dung: this category especially concerns the use of dried dung. Total estimated contribution could be 5–55 EJ worldwide. The low estimate is based on global current use, the high estimate is the technical potential. Utilization (collection) in the longer term is uncertain because this is particularly considered a poor man's fuel (Lysen, 2001).

Organic wastes: this category includes the organic fraction of MSW and waste wood. Estimates are strongly dependent on assumptions on economic development, consumption and the use of biomaterials; the ranges projected for MSW in the longer term (beyond 2040) amount to 5–50 EJ. Higher values are possible when more intensive use is made of biomaterials (Fischer and Schrattenholzer, 2001).

Forest residues: the (sustainable) energy potential of the world's forests is partly uncertain. A recent evaluation of forest reserves and the development of demand for wood products concluded that even in the case of the highest wood demand projections, the demand can be met without further deforestation. The bioenergy potential from forestry can contribute 1 to 98 EJ/y of surplus natural forest growth and 32 to 52 EJ/y harvesting and processing residues in 2050. The most promising regions are the Caribbean and Latin America, the former USSR and partially North America. Key variables are the demand for industrial round wood and fuel wood, plantation establishment rates, natural forest growth and the impact of technology and recycling (Smeets and Faaij, 2006).

Potential for Energy Crops

Dedicated production of crops for energy production, sometimes called 'energy farming', can be achieved with a multitude of crops. Some agricultural (annual) crops such as rapeseed and cereals are presently cultivated for energy purposes in Europe. Rapeseed for biodiesel production may exceed several hundreds of thousands of hectares in Europe, particularly Germany, in 2005 and its use is growing rapidly. Both crops are intermixed with conventional agricultural production and find an application for production of transport fuels. Perennial crops are planted for a longer period of time (15–20 years) and harvest can take place at regular intervals. Sugar cane used for production of bio-ethanol is grown in tropical regions and is the most important crop, covering at present some 2–3 million hectares in

Brazil, which is by far the world's most important producer of bio-ethanol from this crop. The acreage of sugar cane for ethanol production in Brazil, but also in various African and Asian countries, has been growing rapidly in recent years. Willow is a good example of a short rotation coppice (SRC) crop suited for temperate climate zones that is harvested every 2–5 years over a period of some 20–25 years. Most experience with SRC-Willow systems is gained in Sweden where this crop is produced on some 14 000 ha. Poplar and grasses like Miscanthus, which are harvested each year, and Sweet Sorghum are also examples of perennial crops, which are attracting interest in Europe and North America. Commercial use for energy production of perennials is, however, negligible at present (Faaij, 2006).

The potential for energy crops largely depends on land availability, considering that worldwide a growing demand for food has to be met, combined with nature protection and sustainable management of soils and water reserves. Given that a major part of the future biomass resource availability for energy and materials depends on these intertwined, uncertain and partially policy-dependent factors, it is impossible to present the future biomass potential in one simple figure. A review of available studies of future biomass availability carried out in 2002 revealed that no complete integrated assessment and scenario studies were available (Berndes *et al.*, 2003). Subsequently, these issues were addressed in a recent integrated assessment-based modelling approach, that explored the geographic (technical) and economic potential for active biomass production under the different SRES (see Figure 2.1) scenarios (Hoogwijk *et al.*, 2003).

The development over time of the geographical potential as the sum of three main land categories, covering good quality abandoned agricultural land up to marginal lands, is shown in Figure 2.2 for each SRES scenario. Figure 2.2 also shows the total simulated primary energy demand over time for the scenarios. The estimated geographical potential of B1 in 2100 is higher than the simulated total primary energy demand for that scenario. The A2 scenario is the scenario with the highest total energy demand and the lowest biomass energy geographical potential. If we consider the share of biomass in the total energy mixture, this would always be limited (22%) in an A2 world, but may reach 100% in a B1 scenario at the end of the century.

The geographical potential of abandoned agricultural land is found to be the largest for the A1 and B1 scenarios. For these scenarios, the potentials are comparable to the present energy consumption of 418 EJ/y (2001). At a regional level, significant potentials are found in the former USSR, East Asia and South America. A1 and B1 scenarios have the highest potential of abandoned land as both scenarios describe a world with decreasing population growth in the second half of the century and a world in which the technical development is high. In the B1 scenario the world is highly oriented towards environmental, ecological and social values. Therefore, competing land-use options, like nature conservation, are more restrictive than in the A1 scenario. However, there is still a high potential left in this scenario. The A2 scenario has the lowest geographical potential. It is a world with rapid population growth up to 15 billion people in the year 2100, and the pressure on the land-use system is already high.

Biomass production costs are influenced by yield, land rent, management system and labour costs. Increases in productivity are important to reduce production costs. Yields can be improved through crop development, production integration (multi-product plantation), and mechanization. Competition for land use should be avoided to minimize

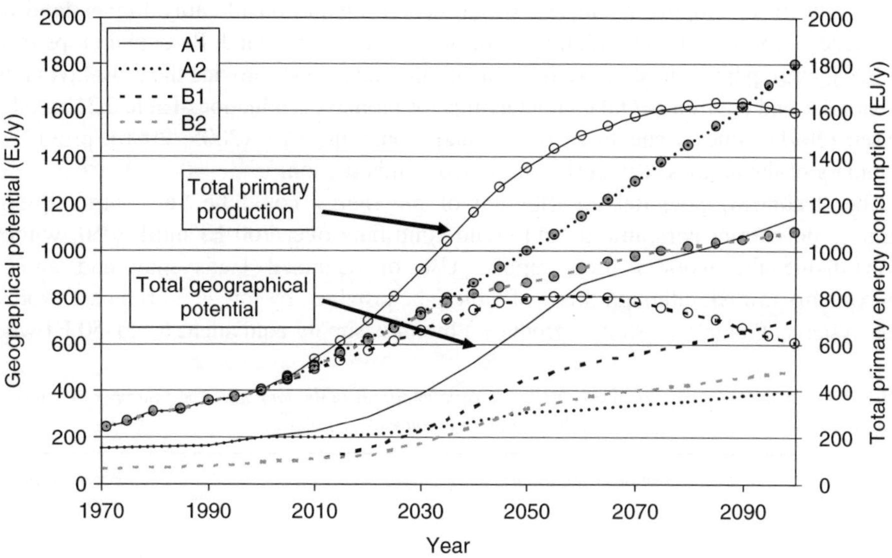

Figure 2.2 Geographical potential of woody biomass energy crops as assessed for the four SRES scenarios over time, as well as the simulated total primary energy consumption
Source: Reprinted Hoogwijk *et al.* (2005b) Copyright (2005), with permission from Elsevier.

inflated land rental rates. Labour costs can be lowered through mechanization. The production costs of plantation biomass are already favourable in some developing countries. Eucalyptus plantations in Brazil supply wood chips for 1.5–2US$/GJ. At present, sugar cane delivers bioethanol at competitive cost levels in Brazil (at around 6–7U$/GJ ethanol produced) and some African countries. In industrialized countries costs of biomass can be much higher (i.e., up to 4US$/GJ) but in the longer run, by about 2020, better crops and production systems are expected to cut biomass production costs in the USA to 1.5–2US$/GJ for substantial land surfaces. Typical cost ranges for perennial woody crops under north-western European conditions are 3–6US$ per GJ (compare to some 1–2US$/GJ for imported coal) (see UNDP (2000, Chapter 5).

Production costs of energy crops develop over time; they may increase due to increased labour costs, and decrease due to productivity increase per ha. Global cost–supply curves have been constructed for the year 2050 based on the SRES scenarios briefly described above (Hoogwijk *et al.*, 2003); they show that in 2050 a significant part (130 (A2)–270 (A1) EJ/y) of the production potential may be realized below 2US$/GJ, which is considered the upper level of the 1998 price for coal. The lowest costs are found at 0.8$ GJ^{-1}, in the A1 scenario in Eastern Africa.

Summarizing, both the technical and economic potential of biomass resources for energy and material use can be very large, up to several times the current global energy demand, without competing with food production, protection of forests and nature. Besides residues from agriculture and forestry, and organic waste, in particular, active production (e.g., energy crops) of biomass is responsible for these potentials. Key to the development of competitive energy cropping systems is the rationalization of agriculture, especially in developing countries, which can result in considerably higher land-use efficiencies for agriculture and, thus, a surplus of productive land. Perennial crops (such as Eucalyptus, poplar, grasses like Miscanthus and sugar cane) provide the most favourable economics and environmental characteristics for biomass production. Table 2.3, based on Lysen (2001), Smeets and Faaij (2004) and Hoogwijk *et al.* (2005; 2006), provides a summary of the biomass categories discussed in this section.

The (technical) potential contribution of bio-energy could be very large. Energy farming on current agricultural land could contribute over 700 EJ until 2050, without jeopardizing the world's food supply. Use of degraded lands may add another (maximum) 150 EJ, although this will largely be provided by low yield biomass production systems. Plantations should produce a biomass supply equivalent to 20–50 EJ when

Table 2.3 *Overview of the global potential of bio energy supply in the long term for a number of categories and the main preconditions and assumptions that determine these potentials*

Biomass category	*Main assumptions and remarks*	*Potential bio-energy supply up to 2050 (EJ)*
Energy farming on current agricultural land	Potential land surplus: 0–4 Gha (more average: 1–2 Gha). A large surplus requires structural adaptation of intensive agricultural production systems. If not feasible, the bio-energy potential could be reduced to zero. On average higher yields are likely because of better soil quality: 8–12 dry tonne/ha*yr is assumed. (*)	0–700 (more average development: 100–300)
Biomass production on marginal lands	A maximum land surface of 1.7 Gha could be involved globally. Low productivity of 2–5 dry tonne/ha*yr. (*) The supply could be low or zero due to poor economics or competition with food production.	(0) 60–150
Bio-materials	Range of land area required to meet additional global demand for bio-materials: 0.2–0.8 Gha. (average productivity: 5 dry tonnes/ha*yr). This demand should come from first two categories in case the world's forests are unable to meet the additional demand.	Minus (0) 40–150
Residues from agriculture	Potential depends on yield/product ratios and the total agricultural land area as well as type of production system: Extensive production systems require re-use of residues for maintaining soil fertility. Intensive systems allow for higher utilisation rates of residues.	Approx. 15–70
Forest residues	The (sustainable) energy potential of the world's forests is unclear. Part is natural forest (reserves). Low value: figure for sustainable forest management. High value: technical potential. Figures include processing residues.	(0) 30–150
Dung	Use of dried dung. Low estimate based on global current use. High estimate: technical potential. Utilization (collection) on longer term is uncertain.	(0) 5–55
Organic wastes	Strongly dependent on economic development, consumption and the use of bio-materials. Figures include the organic fraction of MSW and waste wood. Higher values possible by more intensive use of bio-materials.	5–50 (**)
Total	Most pessimistic scenario: no land available for energy farming; only utilization of residues. Most optimistic scenario: intensive agriculture concentrated on the better quality soils. (between brackets: more average potential in a world aiming for large scale utilization of bio-energy).	40–1100 (250–500)

Notes: (*) Heating value: 19 GJ/tonne dry matter. (**) The energy supply of bio-materials ending up as waste can vary between 20–55 EJ (or 1100–2900 Mtonne dry matter per year). This range excludes cascading and does not take into account the time delay between production of the material and 'release' as (organic) waste.

existing forests are not able to meet the growing demand for biomaterials. Organic wastes and residues could possibly supply another 40–170 EJ, with uncertain contributions from forest residues and potentially a very significant role for organic waste, especially when biomaterials are used on a larger scale. In total, the upper limit the of bio-energy potential could be 1000–1200 EJ/y. This is considerably more than the current global energy use of 418 EJ/y (2001).

2.2.4 Resources for Materials

The most important renewable resources used as materials (including chemicals) are wood for timber products, cellulose/hemicellulose (including paper production), natural fibres (e.g., cotton, flax or hemp), natural rubber, starch, sugar and natural fats/oils. Wood use for timber products and paper/board production and the use of natural rubber are mature applications. Also, many non-food uses of starch (e.g., for paper additives) and natural fats/oils (e.g., for the production of fatty acids and esters) are well established. New (or rather renewed) attention is being paid to the use of natural fibres for composites and, in particular, to the production of biobased polymers and chemical building blocks (e.g., aromatics, alcohols or carboxylic acids). Biobased polymers can be made by treatment (e.g., with heat) of *natural* polymers such as starch, thus leading to so-called thermoplastic starch (Crank *et al.*, 2004). Other options are to produce polymer precursors or building blocks by means of thermochemical processes (e.g., by pyrolysis of biomass including organic waste) or by biotechnological conversions. The use of biotechnology in industrial processes has been gaining momentum in the past few years. Milestones for industrial application are large-scale facilities for the biotechnological production of lactic acid and subsequent polymerization to polylactic acid (PLA) by NatureWorks in the USA (former name: Cargill Dow; 140 000 t p.a. PLA) and the production of 1,3-propanediol by DuPont also in the USA (Crank *et al.*, 2004). In the short to medium term, production facilities based on biotechnology and located in moderate climate zones will use fermentable sugars (dextrose) originating from starch crops. In tropical regions, sucrose from sugar cane is likely to be *the* preferred feedstock next to vegetable oils or glycerol (available as processing by-product from vegetable oils). Research, development and commercialization are ongoing to also make use of lignocellulosics (woody biomass, e.g. corn stover) as a source of fermentable sugars especially for moderate climate zones. If successful, this may revolutionize agriculture, the chemical industry and the energy sector because lignocellulosics are available in large amounts and at low cost.

2.3 Demand for Energy and Materials

2.3.1 The Dynamics of Energy Use

The key drivers for energy demand are population growth, the increase of wealth, efficiency developments (e.g. insulation of buildings, heat recovery and re-use in industry), technology choices (e.g. fuel-based versus electric cars) and consumption patterns (e.g. air transport versus rail). Total final energy demand has been projected by the IPCC to increase from around 300 EJ by the year 2000 to 500–650 EJ in 2030 (Nakicenovic, 2000). Depending on the developments in the longer term, global final energy demand by the year 2100 could amount to 450 EJ (B1) up to more than 1700 EJ (A1). These values for final energy represent the total demand for fuels (including heat) and electricity. The demand for fuels is projected to increase from approximately 250 EJ in 2000 to values between 250 EJ (B1) and 950 EJ (A2) in 2100. This wide range of values – from stagnation to a fourfold increase – reflects the combined effect of the key drivers on final energy demand. In contrast to fuels, it is expected in *all* IPCC scenarios that the demand for electricity will increase substantially (factor 4 in B1 up to factor 18 in A1).

Energy demand in developing countries will grow over proportionality: while total final energy demand in the developing world accounted for somewhat more than 40% in 2000, it is projected to increase to around 70% in 2100 (the extreme cases are 66% in A2 and 75% in A1). Comparable shares are expected for the subtotals for fuel and electricity.

There are various options to make use of renewable resources: They can be used to generate electricity, make hydrogen or raise heat. While various types of renewable resources can be used for the first three options (e.g., power from wind turbines, geothermal heat, solar thermal energy or photovoltaics), biomass is the only option to produce carbonaceous materials from renewable resources. We will discuss the opportunities in this area in the next section.

2.3.2 Materials from Renewable Resources

In mass terms, worldwide, the most important material from renewable resources is nowadays wood (as a construction material), followed by paper. Natural rubber, cotton, other natural fibres and organic chemicals derived from renewable materials (e.g. biobased polymers and surfactants) are nowadays far less important, accounting for only a few percentage of the total of wood and paper (Gielen *et al.*, 2001; Lysen, 2001; Hoogwijk *et al.*, 2003). Instead of fossil fuels, it is also possible to use charcoal (i.e. a biofuel/biomaterial) for iron production in blast furnaces. Being a reducing agent, the carbon input in blast furnaces is sometimes referred to as non-energy use (in contrast to energy use). Iron made by use of charcoal is therefore sometimes listed under biomaterials (Lysen, 2001). However, this is disputable because a considerable part of the carbon input is necessary to provide the process heat and can hence be considered simple fuel combustion. Therefore, the use of charcoal for iron production is further excluded.

Table 2.4 shows demand projections for the three key carbonaceous materials: wood, paper and polymers. The values for polymers represent totals regardless of whether they are produced from fossil or renewable resources. Polymers account for around two-thirds of today's total production of organic chemicals.

Renewable resources can also be used to produce other organic chemicals apart from polymers, namely solvents and surfactants. However, the overall demand trends are less clear for these products (partially stagnating or even decreasing). We therefore limit ourselves to polymers, thereby keeping in mind that we underestimate the total market of petrochemical products. The projections in Table 2.4. are based on historical analyses of the relationships between material use per capita and GDP per capita in numerous countries

Table 2.4 *Global projections for material demand in a 'business-as-usual' scenario*

	Total global demand (Mt/yr)					*Of which from biomass in 2100*	
	2000	*2020*	*2030*	*2050*	*2100*	*(%)*	*Mt*
Wood	306	530	670	1010	1620	100	1620
Paper	320	600	760	1110	1750	100	1750
Polymers	180	370	490	820	1880	10–100	190–1880
Total	806	1500	1920	2940	5250	68–100	3560–5250

Notes: Wood refers to use in construction only. For polymers, the total global demand represents the total amount independent of whether it is produced from petrochemical or renewable feedstock. The 10% estimate for biobased polymers represents a conservative, lower bound for 2100 while the 100% represents an upper bound.

Source: Based on data from Lysen (2001) and Chateau *et al.* (2005).

(Chateau *et al.*, 2005). The historical relationships found for industrialized countries (gradual decoupling of economic growth and material production) have been assumed to apply also for the future development in developing countries. For industrialized countries, the historical relationships were extrapolated into the future. In combination with projections for GDP, this leads to 'top-down' estimates for material consumption according to a 'business-as-usual' scenario which assumes similar dynamics and underlying mechanisms as experienced in the past. The projections presented in Table 2.4. have been prepared assuming a medium development of both population (8.3 billion people worldwide by 2100) and wealth. The assumed per-capita GDP values are higher than in the IPCC scenarios A2 and B2 but generally lower than B1 and A1; global GDP in purchasing power parities is assumed to increase by 2.1% p.a. until 2100 (Chateau *et al.*, 2005). Today, the three materials (Table 2.4) represent about 20% of the total of *all* major bulk materials produced globally. This total includes, in order of decreasing production volumes, cement, steel, bricks/tiles, glass, ammonia and aluminium. In total, their demand is projected to increase from 0.8 Gt globally to 5.3 Gt in 2100 (Table 2.4), i.e. by a factor of 6.5 or by 1.9% per year on average. Since the three materials outpace the other bulk materials in growth, their share is projected to increase in the next decades from approximately 20% today to around 35% in 2100.

The projections according to Table 2.4 are highly uncertain. For all materials, the market prices for fossil fuels and agricultural products and the policies implemented (e.g. on renewable resources, GHG emissions and nature preservation) will decisively influence the future production of materials from biomass, possibly with clearly different directions for the various materials. For wood, the 'building culture' and consumer preferences will additionally play important roles, leading to either lower or higher values. For example, the maximal global use of wood for construction in 2020 was projected to be twice as high under favourable conditions compared to the business-as-usual values in Table 2.4 (Lysen, 2001). For paper, the projected values may turn out to be clearly overestimated if a technology breakthrough towards digital paper can be realized. In contrast to wood and paper, which are mature applications of renewable resources, further developments are particularly uncertain for biobased polymers. This will depend largely on the future level of fossil fuel prices and on technological progress (e.g., in biotechnology) which share of the demand for polymers will be biobased (10–100% in 2100 according to Table 2.4).

Depending on the share of polymers produced from biomass, the total production of bio-based materials could amount to 3.6 to 5.3 billion tonnes (Table 2.4). We use the higher value to estimate the maximum potential biomass input and the attendant land use. As noted above, the real maximum estimate could be clearly larger due to higher future demand for wood but also for paper, polymers and other chemicals from renewable resources. We estimate on this basis, a potential maximum biomass input of 7 Gt per year in 2100. In energy terms, this is equivalent to approximately 100 EJ, which is around 25% of the current total primary energy use worldwide. The results for land use show that the total agricultural land needed to satisfy the demand for biomaterials, food, fodder and biofuels may exceed its availability. Maximization of land use efficiency and subsequent conversion to materials is hence likely to remain an R&D priority.

While the issues just discussed concern the very long term, the main goals for the short to medium term concern process and product improvement, higher product performance and decisively lower cost, while land use will not represent a bottleneck in the next 20 years. According to Crank *et al.* (2004), even a very optimistic projection assumption for biopolymer production in Western Europe would only require 5% of the total land used to grow wheat; this conclusion for Europe is valid also for most other world regions. This is a consequence of the high current market price and the inferior material properties of many biopolymers. The technical substitution potential, which can be derived from the material property set of each biobased polymer and its petrochemical-based equivalent, has been estimated at around 15.4 Mt for EU-15, or 33% of the total current polymer production. A more detailed analysis taking into account economic, social, ecological and technological influencing factors leads to a market substitution potential of 1 to 3 Mt of biobased polymers in EU-15 until 2020 (ibid.). While these are relatively low values compared to today's current *total* polymer production (ca. 45 Mt), the establishment of the first biobased production plants represent key milestones in the agenda of reorienting the chemical sector towards renewables.

2.4 Summary

In this chapter it has been shown that renewables can be used for numerous purposes in energy supply and for material production. Theoretical, geographical, technical and economic potentials are determined for PV, wind and biomass as renewable resources for global energy and material demand. Table 2.1 compares these potentials to the WEA data. Both biomass and wind resources are estimated to have a higher potential in the present study, while for PV this is not the case. Here we would emphasize that several uncertainties exist in the determination of these potentials.

Controversial views exist on the rate of cost reduction of renewable sources, and on their penetration rate. Also, there are large uncertainties about costs and land requirements for large-scale production of commercial biomass energy, and on the interference with food and biomaterials production. The inherent intermittency of solar irradiation may impair the wide use of electricity generated from solar and wind converters until low-cost and reliable storage technologies are available. Large-scale off-shore wind applications would require costly adaptations to the grid, and might suffer from public resistance.

While nature takes care of storage of solar energy by means of photosynthesis in biomass, artificial techniques comprise of, among others, batteries, hydrostorage, flywheels, or compressed air. Thermal storage is used for solar heat. Electricity generated from PV and wind can be used for the generation of hydrogen by means of electrolysis, thereby not only storing the sun's energy, but also producing *true* renewable hydrogen. In a fuel cell hydrogen is used to generate electricity on consumer demand, at the price of lower total chain (sun–electricity–H_2–fuel cell–electricity) efficiency than direct use of PV-generated electricity. In addition, local generation of hydrogen with excess PV electricity could provide self-sufficiency for households in fuelling their electric vehicles.

These new developments ('the hydrogen society') are even more uncertain to predict. Nevertheless, in a future sustainable world it is imperative that mankind benefits from the *mix* of available renewables. Importantly, the transition to such a sustainable future is not merely a matter of cost and technology; also societal developments such as sustainable lifestyles, market development and the balance between decentralized and centralized options are to be addressed. The transition to a sustainable world is *the* challenge of the twenty-first century.

References

Alsema, E. A. and Van Brummelen, M. (1993) *The Potential of PV Systems in OECD Countries* (in Dutch), Utrecht University, Utrecht, the Netherlands.

Berndes, G., Hoogwijk, M. M. and Van den Broek, R. (2003) The contribution of biomass in the future global energy supply: a review of 17 studies, *Biomass and Bioenergy*, **25**, 1–28.

Bucher, K. (1997) Site dependence of the energy collection of PV modules, *Solar Energy*, **47**, 85–94.

Chateau, B., Biberacher, M., Birnbaum, U., Hamacher, T., Lako, P., Martinsen, D., Patel, M., Pospischil, W., Quercia, N. and Smekens, K. (2005) *VLEEM 2 (Very Long Term Energy-Environmental Model)*, ENERDATA, Grenoble, France.

Crank, M., Patel, M., Marscheider-Weidemann, F., Schleich, J., Hüsing, B. and Angerer, G. (2004) *Techno-Economic Feasibility of Large-scale Production of Bio-based Polymers in Europe (PRO-BIP)*, Copernicus Institute, Utrecht University, Utrecht, the Netherlands.

Duffie, J. A. and Beckman, W. A. (1991) *Solar Engineering of Thermal Processes*, Wiley-Interscience, New York.

European Photovoltaic Industry Association (2004) *EPIA Roadmap*, European Photovoltaic Industry Association, Brussels.

European Renewable Energy Council (2004) *Renewable Energy Scenario to 2040*, EREC, Brussels.

European Wind Energy Association (2003) *Wind Energy: The Facts*, European Wind Energy Association, Brussels.

Faaij, A. P. C. (2006) Bio-energy in Europe: changing technology choices, *Energy Policy*, **34**, 322–342.

Fischer, G. and Schrattenholzer, L. (2001) Global bioenergy potentials through 2050, *Biomass and Bioenergy*, **20**, 151–159.

Gielen, D. J., de Feber, M. A. P. C., Bos, A. J. M. and Gerlagh, T. (2001) Biomass for energy or materials?: A Western European systems engineering perspective, *Energy Policy*, **29**, 291–302.

Graßl, H., Kokott, J., Kulessa, M., Luther, J., Nuscheler, F., Sauerborn, R., Schellnhuber, H.-J., Schubert, R. and Schulze, E.-D. (2004) *World in Transition: Towards Sustainable Energy Systems*, German Advisory Council on Global Change (WBGU), Earthscan, London.

Green, M. A., Emery, K., King, D. L., Igari, S. and Warta, W. (2004) Solar cell efficiency tables (Version 24), *Progress in Photovoltaics: Research and Applications*, **12**, 365–372.

Grubb, M. J. and Meyer, N. I. (1993) Wind energy: resources, systems, and regional strategies, in Johansson, T. B., Kelly, H., Reddy, A. K. N. and Williams, R. H. (eds), *Renewable Energy: Sources for Fuels and Electricity*, Island Press, Washington, DC, pp. 157–212.

Hoogwijk, M. M. (2004) On the global and regional potential of renewable energy sources, PhD thesis, Utrecht University, Utrecht, the Netherlands.
Hoogwijk, M. M., de Vries, B. and Turkenburg, W. (2004) Assessment of the global and regional geographical, technical and economic potential of onshore wind energy, *Energy Economics*, **26**, 889–919.
Hoogwijk, M. M., Faaij, A., de Vries, B. and Turkenburg, W. (2006) Global potential of biomass for energy from energy crops under four GHG emission scenarios Part B: the economic potential, *Global Environmental Change*, in press.
Hoogwijk, M. M., Faaij, A., van den Broek, R., Berndes, G., Gielen, D. and Turkenburg, W. (2003) Exploration of the ranges of the global potential of biomass for energy, *Biomass and Bioenergy*, **25**, 119–133.
Hoogwijk, M. M., Faaij, A., Eickhout, B., de Vries, B. and Turkenburg, W. (2005) Potential of biomass energy to 2100, for four IPCC SRES land-use scenarios, *Biomass and Bioenergy*, **29**, 225–257.
Hubbert, M. K. (1971) The energy resources of the earth, *Scientific American*, **225**, 60–84.
IMAGEteam (2001) *The IMAGE2.2 Implementation of the SRES Scenarios: A Comprehensive Analysis of Emissions, Climate Change and Impacts in the 21st Century*, National Institute for Public Health and the Environment (RIVM), Bilthoven, the Netherlands.
International Climate Change Taskforce (2005) *Meeting the Climate Challenge*, Institute for Public Policy Research, London.
Junginger, M., Faaij, A., Koopmans, A., Van den Broek, R. and Hulscher, W. (2001) Setting up fuel supply strategies for large scale bio-energy projects: a methodology for developing countries, *Biomass and Bioenergy*, **21**, 259–275.
Junginger, M., Faaij, A. and Turkenburg, W. C. (2004) Cost reduction prospects for offshore wind farms, *Wind Engineering*, **28**, 97–118.
Junginger, M., Faaij, A. and Turkenburg, W. C. (2005) Global experience curves for wind farms, *Energy Policy*, **33**, 133–150.
Leutz, R., Ackermann, T., Suzuki, A., Akisawa, A. and Kashiwagi, T. (2001) Technical offshore wind energy potentials around the globe, paper presented at European Wind Energy Conference EWEC2001, Copenhagen, Denmark, WIP, Munich, Germany.
Lysen, E. H. (1996) The Trias Energica – Solar energy strategies for developing countries, EuroSun '96, Freiburg, DGS-Sonnenenergie Verlag-GmbH, Munich.
Lysen, E. H. (2001) *Global Restrictions on Biomass Availability for Import to the Netherlands*, Utrecht Centre for Energy Research, Utrecht, the Netherlands.
Martí, A. and Luque, A. (eds) (2004) *Next Generation Photovoltaics, High Efficiency through Full Spectrum Utilization*, Series in Optics and Optoelectronics, Institute of Physics Publishing, Bristol.
Nakicenovic, N. (2000) *Special Report on Emission Scenarios*, Cambridge University Press, Cambridge.
Nakicenovic, N., Grübler, A. and McDonald, A. (1998) *Global Energy Perspectives*, Cambridge University Press, Cambridge.
NASA Atmospheric Sciences Data Center (2005) Available at: eosweb.larc.nasa.gov, NASA Langley Research Center, USA.
New, M., Hulme, M. and Jones, P. D. (1999) Representing twentieth century space–time climate variability. Part 1: development of a 1961–90 mean monthly terrestrial climatology, *Journal of Climate*, **12**, 829–856.
Photovoltaic Technology Research Advisory Council (2004) *A Vision for Photovoltaic Technology for 2030 and Beyond*, European Commission, Brussels.
Schmela, M. (2005) Market survey on global cell and module production in 2004, *Photon International*, **March**, 66–82.
Smeets, E. and Faaij, A. (2004) *A Quickscan of Global Bio-Energy Potentials to 2050: An Analysis of the Regional Availability of Biomass Resources for Export in Relation to Underlying Factors*, Copernicus Institute, Utrecht University, Utrecht, the Netherlands.
Smeets, E. and Faaij, A. Bioenergy potentials from forestry to 2050 *Climatic Change* (in press).
Sørensen, B. (1999) *Long-Term Scenarios for Global Energy Demand and Supply*, Roskilde University, Roskilde, Denmark.
Stainforth, D. A., Aina, T., Christensen, C., Collins, M., Faull, N., Frame, D. J., Kettleborough, J. A., Knight, S., Martin, A., Murphy, J. M., Piani, C., Sexton, D., Smith, L. A., Spicer, R. A., Thorpe, A. J. and Allen, M. R. (2005) Uncertainty in predictions of the climate response to rising levels of greenhouse gases, *Nature*, **433**, 403–406.

UNDP (2000) *World Energy Assessment: Overview*, United Nations Development Programme, United Nations Department of Economic and Social Affairs, World Energy Council, New York.
UNDP (2004) *World Energy Assessment: Overview 2004 Update*, United Nations Development Programme, New York.
Van Wijk, A. J. M. and Coelingh, J. P. (1993) *Wind Power Potential in the OECD Countries*, Utrecht University, Utrecht, The Netherlands.
World Energy Council (2004) *Survey of Energy Resources*, Elsevier, Amsterdam.

3

Sustainability Performance Indicators

Alexei Lapkin

3.1 Introduction

In this chapter we will consider available approaches to assessing the sustainability of new and existing processes or products, and potential approaches to comparing the alternative technologies, such as renewable resource-based technologies. The importance of a reliable method of assessment of technology stems from the need to find an economically viable and socially acceptable path towards a sustainable society. Since its inception in the early 1980s, the concept of sustainable development has undergone a period of maturing of the basic understanding of what *sustainability* implies (the history of development of the term 'sustainability' until the present is well described by Hueting and Reijnders (2004), Laws *et al.* (2004), and Perdan (2004); the historical and philosophical background of the sustainability debate is reviewed by Becker (1997)). It is now widely accepted that the sustainable development concept encompasses three main aspects: (1) ecological balance (including health of natural ecosystems, depleting feedstocks, climate change, etc.); (2) sustained economic stability; and (3) social development and equity, leading to the situation when the activities of current generations do not endanger the opportunities of the future generations (Brundtland, 1987).

The focus of this chapter is on the methods of assessment of *environmental* sustainability of technologies. However, some economic and social aspects will also be discussed. There is a considerable body of research and growing practical experience of the methods of environmental assessment of technology, ranging from very specific narrow methods, targeting a particular problem (e.g., cleaning up of organic syntheses by implementing

Renewables-Based Technology Edited by J. Dewulf and H. Van Langenhove

'green chemistry' solutions (Sheldon, 2000; Curzons *et al.*, 2001), to multi-dimensional indices aiming at capturing the overall concept of sustainability (Sikdar, 2003). Since finding a consensus on the best metrics methodology between all the stakeholders of the sustainable development process is a futile task, in this chapter the very different approaches are structured according to their aims and the user-base, thus providing an overview of the available toolbox of assessment of sustainability of technology.

3.2 The Hierarchy of Sustainability Metrics

Since sustainable development must ultimately be a global effort, it is important that there is a global consensus on the path towards sustainability, the corresponding targets and measures. There should be a continuous debate at the pan-national level about targets for economic development, cultural and social equity, the quality of life and the environmental performance of technological societies, since society, technology, ethics and biosphere are not static. Three major pan-national frameworks of sustainability are currently widely used: (1) the Pressure–State–Response framework, developed by the Organization for Economic Cooperation and Development; (2) the thematic framework developed by the United Nations Commission on Sustainable Development, following the Rio Earth Summit in 1992; and (3) the World Bank's project-oriented framework (see Figure 3.1).

The OECD framework is based on the concept of direct causal links between human activities which generate pressures on the state of the environment, whereas changes in the state of the environment cause societal response in the form of new regulation and policies. In the more recent development of this framework the driving force and impact categories are added to compensate for the apparent oversimplification of the interactions between biosphere, technosphere and society in the linear causality model (Becker, 1997).

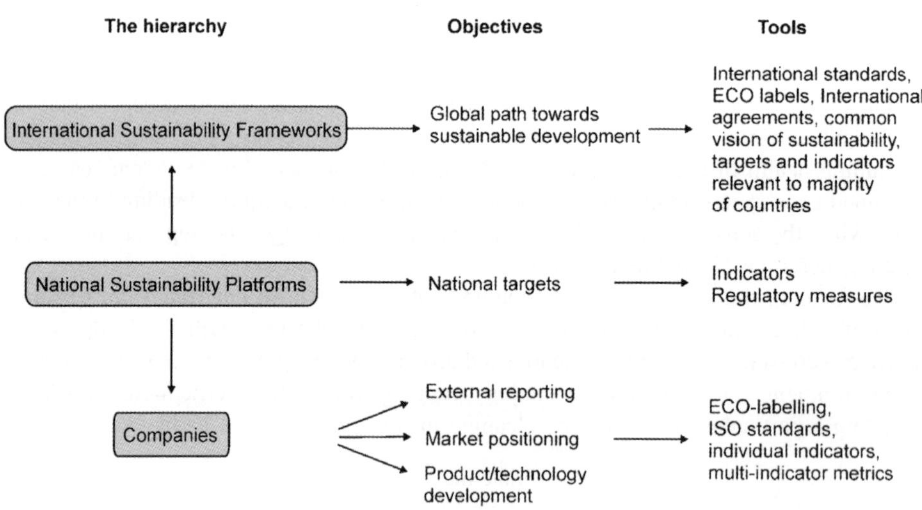

Figure 3.1 A hierarchy of sustainability metrics – from global to local and technology focused players

The framework provides an authoritative set of indicators reflecting the state of the environment, that can be used as a guideline by the national governments.

The UN approach focuses on what is perceived by the majority view of a vision of sustainability, and identifies problem themes, which can later be used at the national and regional levels. In this sense it has two major advantages: (1) since achieving the goal of sustainable development is as much about technology as it is about public engagement and education, it is hugely important to have a shared vision; and (2) the less normative approach provides a framework adaptable to local situations, thus driving the efforts from within the communities (Valentin and Spangenberg, 1999).

The third framework is that developed by the World Bank for evaluating the efficiency of projects aimed at improving sustainability (Segnestam, 2002). This framework allows the assessment of the progress and outcomes of specific projects, such as technology implementation or social infrastructure improvement projects. It is a fairly generic approach, having its roots in good accounting practices. In this framework any project is analysed in terms of inputs, which include all tangible and intangible resources employed within the project, components and outputs, which are the tangible results of the project e.g., services, goods, technology, etc., and outcomes and impacts that measure immediate, mid- and long-term effects of the project's implementation. This framework may potentially be used at the national level for assessment of the effectiveness of investment in science and technology through research and innovation grants.

At a national level, the regulatory frameworks and targets reflect the political reality of each country, however, progress towards global targets is only possible if metrics at the national level is tied up with one or several of accepted pan-national frameworks. Several illustrations of how national agencies use pan-national frameworks of sustainability can be cited. The UK Ministry of Agriculture, Fisheries and Food (now the Department for Environment, Food and Rural Affairs, DEFRA) published a set of indicators of sustainable agriculture (MAFF, 2000), which is based on the OECD's Driving Force–State–Response framework. Implementation of Agenda 21 at the national level can be illustrated using the example of the UK. In 1999 the UK Government published a set of baseline indicators of sustainability which since then have been used to monitor progress annually (DEFRA, 2004). This set of indicators includes 15 headline indicators in the areas of employment, education, health, air quality, etc., and 147 specific indicators, grouped according to issues. There are also 16 additional indicators that are aimed at uncoupling complex interrelations between economic, social and environmental aspects of sustainability. For example, an energy indicator is evaluated as (energy consumption) vs (growth in GDP) as well as (energy consumption) vs (CO_2 emissions).

At the company level, there are at least three goals of sustainability and/or environmental metrics: (1) outside reporting in response to local, national and pan-national regulatory pressures; (2) certification by ISO standards and eco-labelling, leading to better product/process/service marketing; and (3) for internal development. In these three instances the metrics could be different, e.g., a list of indicators corresponding to the national emission and energy efficiency targets for outside reporting, a life cycle assessment (LCA) of key products and processes for ISO certification, and a set of specific technical indicators leading future products development. In order to use the appropriate metrics, it is necessary to align the relevance of specific issues, such as effect of a product/process on the state of the environment, or future costs of emission abatement due to incoming legislation, with the appropriate stakeholders.

Here we utilize the systematic classification of environmental issues, first presented by Graedel (2001), and later adopted for environmental metrics by Lapkin *et al.* (2004). The main stakeholders associated with the societal issues, reflecting the influence of products and processes on society as a whole through their existence in the common environment, are end-users, the population in general, policy-makers, regulators and non-governmental organizations (NGOs). The environmental and/or sustainability issues crossing company and national boundaries, for example, the current practice of trading CO_2 emission quotas, or availability of common strategic feedstocks such as energy, water, land, etc., are related to infrastructure. The relevant stakeholders are company managers, national governments, and pan-national regulators, e.g. the European Commission. The issues relating to the necessary balance between the environmental and economic aspects of products and processes are company-oriented, and the stakeholders are company business managers. Finally, the main stakeholders of the issues of products and processes are scientists and engineers, whose aim is to develop products and processes delivering certain functions characterized in terms of specific performance parameters, i.e. the best possible translation of externally defined function into technical performance characteristics. The function 'brief' is to a significant degree defined by the end-users.

The issues identified and their stakeholders have very different objectives in the use of metrics. It is therefore impractical to develop an indicator or a metric that would be universally applicable. The majority of the methods of assessment of technology can be classified into four types: (1) multidimensional indices that aim to capture all three aspects of sustainability; (2) specific multidimensional indicators that aim to decouple economic-environmental, economic-social and other complex interrelations between sustainability issues; (3) individual indicators in each aspect of sustainability; and (4) the multi-indicator metrics. These methods comprise the 'toolbox' of sustainability metrics.

3.3 Aspects of Methodology

3.3.1 Spatial and Temporal Boundaries of Assessment

The importance of *spatial* boundaries of assessment can be illustrated by the case of hydrogen as 'clean fuel'. Considered within a narrow system boundary of the process itself, also called 'gate-to-gate' boundary, the hydrogen-fed fuel cell has moderate efficiency, but is attractive from the point of view of zero emissions, see Figure 3.2. However, hydrogen is not a source of energy, but its carrier (Shinnar, 2003), and its production requires a substantial amount of energy. Since the only sources of energy for the production of hydrogen that are currently viable are based on non-renewable feedstocks (ca. 75% of hydrogen is currently produced by methane steam reforming), hydrogen does not qualify as a sustainable energy source when considered within the broad, 'cradle-to-gate' system boundaries. Similarly, it has been shown that renewable feedstocks for production of chemicals (as an alternative to the petrochemicals-based supply chain) would lead to more sustainable processes only if no fossil fuels are to be used in the production and transportation of the agricultural produce i.e., in the manufacture of fertilizers, roads, fuel, storage etc. (Curran, 2003). Energy used in transportation was also shown to have a major negative effect on the viability of the end-of-life product re-cycling in the case of mobile

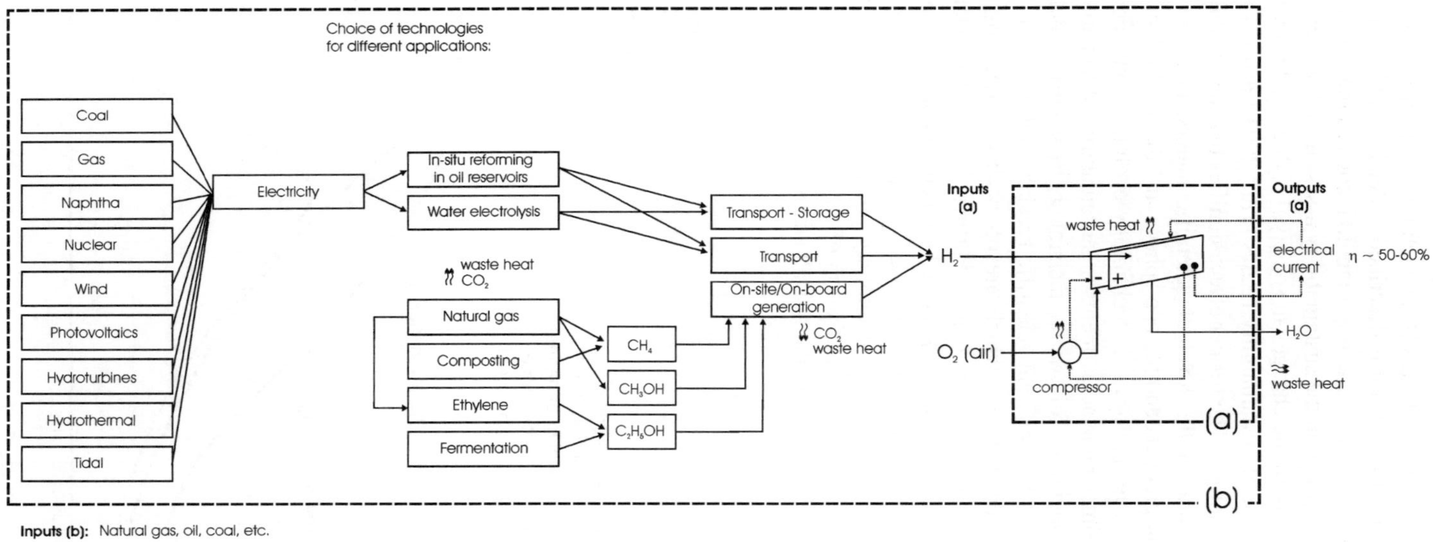

Figure 3.2 *An illustration of spatial boundaries of assessment. An example of hydrogen as an energy carrier: (a) narrow system boundary, gate-to-gate; (b) broad system boundary, cradle-to-gate*

phones (Clift and Wright, 2000). Such problems gave rise to the idea of cradle-to-grave analysis (or 'life-cycle assessment') (LCA), which forms the basis of many indicators and indices. The LCA methodology itself is described in detail elsewhere in this book.

Besides spatial boundaries, it is also important to recognize the transient nature of technology and the effect of time on the sustainability assessment. Two temporal aspects can be identified: (1) relating to the changes in society, i.e. due to education and continuous evolution of the concept of sustainability (Laws *et al.*, 2004); and (2) natural evolution (development) of technology. The former effect signifies that sustainability is a 'moving' target and depends both on the level of scientific understanding of the interactions between technological products and the eco-sphere, and the level of social development. The transient nature of technology, i.e. technology evolution, sets the boundary to the validity of any fixed time-frame assessments of sustainability of technology.

Evolution of any technology follows the technology trend curves such as shown schematically in Figure 3.3. A significant scientific breakthrough leads to the emergence of the new 'pacemaker' technologies which are often characterized by poor efficiency, including poor environmental performance. Further development identifies the 'key' technologies from a variety of 'pacemakers', while their continuous improvement results in a wide acceptance and development into 'base' technologies. This is accompanied by an improvement in the efficiency of technology (environmental performance within a narrow system boundary). However, at the top of an S-curve, the wider adoption of the technology may lead to a decrease in the overall environmental efficiency (calculated for

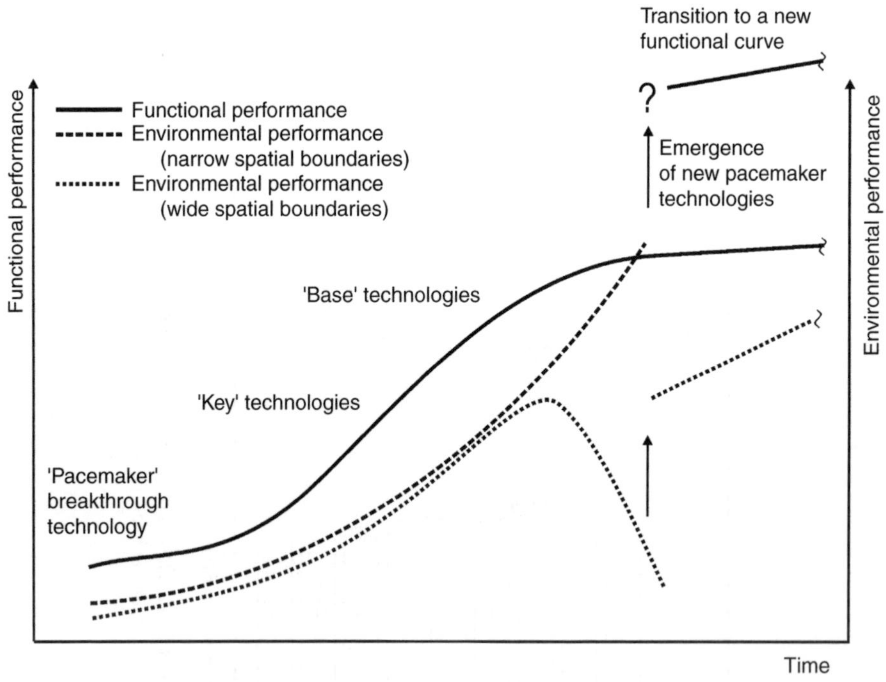

Figure 3.3 Schematic representation of evolution of a single technology

total use, rather than per process or product). The top of an S-shaped curve of the technology evolution trend is indicative of the transition to a new breakthrough technology. One example of such analysis is the gradual evolution of biotechnologies presented by Moser (1994). The evolution of biotechnology has now reached the state of base technology, being used routinely for the production of the ever widening range of chemical intermediates (OECD, 1998).

Bearing in mind the temporal nature of technical systems, it can be argued that a 'static' LCA analysis, i.e. performed at a fixed time-frame in a technology's evolution, can lead to misleading or inconclusive results. Furthermore, if used as a tool for technology development, such analysis would only enable incremental technology improvements, rather than breakthrough solutions. An example is the developments in the window cleaning technology: an LCA is likely to lead to improvements in the performance of washing liquids, an increased use of biodegradable surfactants and renewable feedstocks in the manufacture of surfactants. These are incremental improvements. The technology breakthrough, however, makes all cleaning infrastructure redundant by developing new types of glass surfaces e.g., hydrophobic surface utilizing the Lotus leaves bio-mimetic principle, or a super-hydrophilic surface with the photo-active layer. The breakthrough solutions are the beginnings of new technologies, with their own S-shape trends.

3.3.2 Specific Aspects of Indicator and Indices Development

It is useful to record the differences between indicators, indices and metrics. An indicator is a tool for simplifying, quantifying and communicating information, and is the first level of analysis of the basic data. Because indicators should ideally fulfil the three functions simultaneously, the construction and selection of indicators are not always straightforward and have been a subject of great many publications, see e.g., MacGillivray (1995), Segnestam (2002). The criteria of a good indicator include: technical validity, relevance to stakeholders, manageable cost of collection, reliability, suitable boundaries (spatial and temporal), ease of interpretation, having a standard for comparison, and ability to show trends over time. However, a technically robust indicator may also be difficult to interpret, thus failing in its communication function.

In many instances, evaluation of indicators involves either normalization or comparison with a pre-defined value. Normalization often enables an alternative representation of a very technical indicator. Thus, when an indicator is normalized and expressed as a percentage (e.g., a percentage of sustainable level of emission of a substance) or compared with a pre-defined standard or a target (e.g., a percentage of renewable energy used compared to a national average), the interpretation of such an indicator becomes simpler. Normalized or compared indicators are also more useful for the purposes of monitoring, target setting and eco-labelling. Selection and justification of normalization values (e.g., targets) remain a highly contentious problem.

In order to characterize various aspects of a complex phenomenon, cumulative indices or collection of indicators into a metric are used. Usefulness of a metric depends on the number of indicators: too few may not provide an adequate description of a phenomenon, too many would make the cost of completing the metric prohibitively high. In a comparatively simple case – environmental performance of an organization (this may include the use of various resources e.g., paper, energy, water, as well as emissions), the total number of seven indicators has been recommended (Schmid-Schonbein and Braunschweig, 2000).

Krotscheck and Narodoslawsky (1996) suggested that a distinction should be made between indicators, as tools for the characterization of a certain phenomenon, and measures (or indices), as cumulative numerical parameters suitable for strategic planning. The advantage of a single index, rather than the collection of indicators, is the ease of communication (MacGillivray, 1995). However, there are many disadvantages, such as loss of detail, loss of accuracy due to the combination of parameters with varied accuracy and magnitude, and the need to provide conversion coefficients to express all variables in the same units. Four types of cumulative indices have been proposed based on the area of land, cost, mass and energy.

3.4 Examples of Sustainability Metrics for Technology Assessment

3.4.1 Environmental Sustainability Assessment by Process-Oriented Metrics

One of the roles of metrics is to guide future process and product development. An example of the specific focus on the issue of sustainability in the product/process development is the 'green chemistry' approach, aimed at avoiding emissions and enabling minimization of the energy use at the stage of chemical processes and products design (Anastas and Warner, 1998). The indicators that are being used in the area of green chemistry deal with the resource efficiency of chemical synthesis (atom efficiency (Trost, 1991) and reaction mass intensity (Constable *et al.*, 2001)), energy efficiency of manufacturing processes, use of solvents, inventory of toxic and hazardous compounds, etc. Several companies have developed industry-sector specific indicators, since the problems of different chemical industries, e.g. pharmaceuticals and petrochemials, are quite different. Thus, GlaxoSmithKline developed a set of indicators to guide the development of new routes to complex organic molecules, while adhering to the principles of green chemistry (Constable *et al.*, 2001). This approach is also being used by other companies (Dunn *et al.*, 2004). In the case of bulk chemicals manufacture, energy efficiency can be used for comparison between alternative technologies, as well as for process optimization. One example is the pilot study of the energy efficiency of several chemical processes, which adopted a hierarchy of process improvements from incremental to a complete re-design as a guide to decision-making (CWRT, 2001). As well as individual indicators and collection of indicators, LCA-based methods can be used for comparative assessment of competing technologies.

3.4.2 Environmental Sustainability Assessment by Environmental Pressure-Oriented Metrics

Environmental pressure-oriented metrics are based on the concepts of limited availability of resources that are being depleted as a result of technological development. Such resources are energy, land, clean air, fresh water and non-renewable feedstocks. Several methods of assessment can be classified according to the employed normalization of environmental information: to the area of land, energy or mass, or using normalization to non-physical parameters.

Area of land as a limiting factor is the basis of the ecological footprint sustainability index (Wackernagel and Yount, 1998), and the Sustainable Process Index (SPI) (Krotscheck and Narodoslawsky, 1996; Narodoslawsky and Krotscheck, 2000). The ecological footprint index is based on the concept of Earth's limited carrying capacity: manufacturing processes, agriculture, recreation, preservation of biodiversity, biological mechanisms of assimilation of waste, etc., all require a certain area of land. Comparison of the area of land per capita used within a particular region with (1) the available biotically productive area per capita within the same region, or (2) with the world average allows us to qualify the extent to which a particular region is self-sustainable and supporting other regions. The main advantage of the ecological footprint index is the simplicity and power of its message: many industrialized societies are living at the expense of poorer nations (Wackernagel and Yount, 1998).

The SPI is the translation of the ecological footprint concept to the problem of assessing technologies. Any process considered within the broad – cradle-to-grave – system boundary requires energy, transport, feedstocks, which can be expressed in terms of the area of space, specific to the energy generation technology and the type of feedstocks. Other space requirements include manufacturing facilities, human habitat (including recreation and waste disposal), and systems for industrial and natural processes for treating waste and emissions, etc., which can be normalized to the required land area. The SPI is then a fraction of total surface area available for one person within one year, which is required for delivery of a service or manufacture of a product.

The index is sensitive to the specific geographical location and some aspects of social development. Arguably, the main strength of the SPI method of assessment is its ability to account for specific local conditions: social as well as environmental. Difficulties arise from the need to develop a large number of conversion coefficients, as well as the need to define such standards as the required area of land for habitat, etc. In the case of defining the area of sustainable assimilation of chemicals, the complexity of the network of interactions between the components of natural ecosystems and of the fate of chemicals in the environment is such that establishment of quantitative exposure limits with any degree of accuracy is problematic. However, this approach may be particularly relevant for the assessment of technologies based on renewable raw materials and energy resources. Production of biomass for conversion into chemical products and/or energy requires fertile arable land, thereby creating a competition between land for food production and land for energy and chemicals production, as well as increasing demand for fertilizers. A case study of ethanol production from sugar beet for energy generation is presented by Krotscheck and Narodoslawsky (1996). The analysis of all contributions to the total required area of land showed that the two most important contributions are production of beet crop (20%) and sustainable dissipation of by-products (63%). It is very likely that a much better performance in terms of lower SPI could be obtained if all by-products were converted into benign emissions i.e., water and carbon dioxide.

Since all technological processes depend on the use of energy, it is a convenient normalizing parameter. Production can be defined as 'anything that happens to an object or set of objects that increases its value... The basic physical condition necessary (to effect any such changes) is that energy must be applied' (Chenery, 1953). Because most energy is currently produced from depleting resources, there is a natural limit to how much energy can be produced. Therefore, less energy-intensive processes are more sustainable. A comprehensive analysis of the contributions of energy to transport, the manufacturing process itself, support of labour, and production of energy is given in Slesser (1979) and recent developments of the energy analysis method are described in Slesser *et al.* (1997).

Based on the analysis of the exergy flows in the technosphere (where exergy is the amount of energy that can be converted to work in an open system), a thermodynamically sound sustainability criterion contrasting the use of renewable to non-renewable energy sources and process efficiency was developed (Dewulf *et al.*, 2000). The exergy-based sustainability index does not account for any aspects of social development. However, it is able to reflect the inevitable loss of energy to entropy, and has the potential to reflect a more complex phenomena, such as the integrated material and energy flows within a highly structured hierarchy of industrial ecology (a trivial point would be the possibility of accounting for energy demand across process/product system boundaries, which is a variation on the Life Cycle Inventory; a non-trivial point is that any organized system requires additional energy to sustain the organization itself, the area where further development based on the principles of non-equilibrium thermodynamics is called for). The most significant drawback of the energy-based indices, despite their technical accuracy, is the difficulty in interpretation outside the science and technology communities. Further analysis and comparison of highly aggregated indices can be found in several earlier reviews (Krotscheck, 1997; Hertwic *et al.*, 1997; Brunner and Starkl, 2004; Wrisberg *et al.*, 2002).

Material intensity indicators are often used to highlight the depletion of non-renewable feedstocks. The material intensity per unit service (MIPS) index (Schmidt-Bleek, 1993) reveals the resource efficiency of a process, although it does not take into account the stages of product use and afterlife, nor waste disposal and thus can easily lead to erroneous conclusions if applied inappropriately.

A series of methods based on life cycle assessment have been developed, converting physical measures of environmental damage, such as energy and resource intensity, emissions, etc. in the form of non-physical parameters. The two most important methods are the Eco-Indicator 95 and 99, and the Eco-points method. In both cases the data from either broad (in the case of Eco-indicator) or narrow (in the case of Eco-points) life-cycle inventory for a product of a process are compared against a set of targets, the main difference being the process of setting the targets. In the case of the Eco-indicator, the targets correspond to the lowest allowed environmental or health damage based on agreed criteria e.g., percentage degradation of an ecosystem over given number of years, or number of deaths per million of population per year. In the case of the Eco-points method, the targets are set by the government as industry targets. In both cases, the result of the analysis is a single number.

In addition to the overall indexes of sustainability described above, there are a number of approaches to decoupling the interrelations between different aspects of sustainability through use of multidimensional indicators, see Figure 3.4 illustrating the concept of single-issue and multidimensional indicators. The economic vs environmental boundary of this problem is dealt with more frequently than anything else, since it is of huge

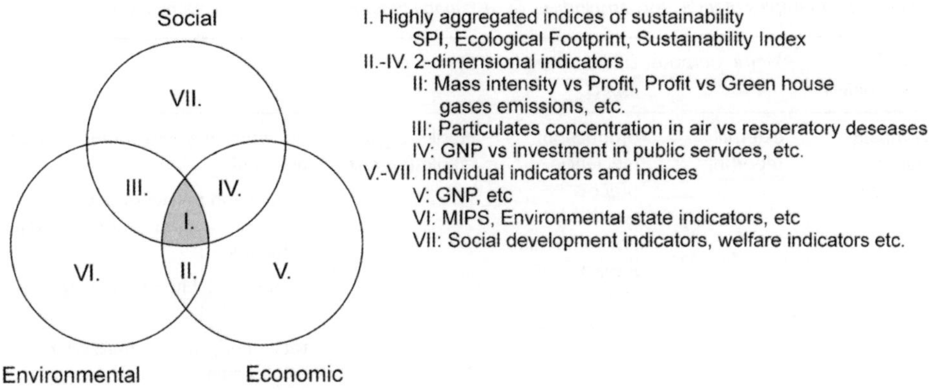

Figure 3.4 *Illustration of different approaches to sustainability metrics*
Source: Reproduced from Moser (1994). *Acta Biotechnol.*, 14, p. 315–335 with permission from Wiley-VCH.

importance for the development of new clean processes to ensure that such processes are also economically viable.

3.4.3 Environmental-Economic Sustainability Assessment

One of the major barriers to the implementation of new, more sustainable, technologies is the entrenched opinion that all such technologies are inevitably less economically attractive than conventional technologies (Crystal, 2003). Therefore, there is a particular need for environmental vs economic analysis. A number of approaches decoupling the economic and environmental aspects of sustainability were developed specifically for screening of new technologies. Thus, a three-dimensional or a spider diagram could be built based on the indicators of reaction mass efficiency, the overall mass efficiency (using e.g., MIPS index) and cost (Heinzle *et al.*, 1998). Combining the mass intensity index with a cost indicator, total annualized profit per service unit (TAPPS), enables the plotting of the economic process performance against its material intensity (Hoffmann *et al.*, 2001).

The economic vs environmental balance is further represented in the concepts of eco-efficiency and environmental cost accounting. The eco-efficiency concept is largely promoted by the business community with a significant contribution by the World Business Council for Sustainable Development (Verfaille and Bidwell, 2000). The basic idea of the eco-efficiency concept is to break the trend that has persisted until now, that economic growth and improvement in the welfare are directly linked to the increase in the environmental degradation (AniteSystems, 1999). Eco-efficiency is defined as the ratio of economic output to environmental influence, the latter being assessed using an appropriate set of indicators, such as shown in Table 3.1.

Environmental cost accounting is a more complex concept, when damage to the environment due to industrial activity is evaluated in monetary terms. The attractiveness of such an approach is due to its direct link with the well-developed economic tools of assessment and management. At the level of processes or products, the environmental cost accounting methods require the assigning of monetary values to specific environmental

Table 3.1 *Examples of indicators and indices of sustainability applicable at the company level*

Level/ stakeholders	*Metric purpose*	*Environmental aspect*	*System boundary*	*Possible indicators*
company/ business managers	external reporting	production efficiency, process economics vs environmental impact	cradle-to-gate or cradle-to-grave	eco-efficiency, environmental cost accounting production efficiency (e.g. kg product per annum) normalized to value added total material input normalized to value added total energy use normalized to value added total green house emissions as % of Kyoto quota
	better marketing	energy and material efficiency responsible manufacturing and agriculture practices		ISO standardization, eco-labels
products and services/ scientists and engineers	product, process or service development	resource efficiency	gate-to-gate	reaction mass intensity (RMI) $\frac{\text{kg raw materilas}}{\text{kg product}}$ $\frac{\text{kg net water consumed}}{\text{kg product}}$ $\frac{\text{kg solvent loss}}{\text{kg solvent used in the process}}$ $\frac{\text{kg product recyclable}}{\text{kg product}}$ $\frac{\text{kg recycled feed}}{\text{kg feed}}$
		energy efficiency		total energy required for a process/ product normalized to the theoretical minimum
		toxicity to human life and the environment		eco-toxicity to aquatic life photochemical ozone formation human health, exposure limits and indices aquatic oxygen demand aquatic acidification eutrophication

damages. Such assignment is highly problematic, since money is an arbitrary concept and has no relation to the physical reality. Furthermore, there is a powerful moral argument against monetary methods of assessment.

The economic vs environmental argument is also the main focus of the two industry-oriented metrics published recently by the Centre for Waste Reduction Technologies (CWRT) at AIChE and by the Institute of Chemical Engineers (IChemE, 2003). The main aim of the IChemE and CWRT metrics is year-on-year internal company assessment as a guide towards the improving of its existing processes and products. The CWRT metric is developed on the basis of the eco-efficiency concept and the indicators introduced by the World Business Council for Sustainable Development. The important aspects of the CWRT metrics are: (1) normalization to sales monetary value or value added; and (2) consideration of relative environmental impacts of different pollutants. The main categories included in CWRT metrics are: energy, mass, water usage, pollutant, human health and eco-toxicity. Indicators of human health and eco-toxicity are based on the parameters already widely used in the assessment of chemical hazards, i.e. permissible exposure limits and 50% of lethal concentration. The indicators also take into account the lifetime of chemical pollutants in various media of the environment.

The IChemE metrics of sustainability consist of 49 indicators classified into three main categories: economic, environmental and social (IChemE, 2003). The IChemE metrics include the area of land as an environmental indicator. The actual indicators are: (1) the sum of directly occupied and affected land per value added; and (2) the rate of land restoration. Other differences relate to the assessment of the relative impacts of pollutants on the environment and human health. The IChemE indicators do not take into account the lifetime of chemicals in various media of the environment. The human health indicator is limited to carcinogenic effects and is normalized to benzene.

3.5 Summary

This chapter provided a broad overview of the topic of assessment of sustainability of industrial processes, products and services. The framework developed in this work starts with the pan-national programmes on sustainability, which define problems and themes for specific projects, or develop sets of indicators for use on a national level. These large programmes promote wide discussions of the topics of sustainability, which are essential if society as a whole is to achieve reasonable progress in this area. A very different problem is the assessment of technologies that is undertaken by companies. Regulatory drivers lead to the use of 'defensive' indicators, which usually correspond to the pressure indicators developed at a pan-national level, whereas commercial drivers promote the use of eco-labelling, a more pro-active approach to developing better processes and products. This chapter also focused on the problems of spatial and temporal boundaries of environmental assessment, largely adopting a functional approach to technology with technical evolution. Although the functional approach has its limitations, it enables one to take into account technology breakthroughs. Finally, a considerable amount of literature and a significant international effort have led to the effective discussion of the topic of sustainability metrics. The next few years should see the convergence of many available methods and indicators through user selection to a small set of widely recognized approaches, that would require the development of new international standards.

References

Anastas, P. T. and Warner, J. C. (1998) *Green Chemistry: Theory and Practice*, Oxford University Press, Oxford.

AniteSystems (1999) A first set of eco-efficiency indicators for industry: pilot study. Eco-efficiency Indicators Project. Available at: http://europa.eu.int/comm/enterprise/environment/reprots_studies/studies/anite_final.pdf

Becker, B. (1997) Sustainability assessment: a review of values, concepts, and methodological approaches, *Issues in Agriculture 10*, The World Bank, Washington, DC.

Brundtland, G. H. (1987) *Our Common Future: Report of the World Commission on Environment and Development*, United Nations, New York.

Brunner, N. and Starkl, M. (2004) Decision aid systems for evaluating sustainability: a critical survey, *Environmental Impact Assessment Review*, **24**, 441–469.

Chenery, H. B. (1953) Process and production functions from engineering data, in Leontieff, W. W. (ed.), *Studies in Structure in the American Economy*, Oxford University Press, Oxford.

Clift, R. and Wright, L. (2000) Relationships between environmental impacts and added value along the supply chain, *Technological Forecasting and Social Change*, **65**, 281–295.

Constable, D. J. C., Curzons, A. D., Santos, L. M. F. D., Geen, G. R., Hannah, R. E., Hayler, J. D., Kitteringham, J., McGuire, M. A., Richardson, J. E., Smith, P., Webb, R. L. and Yu, M. (2001) Green chemistry measures for process research and development, *Green Chemistry*, **3**, 7–9.

Crystal (2003) *Facilitating the Uptake of Green Chemical Technologies*, Report by Crystal Faraday Ltd (UK), Keighley.

Curran, M. A. (2003) Do bio-based products move us towards sustainability? A look at three USEPA case studies, *Environmental Progress*, **22**, 277–292.

Curzons, A. D., Constable, D. J. C., Mortimer, D. N. and Cunnigham, V. L. (2001) So you think your process is green, how do you know? Using principles of sustainability to determine what is green: a corporate perspective, *Green Chemistry*, **3**, 1–6.

CWRT (2001) *A Pilot Study of Energy Performance Levels for the U.S. Chemical Industry*, BRIDGES to Sustainability, Houston, TX.

DEFRA (2004) *Quality of Life Counts: Indicators for a Strategy for Sustainable Development for the UK*, Office of National Statistics, London.

Dewulf, J., Langenhove, H. V., Mulder, J., Berg, M. M. D., Kooi, H. J. V. D. and Arons, J. D. S. (2000) Illustrations towards quantifying the sustainability of technology, *Green Chemistry*, **2**, 108–114.

Dunn, P. J., Galvin, S. and Hettenbach, K. (2004) The development of an environmentally benign synthesis of sildenafil citrate (Viagra™) and its assessment by Green Chemistry metrics, *Green Chemistry*, **6**, 43–48.

Graedel, T. E. (2001) Green chemistry as systems science, *Pure and Applied Sciences*, **73**, 1243–1246.

Heinzle, E., Weirich, D., Brogli, F., Hoffmann, V. H., Koller, G., Verduyn, M. A. and Hungerbuhler, K. (1998) Ecological and economic objective functions for screening in integrated development of fine chemical processes. 1. Flexible and expandable framework using indices, *Industrial and Engineering Chemistry Research*, **37**, 3395–3407.

Hertwic, E. G., Pease, W. S. and Koshland, C. P. (1997) Evaluating the environmental impact of products and production processes: a comparison of six methods, *The Science of Total Environment*, **196**, 13–29.

Hoffmann, V. H., Hungerbuhler, K. and Mcrae, G. J. (2001) Multiobjective screening and evaluation of chemical process technologies, *Industrial and Engineering Chemistry Research*, **40**, 4513–4524.

Hueting, R. and Reijnders, L. (2004) Broad sustainability contra sustainability: the proper construction of sustainability indicators, *Ecological Economics*, **50**, 249–260.

IChemE (2003) *The Sustainability Metrics*. Rugby, The Institute of Chemical Engineers.

Krotscheck, C. (1997) Measuring eco-sustainability: comparison of mass and/or energy flow based highly aggregated indicators, *Environmetrics*, **8**, 661–681.

Krotscheck, C. and Narodoslawsky, M. (1996) The sustainable process index: a new dimension in ecological evaluation, *Ecological Engineering*, **6**, 241–258.

Lapkin, A., Joyce, L. and Crittenden, B. (2004) A framework for evaluating the 'greenness' of chemical processes: case studies for a novel VOC recovery technology, *Environmental Science and Technology*, **38**, 5815–5823.

Laws, D., Scholz, R. W., Shiroyama, H., Susskind, L., Suzuki, T. and Weber, O. (2004) Expert views on sustainability and technology implementation, *International Journal of Sustainable Development and World Ecology*, **11**, 247–261.

MacGillivray, A. (1995) Indicators indicators everywhere, in *Accounting for Change*. Toynbee Hall, New Economics Foundation, London.

MAFF (2000) *Towards Sustainable Agriculture: Pilot Set of Indicators*, Ministery of Agriculture, Fisheries and Food, London.

Moser, A. (1994) Trends in biotechnology: sustainable technology development: from high tech to eco tech, *Acta Biotechnologica*, **14**, 315–335.

Narodoslawsky, M. and Krotscheck, C. (2000) Integrated ecological optimization of processes with the sustainable process index, *Waste Management*, **20**, 599–603.

OECD (1998) *Biotechnology for Clean Industrial Products and Processes: Towards Industrial Sustainability*, OECD, Working Party on Biotechnology, Paris.

Perdan, S. (2004) Introduction to sustainable development, in Azapagic, A., Perdan, S. and Clift, R. (eds), *Sustainable Development in Practice*, John Wiley and Sons, Ltd., Chichester.

Schmid-Schonbein, O. and Braunschweig, A. (2000) *Environmental Performance Indicators for the Financial Industry*, EPI-Finance, Zurich.

Schmidt-Bleek, F. (1993) MIPS re-visited, *Fresenius Environmental Bulletin*, **2**, 407–412.

Segnestam, L. (2002) *Indicators of Environment and Sustainable Development: Theories and Practical Experience, Environmental Economics Series*, The World Bank, Washington, DC.

Sheldon, R. A. (2000) Atom utilisation, E factors and the catalytic solution, *C.R. Acad. Sci. Paris, Serie IIc, Chimie*, **3**, 541–551.

Shinnar, R. (2003) The hydrogen economy, fuel cells and electric cars, *Technology in Society*, **25**, 455–476.

Sikdar, S. K. (2003) Sustainable development and sustainability metrics, *AIChE Journal*, **49**, 1928–1932.

Slesser, M. (1979) *Energy in the Economy*, Macmillan, London.

Slesser, M., King, J. and Crane, D. C. (1997) *The Management of Greed*, Resource Use Institute, Ltd., Dunblane.

Trost, B. M. (1991) The atom economy – a search for synthetic efficiency, *Science*, **254**, 1471–1477.

Valentin, A. and Spangenberg, J. H. (1999) *Indicators for Sustainable Communities*, Wuppertal Institute, Wuppertal.

Verfaille, H. A. and Bidwell, R. (2000) *Measuring Eco-Efficiency: A Guide to Reporting Company Performance*, World Business Council for Sustainable Development, Geneva.

Wackernagel, M. and Yount, J. D. (1998) The ecological footprint: an indicator of progress towards regional sustainability, *Environmental Monitoring and Assessment*, **51**, 511–529.

Wrisberg, N., Haes, H. A. U. D., Bilitewski, B., Bringezu, S., Bro-Rasmussen, F., Clift, R., Eder, P., Ekins, P., Frischknecht, R. and Triebswetter, U. (2002) Demand and supply of environmental information: analytical tools, in Wrisberg, N., Haes, H. A. U. D., Triebswetter, U., Eder, P. and Clift, R. (eds), *Analytical Tools for Environmental Design and Management in a Systems Perspective*, London, Kluwer, Dordrecht, pp. 45–73.

Part II
Relevant Assessment Tools

4

Life Cycle Inventory Analysis Applied to Renewable Resources

Niels Jungbluth and Rolf Frischknecht

4.1 Introduction

A comprehensive environmental assessment of products and services requires solid data on energy requirements, semi-finished products inputs, transport and waste management service requirements and emissions to air, water and soil of all the individual industrial processes that are necessary to deliver the product or provide the service. The compilation of these data and the modelling of all individual processes are performed in the so-called life cycle inventory (LCI) analysis. In that sense, life cycle inventories form the backbone of many different analytical methods that apply a life cycle thinking approach such as net energy balancing (NEB) or the MIPS concept (material intensity per service unit). At the level of individual processes, data requirements of the life cycle inventory analysis as defined in the context of life cycle assessment and described in this chapter has similarities with the data requirements of other methods such as material and substance flow analysis or exergy analysis. However, the methodology of life cycle inventory analysis has been discussed and described most extensively in the context of life cycle assessment (LCA).

In the past, some authors have made a distinction between a life cycle inventory and an eco balance (Van Berkel *et al.*, 1997). These authors make the distinction between a product-oriented (per functional unit of product, life cycle inventory) and a process-oriented (per physical unit of production, eco-balance) approach. The coverage of life cycle inventory analysis and eco-balance is the same and there is no difference in the procedure nor in modelling approaches and methodological challenges between the standardized

Renewables-Based Technology Edited by J. Dewulf and H. Van Langenhove

life cycle inventory analysis (including goal and scope definition) and the eco-balance as described in van Berkel *et al.* (1997). We assume that the English term 'eco-balance' or 'ecobalance' is derived from the German 'Ökobilanz', which in turn is the German synonym for life cycle assessment. The LCA aims at investigating and comparing environmental impacts of products or services that occur from cradle to grave. This means the whole lifecycle from resource extraction to final waste treatment is investigated.

The International Organization for Standardization (ISO) (1997–2000) has standardized the basic principles. The following description is based on the ISO standards 14040 on general principles and 14041 on goal and scope definition and inventory analysis (International Organization for Standardization 1997, 1998) and the most recent and complete LCA guide published by Guinée *et al.* (2001a, 2001b).

4.2 Conceptual Background to LCA in ISO 14040ff

LCA studies systematically and adequately address the environmental aspects of product systems, from raw material acquisition to final disposal (from 'cradle to grave'). Hence, the analysis normally includes also the life cycle of all semi-finished products, services and energy carriers used. Many kinds of environmental interventions, e.g. emissions into water, air and soil as well as resource uses (primary energy carriers, land, etc.) are accounted for. Some authors also include additional effects, e.g. the direct health hazards for employees in the production facilities.

Normally LCA aims at analyzing and comparing different products, processes or services that fulfil the same utility (e.g. one kg of synthetic ethanol against one kg of ethanol from sugar beets, or different technologies to produce one kg of synthetic ethanol). The ISO standards are not mandatory in any way for conducting LCA studies. Many published studies deviate in certain points from this procedure. However, following the guidelines of these standards as far as possible is recommended in order to increase the credibility of LCA studies. This is especially important if such studies are disclosed to the public and contain comparative assertions of competing products.

The method distinguishes four main phases, namely, (1) goal and scope definition; (2) inventory analysis; (3) impact assessment; and (4) interpretation. The goal and scope definition describes the underlying questions, the system boundaries and the definition of a functional unit for the comparison of different alternatives. Furthermore, the environment to be protected is defined by selecting impact categories or impact assessment methods. The inputs of resources, materials and energy as well as the outputs of products and emissions are investigated and recorded and the inventory analysis results are computed in the inventory analysis. The life cycle inventory results (cumulative emissions and resource consumptions) are classified, characterized and aggregated during the impact assessment. Final conclusions are made in the interpretation phase. This chapter focuses on the first two steps, 'goal and scope definition' and 'lifecycle inventory analysis', and within the goal and scope definition on the aspects relevant for the inventory analysis. Chapter 6 in particular focuses on the overall life cycle analysis, paying attention predominantly to the impact assessment and interpretation stages.

4.3 Goal and Scope Definition

The ISO 14041 (International Organization for Standardization (ISO), 1998) describes the procedure for the goal and scope definition. Some key aspects are described in the following section. The goal of an LCA study shall unambiguously describe the intended application, the reasons for carrying out the study and the intended audience, i.e. to whom the results of the study are intended to be communicated (ibid.). An important question not treated in the ISO standards is whether the study should evaluate the environmental impacts of an existing system (attributional LCA) or if consequences due to the change of production patterns should be analysed (see Section 4.6 for a further explanation).

It should be clearly specified which functions the system under study has. Goods or services are defined as a functional output. The functional unit is a measure of the performance of the functional output of the product system. The purpose of the functional unit is to provide a reference, quantified in terms of the reference flow, to which the inputs and outputs of the product system are related (ibid.).

The system boundaries define the unit processes to be included in the system to be modelled. The analysis of technical processes required to manufacture products and deliver services are based on pure environmental process chain analysis. In many cases there will not be sufficient time, data, or resources to conduct a fully comprehensive study. According to ISO 14041 (ISO, 1998) several criteria are used to decide which inputs to be studied, including: (1) mass; (2) energy; and (3) environmental relevance. Any decisions to omit life cycle stages, processes or inputs/outputs should be clearly stated and justified. The criteria used in setting the system boundaries dictate the degree of confidence in ensuring that the results of the study have not been compromised and that the goal of the study will be met.

One important question for agricultural products is the definition of system boundaries between the technosphere system (agricultural production) and nature (e.g. agricultural soil and water). Here it has to be clearly defined which part of soil and water belongs to the technical system and which to the natural system.

4.4 Inventory Analysis

The second stage of an LCA is the life cycle inventory analysis standardized in ISO 14041 (ISO, 1998). This involves the data collection and calculation procedures to quantify relevant inputs and outputs of a product system. An intermediate result of an LCA is the life cycle inventory analysis result with data on the emission of hundreds of individual substances and on many resource uses caused during the complete life cycle of the product at issue. These data constitute the input to the life cycle impact assessment (International Organization for Standardization (ISO), 1997).

Normally, data investigation is the most time-consuming step of an LCA. In the past few years, the situation has been continuously improving due to the set-up of standardized background databases (e.g. ecoinvent Centre 2004) and LCA software products (Ifu and Ifeu, 2004; IKP and PE Europe, 2004; PRé Consultants, 2004) that include these background data.

The agricultural production stage is more difficult to model in LCA than technical systems such as, for example, coal power plants due to a number of specific methodological problems and less frequent measurements of emissions. A cow produces, for example, milk, meat, fertilizer, leather, etc. and it is difficult to assign or allocate emissions due to fodder production to the single products. Agricultural products are produced by thousands of producers while technical products are often produced in a few facilities. Thus it is difficult to determine the average production parameters such as fertilizer use for so many actors, because they are not monitored sufficiently (Cowell *et al.*, 1999).

The following sections will look further into some challenges of the life cycle inventory analysis.

4.4.1 Product System and Unit Process

The unit process describes the smallest portion of a product system for which data are collected when performing an LCA. Figure 4.1 shows the main processes of the product system for the transport service with a truck that uses ethanol. The product system is divided into unit processes, e.g. potatoes production or fermentation to ethanol, in order to facilitate and structure the further analysis.

4.4.2 Unit Process Inventory

The unit process inventory is an inventory of energy and material flows (inputs and outputs) that are used and emitted by an unit process. It is also termed as unit process raw data. There are two classes of inputs and outputs: technosphere flows and elementary

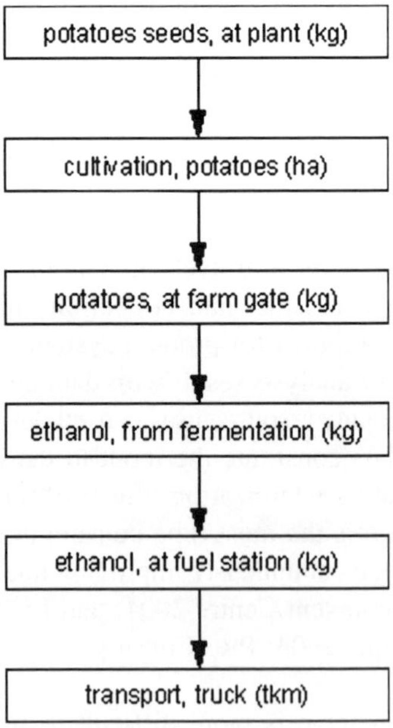

Figure 4.1 *Product system from well to wheel of a truck fuelled with ethanol from potatoes*

flows. Technosphere flows take place between different processes which are controlled by humans, e.g. the delivery of ethanol from the plant to the fuel station. They can be physical inputs (e.g. fertilizer or seeds) or physical outputs (e.g. the product or wastes that have to be treated). Elementary flows in this context are all emissions of substances to the environment (outputs) and resource uses (inputs, e.g. of water or land). An emission is a single output of a technical process to the environment, e.g. the emission of one kg SO_2. Figure 4.2 shows the unit process for potatoes cultivation as an example. Potato seeds are the direct input, potatoes are the major output of this unit process. Also further inputs, e.g. fertilizer, machinery hours or pesticides are necessary. The unit process causes also some emissions, e.g. pesticides to water or N_2O to air.

Table 4.1 shows some unit process raw data for the production of 1 kg of potatoes in Switzerland with integrated production technology (excerpt from Nemecek *et al.*, 2004). Only a part of the recorded 67 inputs and outputs is shown in Table 4.1. One can first see some examples for the input of fertilizers (ammonium nitrate, as N, at regional storehouse), pesticides ([sulfonyl]urea-compounds, at the regional storehouse) and transport services (transport, lorry 28t). These technosphere inputs are linked to similar tables for these unit processes. Then resource uses of carbon dioxide and land are recorded (input flows from nature). Emissions are distinguished according to the compartments (air, water, soil) and sub-compartments (such as industrial soil or agricultural soil). Finally, the technosphere output (reference flow quantifying the functional unit) of the process is defined as 1 kg of potatoes from integrated production (IP) in Switzerland. The integrated production is an agricultural technique in which biological, technical and chemical methods are balanced carefully, taking into account the protection of the environment, profitability and social requirements.

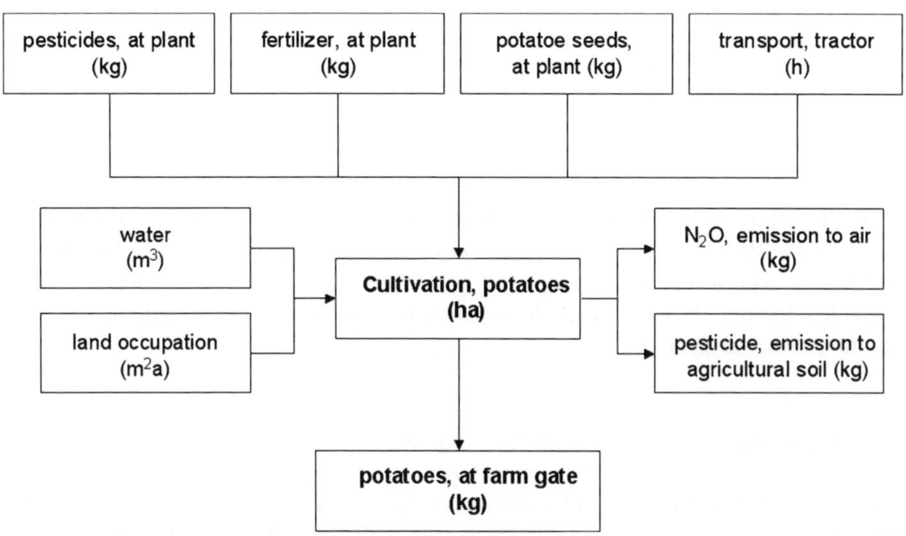

Figure 4.2 The unit process of the cultivation of potatoes and selected inputs from technosphere and from nature and selected emissions

Table 4.1 *Selected unit process raw data of the production of 1 kg of potatoes in Switzerland with integrated production technology*

Inputs and outputs	*Name*	*Unit*	*Potatoes IP, at farm/CH kg*	σ_g^2
inputs				
technosphere input	ammonium nitrate, as N, at regional storehouse/RER	kg	4.35E-04	1.07
	[sulfonyl]urea-compounds, at regional storehouse/CH	kg	2.69E-07	1.13
	potato seed IP, at regional storehouse/CH	kg	0.0678	1.07
	fertilizing, by broadcaster/CH	ha	8.08E-05	1.07
	harvesting, by complete harvester, potatoes/CH	ha	2.69E-05	1.07
	transport, lorry 28t/CH	tkm	0.00157	2.71
resource, in air	carbon dioxide, in air	kg	0.342	1.07
resource, biotic	energy, gross calorific value, in biomass	MJ	3.87	1.07
resource, land	occupation, arable, non-irrigated	m^2a	0.127	1.77
	transformation, from arable, non-irrigated	m^2	0.269	2.67
	transformation, to arable, non-irrigated	m^2	0.269	2.67
outputs				
emission to air, low population density	ammonia	kg	4.36E-04	1.3
	dinitrogen monoxide	kg	1.29E-04	1.61
emission to agricultural soil	cadmium	kg	2.62E-08	1.77
	chlorothalonil	kg	8.83E-05	1.32
emission to groundwater	nitrate	kg	0.00936	1.77
	phosphate	kg	3.06E-06	1.77
emission to river	phosphate	kg	1.06E-05	1.77
product output	potatoes IP, at farm/CH	kg	1.0	

Source: Nemecek *et al.* (2004).

This inventory table provides also information on the uncertainty for the recorded amount of the flows. In this case the uncertainty distribution is lognormal. The standard deviation σ_g^2 records the square value for the 95% percentile. The mean value multiplied or divided by the 95% standard deviation gives the 97.5% maximum or the 2.5% minimum value, respectively.

4.4.3 Multi-Output Processes and Allocation Rules

Some processes not only have one single technosphere output, but several outputs which can have different uses in the technosphere. The seeding of wheat on an agricultural area leads to two products: wheat grains and wheat straw. During one year 6420 kg grain and 3910 kg straw are produced in Switzerland per hectare (Nemecek *et al.*, 2004).

Multi-output processes are ubiquitous in LCA product systems. They are present in the energy industry (e.g., combined oil and gas production, oil refineries, combined heat and power production), in the mining industry (e.g., platinum group metals), in agriculture (e.g. production of wheat and straw), in the chemical industry (e.g., phosphoric acid production), in forestry (e.g., sawing of timber) or in the electronics industry (silicon purification).

Principles According to ISO 14041

The environmental impacts of the multi-output process have to be shared between the different products (allocation). In the LCI of multi-output processes the following stepwise procedure should be applied according to ISO 14041 (ISO, 1998):

- wherever possible, allocation should be avoided by dividing the subprocesses to be allocated into two or more subprocesses and collecting the data related to these subprocesses or,
- expanding the product system to include the additional functions related to the co-products.The ISO standard does not specify how the system expansion has to be made and two possibilities can be distinguished:
  - subtraction of avoided burdens for the co-products which are not of interest for the study at hand (avoided burden approach);
  - expansion of the functional unit in order to include more benefits of the systems under investigation into the analysis (basket of benefits approach).

In principle, there are four possibilities for the choice of the additional products or services included in the system (see Section 4.6):

- Average products can be assumed to be included for determining further functions of the system (attributional LCA), or
- The system expansion is based on the principle that marginal products are identified and included in the production system (consequential LCA).
- If allocation cannot be avoided, the inputs and outputs of the system should be partitioned between its different products or functions in a way that reflects the underlying physical relationships between them.
- If a physical relationship cannot be established, the inputs should be allocated between the products and functions in a way that reflects other relationships between them. For example, input and output data might be allocated in proportion to the economic value of the products.

Whenever several alternative allocation procedures seem applicable, a sensitivity analysis should be conducted to illustrate the consequences of the departure from the selected approach.

Avoiding Allocation by Further Detailing the Analysis

Most farms produce more than just one product. They may produce corn, milk and meat, vegetables, etc. If only inputs and outputs at the level of the entire farm are known, an allocation key is required to attribute emissions and energy and working materials inputs to the various products. To avoid this allocation step, one could further analyse the energy and material flows of the farm and identify processes and activities that are only required for cattle breeding, corn growing or vegetables cultivation. This would help to avoid (at least partly) the application of allocation factors.

System Expansion with the Avoided Burden Approach

The following description for the procedure of system expansion with the avoided burden approach (in a consequential LCA) is based on a case study for rapeseed methyl ester (Calzoni *et al.*, 2000). It is assumed that extracted rapeseed meal is used as protein component in livestock feed and replaces soy meal. The system expansion is based on the preconditions that:

1 Soy meal is the marginal protein fodder and rapeseed oil is the marginal edible oil on the market.
2 Rapeseed contains 40% oil and 20% raw protein in the dry matter and that soy bean contains 17% oil and 34% raw protein in the dry matter.
3 The raw protein and the oil in both rapeseed and soy bean are substitutable in the marginal application.

Per 5 kg rapeseed produced an additional production of 1.66 kg rape seed is added (see Figure 4.3). Then a system expansion is made by subtracting the production of 3.91 kg soy bean. The net output of system is then 2 kg of oil and 0 kg of protein. The resulting resource consumptions and emissions (not shown in Figure 4.3) of the balance are applied on the functional unit of interest, which is 2 kg of rapeseed oil.

System Expansion with the Basket of Benefits Approach

The basic idea for the system expansion with a basket of benefits for the functional unit is quite similar to the example above (Figure 4.3). In this case the comparison is made between a system, which delivers several benefits with one multi-output process, and a

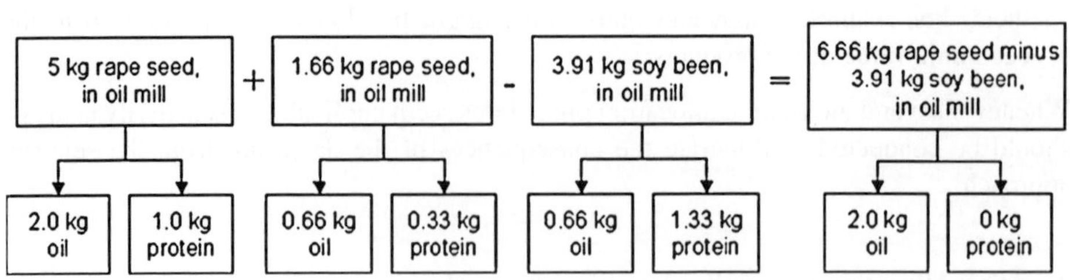

Figure 4.3 *Example of system expansion with the avoided burden approach for rape seed with the purpose of avoiding allocation regarding soy bean oil and protein (Calzoni et al., 2000). All inputs and outputs except the co-products are not shown for the sake of clarity*

system, which delivers these benefits with different separate production processes. The results show the impacts for the whole expanded product system and not for the individual products. Applied to the example in Figure 4.3, one could define the functional unit of rapeseed production as '2 kg oil and 1 kg protein'. This could be compared to an alternative system which produces the same amount of oil and protein in a different way (e.g. oil from crude oil and protein from biomass).

Allocation by Partitioning of Inputs and Outputs Based on Physical Relationships

In some cases, production processes deliver several products where the amount of output can be varied independently. In such cases, physical relationships can be identified by varying one output and analysing the changes in inputs and emissions. A truck may transport several different goods at the same time. Because fuel consumption and emissions are dependent on the payload, one could allocate fuel consumption, emissions, etc. on the basis of the mass of the transported goods. Waste incineration plants burn various types of wastes and use advanced flue gas treatment facilities. In this case, the chemical composition of the waste (e.g. chlorine content) could be used to allocate the consumption of working materials (such as caustic soda).

Allocation by Partitioning of Inputs and Outputs Based on Other Relationships

The procedure for an allocation by partitioning on the basis of other relationships is explained with an example from the ecoinvent Database (Frischknecht *et al.*, 2004). According to ISO 14041, 'the sum of the allocated inputs and outputs of a unit process shall equal the unallocated inputs and outputs of the unit process' (ISO, 1998).

The allocation procedures should be uniformly applied to similar inputs and outputs of the systems under consideration (ibid.). This is especially important if a product is an output for one process and an input for another process. Residues without value that are used by other processes have to be treated in a consistent way for the origin processes and the processes that make use of the residues.

Table 4.2 shows an excerpt of the inputs and outputs of the wheat production process and the allocation factors as modelled in the ecoinvent Database. First, some examples of inputs from technosphere and elementary flows are shown. The column 'wheat IP' gives the amounts used or emitted per hectare. In this example, 67 kg of nitrogen in ammonium nitrate are required and 3.9 grams of cadmium are emitted to agricultural soil per hectare and year. The field is used during 290 days of the year (from tillage to harvest) which results in a land occupation of 7 940 m^2a per hectare wheat cultivation. The allocation factors for the two products shown in the columns 'wheat grains IP, at farm' and 'wheat straw IP, at farm' define the share of this total which is allocated to the specific product. These shares (allocation factors) can and should be determined based on different properties, e.g. product value, carbon or energy content. For carbon dioxide uptake, 61% of the total amount is allocated to the wheat grains because the shares 61/39 equal the shares of carbon found in the two products wheat grain and wheat straw. The energy resource extraction is allocated on the basis of the energy content of the two products and the land occupation and pesticide emission to soil are allocated based on the product values, the

Table 4.2 Selected multi-output process raw data of the cultivation of wheat (per hectare) and allocation factors (%) used for wheat grains and straw

Inputs and outputs	*Name*	*Unit*	*Cultivation wheat, integrated production (IP)/CH ha*	*Wheat grains IP, at farm/CH (%)*	*Wheat straw IP, at farm/CH (%)*
inputs					
technosphere input	ammonium nitrate, as N, at regional storehouse/RER	kg	67.1	92.49	7.51
	grain drying, low temperature/CH	kg	76.4	100.00	0.00
resource, in air	carbon dioxide, in air	kg	13900	61.27	38.73
resource, biotic	energy, gross calorific value, in biomass	MJ	167000	59.13	40.87
resource, land	occupation, arable, non-irrigated	m^2a	7940	92.49	7.51
outputs					
emission to	cadmium	kg	0.00391	42.15	57.85
agricultural soil	chlormequat	kg	0.23	92.49	7.51
co-product	wheat grains IP, at farm/CH	kg	6420	100.00	
output	wheat straw IP, at farm/CH	kg	3910		100.00

Source: Nemecek *et al.* (2004).

default allocation factor of this process. Finally, the cadmium emissions to soil are allocated based on the amounts in the two products, wheat grains and wheat straw.

The data are fed into the database to calculate the unit process raw data. This is shown for the wheat production example in Table 4.3. For instance, the input of 67 kg 'ammonium nitrate' is multiplied by the allocation factor 92.5% and divided by 6.4E+3 kg (the amount of wheat grains per hectare). Hence, 9.7 g ammonium nitrate input is attributed to the production of 1 kg of wheat grains. Only 1.3 g is attributed to the production of 1 kg of wheat straw.

Summary

The ISO methodology for modelling multi-output processes still leaves a range of possible choices. These choices can have an important influence on the final results and they have a subjective component in any case. Regarding the choice of the appropriate approach, the specific goals of the study have to be considered.

4.4.4 Uncertainty Considerations in LCI

In the life cycle inventory of a unit process the amounts of the inputs and outputs are described in single figures (the mean values). This quantitative description of the unit process includes uncertainty because the mean values are uncertain. In reality, there might be a difference between the value that has been investigated (or measured and reported) and the 'real' value.

Table 4.3 *Example of selected derived unit process raw data of the two co-products of the cultivation of wheat ("cultivation wheat, integrated production (IP)"). Input and output flow of the multi-output process times allocation factor divided by co-product output equals input and output flows of the derived unit processes*

Inputs and outputs	*Name*	*Unit*	*Wheat grains IP, at farm/CH kg*	*Wheat straw IP, at farm/CH kg*
inputs				
technosphere input	ammonium nitrate, as N, at regional storehouse/RER	kg	0.00967	0.00129
	grain drying, low temperature/CH	kg	0.0119	
resource, in air	carbon dioxide, in air	kg	1.33	1.38
resource, biotic	energy, gross calorific value, in biomass	MJ	15.4	17.5
resource, land	occupation, arable, non-irrigated	m^2a	1.14	0.153
outputs				
emission to agricultural soil	cadmium	kg	2.57E-07	5.79E-07
	chlormequat	kg	3.31E-05	4.42E-06
co-product output	wheat grains IP, at farm/CH	kg	1.0	
	wheat straw IP, at farm/CH	kg		1.0

Source: Nemecek *et al.* (2004).

Different types of uncertainty are present in the lifecycle inventory data of a process:

- Variability and stochastic error of the figures that describe the inputs and outputs due to measurement uncertainties, process specific variations, temporal variations, etc.
- Appropriateness of the input or output flows. Sometimes an input or output does not perfectly match with the input or output observed in reality. This may be due to temporal and/or spatial approximations. For instance, the electricity consumption of a process that takes place in Nigeria may be approximated to the dataset of the electricity supply mix of the European network.
- Model uncertainty: the model used to describe a unit process may be inappropriate (using, for instance, linear instead of nonlinear modelling).
- Ignoring important flows. Sometimes not all the relevant information is available to completely describe a process. Such unknown inputs and outputs are missing in the inventory.

So far there is no standardized procedure how to document and analyze different types of uncertainties in the LCI. It has to be noted that the impact assessment introduces further uncertainties to the analysis which might be even more important than the inventory uncertainties.

4.4.5 Lifecycle Inventory Analysis Result

The LCI result is the outcome of a lifecycle inventory analysis. It includes all the basic flows crossing the boundaries of the whole product system under investigation. All inputs of resources and outputs of emissions are summed up over the entire life cycle, which

Table 4.4 Example of selected LCI results of the production of 1 kg of potatoes

Inputs and outputs	Name	Unit	potatoes IP, at farm/ CH kg
inputs			
resource, land	occupation, arable, non-irrigated	m^2a	0.33
resource, in air	carbon dioxide, in air	kg	0.37
outputs			
emission to air, high population density	carbon dioxide, fossil	kg	0.013
emission to air, low population density	carbon dioxide, fossil	kg	0.02
emission to air, low population density	nitrogen oxides	kg	0.00028
emission to agricultural soil	cadmium	kg	2.8E-08
emission to soil, unspecified	cadmium	kg	4.8E-11
product output	potatoes IP, at farm/CH	kg	1.0

Source: Nemecek *et al.* (2004).

includes not only the potato seeds but also machinery, agricultural cultivation processes, fertilizer and pesticide inputs.

Table 4.4 shows an excerpt of the cumulative LCI results for the integrated production of 1 kg of potatoes (Nemecek *et al.*, 2004). Only a few of more than 1000 elementary flows are shown in Table 4.4. One can see that the cumulative results are higher than for the unit process raw data in Table 4.1 due to further inputs from the lifecycle. The direct uptake of carbon dioxide for potatoes growing is about 340 grams (see Table 4.1). Another 30 grams are added in the lifecycle, e.g. for the growing of potato seeds, which results in a total cumulative CO_2 update of 370 grams per kg of potatoes (see Table 4.4). Emissions to air are recorded and reported by distinguishing their point of release, i.e., 'low population density', 'high population density', 'lower stratosphere/upper troposphere' and 'unspecified'. It has to be noted that the LCI results do not show any inputs and outputs to the technosphere (except the product output), but only the flows between technosphere and the environment. This LCI table provides the starting point for the life cycle impact assessment phase of an LCA.

4.5 LCI Data Documentation and Exchange Format

ISO/TS 14048 is a technical specification for life cycle inventory (LCI) data documentation format (ISO, 2002). The specification aims to support a transparent reporting, interpretation and review of data collection, data calculation and data quality, as well as facilitating data exchange. This format can serve as a guide to collecting and reporting data according to ISO 14041, independent of the media used (paper or electronic), independent of any specific software, and independent of the industrial context. So far no electronic format for data exchange has been standardized by ISO. ISO 14048 as a technical specification is not mandatory for an LCA according to ISO 14040ff.

In practice, the EcoSpold data format is widely being used. Major LCI databases such as the Swiss ecoinvent Database provide data in this format and all main LCA software tools implemented an EcoSpold interface to at least import EcoSpold datasets. The EcoSpold data format is based on the ISO/TS 14048 format and allows extensive documentation of LCI datasets (see also Frischknecht, 2001).

4.6 Consequential versus Attributional LCI

One of the most controversial themes in LCA methodology development at the moment is the distinction between consequential and attributional LCI and LCA. So far this has not been dealt with in detail in the ISO standards.

The attributional methodology, which has been used so far for most of the LCA studies, aims to describe the environmentally relevant physical flows to and from a product system and its subsystems. Thus it considers only impacts of the running process and not what would have happened if the process had not taken place. Quite often it is used to predict environmental improvements, but it has been shown that it is not fully possible to derive conclusions on future changes from the analysis of existing systems (Ekvall *et al.*, 2004).

Consequential LCI methodology, in contrast, aims to describe how the environmentally relevant physical flows to and from the technosphere will change in response to possible changes in the life cycle of a product or service. Thus, the natural starting point for a consequential LCA is the decision itself. It might not be necessary to follow up the full life cycle of the product, but only the part of the world product system which is affected by the decision (Ekvall *et al.*, 2004; Ekvall and Weidema, 2004).

From an analytical point of view it might, for example, be intended to consider side-effects. Today straw is quite often burned in the field. This causes air emissions but, on the other hand, ashes remain in the soil and serve as fertilizer. If straw is withdrawn as an energy resource, it seems important to consider these effects. On the one hand, avoided air emissions can be subtracted from the environmental impacts of the production system. On the other, surplus production of fertilizers has to be added to the product system, causing additional emissions and resource consumption. Therefore, the system under investigation has to be expanded to take in the consequences due to changing some production patterns outside of the life cycle investigated. The following might, for example, be considered in a consequential LCI and LCA:

- increased or decreased demand for a product or production technology caused by increasing the demand for the product under investigation;
- alternative use of products, residues, resources (e.g. agricultural land, water), etc. that are needed in the life cycle of the product system;
- consequences of not using certain products, residues, resources, etc.;
- consequences of emitting more or less of a pollutant in comparison to the today average.

So far there is no standardized methodology for consequential LCI and LCA. In particular, there are no clear guidelines how to define cut-off rules for consequences to be included or excluded. A proposal how to deal mainly with different aspects of allocation and indirect effects has been elaborated by Ekvall and Weidema (2004).

The idea of modelling consequences has mainly been developed to solve the allocation problem with system expansion (see Section 4.6). Here marginal products or technologies are considered for the system expansion instead of assuming average production patterns. Marginal products are defined as the specific products that are most likely to come to the market if the demand is increased (or which disappear from the market due to decreased demand). The same view can be applied to production technologies. The marginal production technologies are defined as the specific technologies that are most likely to increase their production if the demand is increased. To define these products or production technologies, a very good knowledge of economic options and restrictions is necessary.

Other phases of the LCA are so far seldom modelled in a consequential manner. In order to model the full consequences, it would be necessary to develop this idea also for the goal and scope definition, the modelling of all inventory data and for the impact assessment. A recent comparison of the two approaches concludes that at the moment it is not possible to decide which methodology will point more often in the 'right' direction. In the short-term view, consequential LCA seems to be more easy to misuse because of lack of methodological guidelines and established good practice. In the long term, with improved guidelines, modelling consequences might be the more promising approach (Ekvall *et al.*, 2004). A choice between these two general approaches is necessary in order to make a consistent decision on the models used in LCI and LCA. This decision also depends on the type of questions which will be answered by the LCA study.

4.7 Summary

The main strengths of the life cycle inventory analysis are the holistic approach and the structured procedure for the goal definition, data investigation and systems comparison. The method considers more environmental impacts than other common methods such as environmental risk assessment. Thus, it fits in well with the detailed and systematic comparison of products that cause completely different environmental impacts along the different life cycle phases. Life cycle inventory is the backbone of any LCA and helps to identify hot spots with regard to key pollutants in the life cycle of biomass products. Life cycle inventory data on a unit process level can not only serve LCA but also Material and Substance Flow Analysis (MFA, SFA), energy and exergy analysis but also MIPS and ecological footprint analyses.

Its main weaknesses are the time-consuming acquisition of data and the belief that life cycle inventory analysis is a purely natural science-based method and free of value choices. The application of LCI in a decision-making process is restricted by the specific object and the initial assumptions. For biomass products some areas of concern, e.g. pesticide use, use of soil and water resources are difficult to model. It is still not possible to quantify all the known environmental impacts in the inventory analysis. Noise emissions are, for example, only seldom quantified because the impacts cannot yet be assessed within the LCA framework.

Summarizing the goal and scope definitions of LCA case studies for biomass products shows that they focused on different themes:

- environmental comparison of fossil-based energy and materials and comparable products made from biomass and identification of environmental advantages and disadvantages of biofuels in comparison to conventional fuels;
- estimation of the reduction potential for greenhouse gas emissions due to the increased use of biomass;

- environmental comparison of different types of biomass uses;
- environmental comparison of different production routes and biomass sources for products;
- environmental improvement of production routes for biomass products.

LCA case studies show that major methodological differences in biomass product LCAs are in the allocation approaches chosen in agriculture and further processing (e.g., cattle raising, rapeseed oil production). It will not be possible to standardize this modelling aspect because of the value choices involved and hence its subjective nature.

The distinction between attributional and consequential LCI and LCA helps to better classify the different scopes of LCA work. However, the operationalization of the consequential approach in LCI but also in life cycle impact assessment still needs considerable research and its usefulness in daily LCA practice needs further proof.

References

Calzoni, J., Caspersen, N., Dercas, N., Gaillard, G., Gosse, G., Hanegraaf, M., Heinzer, L., Jungk, N., Kool, A., Korsuize, G., Lechner, M., Leviel, B., Neumayr, R., Nielsen, A. M., Nielsen, P. H., Nikolaou, A., Panoutsou, C., Panvini, A., Patyk, A., Rathbauer, J., Reinhardt, G. A., Riva, G., Smedile, E., Stettler, C., Weidema, B. P., Wörgetter, M. and van Zeijts, H. (2000) *Bioenergy for Europe: Which Ones Fit Best? A Comparative Analysis for the Community*, Institut für Energie- und Umweltforschung, Heidelberg.

Cowell, S., Audsley, E., Brentrup, F., Cederberg, C., Gaillard, G., Goldhan, G., McKeown, P., Jolliet, O., Lindeijer, E. and Satter, I. (1999) *Theme Report: Methodology Working Group*, Fair No. 97-3079, LCAnet Food. Available at: www.sik.se/sik/affomr/miljo/lcanetf.html

Ecoinvent Centre (2004) *Ecoinvent Data v1.1, Final Reports Ecoinvent 2000 No. 1-15*, Swiss Centre for Life Cycle Inventories. Available at: www.ecoinvent.ch.

Ekvall, T., Ciroth, A., Hofstetter, P. and Norris, G. (2004) Evaluation of attributional and consequential life cycle assessment, Working paper, Chalmers University of Technology, Göteborg.

Ekvall, T. and Weidema, B. (2004) *System Boundaries and Input Data in Consequential Life Cycle Inventory Analysis, International Journal LCA* **9**(3), 161–171. Available at: www.scientificjournals.com/sj/lca/welcome.htm.

Frischknecht, R. (2001) Life cycle inventory modelling in the Swiss national database Ecoinvent 2000, in Hilti, L. M. and Gilgen, P. W., *Sustainability in the Information Society, 15th International Symposium Informatics for Environmental Protection, ETH Zürich*, Metropolis-Verlag, Marburg, pp. 699–708.

Frischknecht, R., Jungbluth, N., Althaus, H.-J., Doka, G., Dones, R., Heck, T., Hellweg, S., Hischier, R., Nemecek, T., Rebitzer, G. and Spielmann, M. (2004) *Overview and Methodology*, Final report Ecoinvent 2000, No. 1, Swiss Centre for Life Cycle Inventories. Available at: www.ecoinvent.ch.

Guinée, J. B., Gorrée, M., Heijungs, R., Huppes, G., Kleijn, R., de Koning, A., van Oers, L., Wegener Sleeswijk, A., Suh, S., Udo de Haes, H. A., de Bruijn, H., van Duin, R., Huijbregts, M. A. J., Lindeijer, E., Roorda, A. A. H. and Weidema, B. P. (2001a) *Life Cycle Assessment; An Operational Guide to the ISO Standards; Part 3: Scientific Background*, Ministry of Housing, Spatial Planning and Environment (VROM) and Centre of Environmental Science (CML), Den Haag and Leiden, The Netherlands.

Guinée, J. B., Gorrée, M., Heijungs, R., Huppes, G., Kleijn, R., de Koning, A., van Oers, L., Wegener Sleeswijk, A., Suh, S., Udo de Haes, H. A., de Bruijn, H., van Duin, R., Huijbregts, M. A. J., Lindeijer, E., Roorda, A. A. H. and Weidema, B. P. (2001b) *Life Cycle Assessment; An Operational Guide to the ISO Standards; Parts 1 and 2*, Ministry of Housing, Spatial Planning and Environment (VROM) and Centre of Environmental Science (CML), Den Haag and Leiden, The Netherlands.

Ifu and Ifeu (2004) *Umberto; Software für das betriebliche Stoffstrommanagement; Version 4.2*, Programm, Beschreibung und Anhang, Hamburg, Heidelberg.

IKP and PE Europe (2004) *GaBi 4: The Software System for Life Cycle Engineering*, IKP Universität Stuttgart, Institut für Kunststoffprüfung und Kunststoffkunde, PE Europe GmbH. Available at: www.pe-europe.com.

International Organization for Standardization (ISO) (1997) *Environmental Management – Life Cycle Assessment: Principles and Framework*, ISO 14040, ISO, Geneva.

International Organization for Standardization (ISO) (1997–2000) *Environmental Management – Life Cycle Assessment*, ISO 14040ff, ISO, Geneva.

International Organization for Standardization (ISO) (1998) *Environmental Management – Life Cycle Assessment: Goal and Scope Definition and Inventory Analysis*, ISO 14041, ISO, Geneva.

International Organization for Standardization (ISO) (2002) *Environmental Management – Life Cycle Assessment: Data Documentation Format*, Technical Specification ISO/TS 14048, ISO, Geneva.

Nemecek, T., Heil, A., Huguenin, O., Meier, S., Erzinger, S., Blaser, S., Dux., D. and Zimmermann, A. (2004) *Life Cycle Inventories of Agricultural Production Systems*, Final report Ecoinvent 2000 No. 15, Agroscope FAL Reckenholz and FAT Taenikon, Swiss Centre for Life Cycle Inventories. Available at: www.ecoinvent.ch

PRé Consultants (2004) *SimaPro 6.0*. Available at: www.pre.nl

van Berkel, R., Willems, E. and Lafleur, M. (1997) Development of an industrial ecology toolbox for the introduction of industrial ecology in enterprises – I *Journal of Cleaner Production*, **5**(1–2), 11–25.

5

Net Energy Balancing and Fuel-Cycle Analysis

Hosein Shapouri, Michael Wang and James A. Duffield

5.1 Introduction

Net energy balancing has been a popular method of evaluating the sustainability of biofuels since the energy crisis of the 1970s (US Department of Energy, 1980). During this period, energy shortages and price volatility reminded the world that petroleum reserves were finite and alternative energy sources would someday be needed to sustain future economic growth. National energy security concerns stimulated research efforts aimed at developing alternative transportation fuels, such as methanol, ethanol, and natural gas. By 1990, much of this research had shifted to fuel-cycle analysis that showed how biomass-derived alcohol fuels can reduce greenhouse gases (Delucchi, 1991; Keeney and DeLuca, 1992; Turhollow and Perlack, 1991). Using net energy balancing in combination with fuel-cycle analysis has become the standard procedure for measuring energy and environmental sustainability of biofuels. The aim of this chapter is to describe the procedures used to estimate net energy values and greenhouse gas emissions and show the environmental advantages of biofuels versus petroleum fuels.

5.1.1 Background

Generally, when converting one form of energy into a more useful form of energy, e.g., converting crude oil into transportation fuels, by the second law of thermodynamics, there is always an energy loss. Thus, all fossil fuels have a negative net energy balance (NEB), where NEB is defined as the energy content of a fuel, minus the energy content of

Renewables-Based Technology Edited by J. Dewulf and H. Van Langenhove

the petroleum and other energy sources required to produce it. An exception to this rule is in the case of biofuels because most agricultural energy analysts view the solar energy captured by biomass to be energetically free (Stout, 1990). When solar energy is not included in the NEB calculations, it is possible for a biomass fuel to have a positive NEB, i.e., the conversion process results in a net energy gain. The net energy balancing method is often used to make energy-efficiency comparisons between fuels. The fuel with the higher NEB is said to be more energy efficient. Another measurement that is often used to estimate the net energy value of fuels is net energy ratio (NER), where NER is energy output divided by energy input. A fuel with a NER greater than 1 indicates a net energy gain. On the other hand, a fuel with a NER less than 1 indicates a net energy loss.

5.1.2 Environmental Sustainability

Biofuels are generally regarded as more environmentally sustainable than their fossil fuel counterparts because much of the energy content of biofuels is derived from renewable sources. However, the consumption of fossil fuels required for renewable fuel production results in resource depletion and environmental degradation. NEB can be used as an indicator to measure the environmental effects of biofuels. For example, a low NEB value for a biofuel could imply that the efficiency of its production is low. This, in turn, implies a greater environmental burden and more resource consumption for fuel production. Thus, NEB can be used as a first approximation in measuring the environmental sustainability of a given biofuel.

However, as energy and environmental analyses of biofuels have advanced in the past 15 years, researchers and practitioners have begun to realize the limitations of the NEB method. It has been found that energy use may not have a strict linear relationship with environmental burden. For example, the petrochemicals used by farmers for fertilization often emit significant amounts of nitrous oxide emissions. This potent greenhouse gas would not be included in most NEB analyses. Thus, in some instances, NEB may not be a good indicator of environmental sustainability of a given fuel. Environmental sustainability of a fuel should be analyzed directly using methods such as fuel-cycle analysis (FCA).

5.1.3 Domestic Energy Security

In the USA and many other countries, biofuels are being introduced to replace petroleum use in order to reduce oil imports. It is often argued that fuels with negative NERs cannot reduce imports because they require fossil energy for their production (Pimentel, 1991, 2003). Although it may require fossil energy to produce a biofuel, as long as domestic sources of energy are used, there will be a replacement of oil imports. For example, the USA has an abundant coal supply and many ethanol plants use domestic coal as their primary energy source. When coal is used to increase US ethanol production, the US supply of liquid transportation fuel increases, helping to reduce its dependence on foreign oil. In order to fully address energy security benefits of ethanol, NEB calculations need to distinguish between the energy derived from liquid fuels and the energy derived from non-liquid fuels (Shapouri *et al.*, 1995; Wang *et al.*, 1999).

5.2 Methodology

Estimating NEB begins by defining the entire production system of the biofuel under study. The production system for biobased-derived fuels generally includes four subsystems; (1) feedstock production; (2) feedstock transportation; (3) energy conversion; and (4) product distribution. An inventory is then developed that identifies and quantifies all the nonrenewable energy inputs used in each subsystem. The boundaries of the total system should be broadly defined to ensure that all significant sources of energy are included in the inventory. The analyst should first identify the primary energy inputs, for example, the liquid fuel and electricity used to directly power equipment in the system. The energy content of materials that are made from energy resources, such as fertilizers, pesticides, and other petrochemicals should also be included in the inventory.

Most NEB calculations include activities related to operations (fertilizer production, farming, ethanol production, etc.). Some studies include energy embodied in facilities as well (Pimentel, 1991, 2003). In this case, energy embodied in farming equipment, ethanol plant construction materials, roads, etc. is taken into account. This is a measurement of the energy required to produce raw materials found in the system, e.g., the energy content of the metal used to make tractors and irrigation pumps. Several tiers of these auxiliary energy sources exist, extending further and further from the system. Theoretically, an inventory could include all embodied energy in a system but the enormous data and computational requirements make this task impractical. For embodied energy calculations, the assumption of facility lifetime and multiple usage purposes of facilities are key technical issues that must be addressed. As of now, no studies attempting to include embodied energy have adequately addressed these issues. Thus, most analysts ignore secondary energy sources, arguing that they are distantly related to the fuel production system and maintain that the energy content of embodied energy is insignificant. It has been shown, that if analyzed accurately, the share of embodied energy relative to total energy, particularly for ethanol, is small, and the differences in embodied energy among different fuels are even smaller (Delucchi, 2003).

The energy values of all fossil resources used in the system are adjusted by energy efficiency coefficients to take into account the energy used to convert fossil resources into usable energy. These coefficients include the energy required to mine, extract, and manufacture the raw energy materials that go into the final energy product. Estimates of electricity generation are based on the weighted average of all sources of power, including coal, natural gas, nuclear, and hydroelectric. All electricity used in the system is increased to account for transmission loss by a factor of 1.087 (Energy Information Administration, 2004).

The feedstock subsystem accounts for the energy required to produce the biomass, e.g., growing corn used to make ethanol. Starting with seed production, this phase includes the energy value of the liquid fuels, such as gasoline, diesel and liquefied petroleum gas (LPG) used to operate farm machinery. It includes any energy required to irrigate or to reduce the moisture content of crops. The energy required for manufacturing, packaging, and transporting fertilizers and pesticides to the farm is also included in the feedstock phase. The final step in constructing an inventory for the feedstock subsystem involves loading the feedstock and delivering it to a local storage facility. The feedstock transportation subsystem accounts for the energy used to deliver the feedstock to a regional conversion facility.

The conversion subsystem processes the biomass feedstock into an energy product. Generally, this phase would include energy use for generating the thermal and electric power used in a production facility, such as an ethanol or biodiesel plant. The distribution subsystem accounts for the energy used for transporting the final energy product from the plant to a distribution center, generally by rail, barge, or truck. The next step is to estimate the energy required to transport the energy product to its final retail destination. The inventory for this subsystem ends where the energy is dispensed into a vehicle fuel tank or other end use point just prior to fuel combustion. Adding up all the energy used across the subsystems gives the total energy required over the entire production system of the fuel.

5.2.1 Co-Product Allocation

Many energy conversion processes generate multiple products, called co-products. For example, wet milling operations that produce fuel-ethanol are designed to also process corn into various animal feeds and corn oil. When ethanol is the primary product of interest, the other co-products and the energy used in their production must be identified, so they can be excluded from the inventory. If sufficient details about the production system are available, allocation of energy burdens among co-products can be very straightforward. In many cases there are individual co-product subprocesses that clearly have their own energy requirements. In the case of wet milling, a clear delineation can be made between ethanol and the other co-products by tracking the starch through the conversion subsystem. The starch, which is the primary ingredient for ethanol production, is separated from the grain in the pre-treatment phase, with the remaining non-starch components being directed to other co-product subprocesses. The energy associated with producing the non-starch coproducts are outside the ethanol production system and should not be included in the NEB calculations. This method is called process-based allocation.

When detailed information is not available on co-products or there is no clear method to separate them into subprocesses, allocation methods must be used to assign co-product values. This is generally the case for feedstock production subsystems because they proceed the processing phase before the crop is separated into co-products. Continuing with the ethanol example, corn is a single product on the farm, however, it is converted into multiple products at the mill, therefore the energy used in the feedstock subsystem must also be allocated among the co-products. There are several allocation methods that can be used to estimate the energy value of co-products. A discussion of the pros and cons of each allocation rule would be too lengthy to include here, so we shall only briefly discuss four allocation methods. In general, no allocation rule is always applicable and the appropriate method should be chosen on a case-by-case basis.

Perhaps the most common method is the mass allocation rule. It is favored by some researchers because it is easy to apply and provides very reasonable results (Vigon *et al.*, 1993). This method simply allocates energy to the primary product by its share of the total co-product weight. Since ethanol is about 48% of the total co-product weight, then 48% of the energy used in corn production is allocated to ethanol. This can also be done based on the weight of the starch of the corn. The relative market value of each of the co-products is another method of allocating co-product energy. For example, if energy used to produce ethanol is allocated between ethanol and the other co-products based on their 10-year average market values, about 72% of the energy used in the feedstock subsystem would be assigned to ethanol. When co-products have a food value, their caloric content can be used to allocate energy credits. For example, the co-product energy value of a pound of corn gluten feed or corn gluten meal would be about 8.44 MJ based on their caloric values. This method results in 58% of the energy allocated to ethanol.

The final energy allocation method presented in this section is based on the replacement value of the primary product. For ethanol, the replacement value is based on the energy required to produce a substitute for each coproduct. This hypothetical approach relies on the assumption that co-products replace other products, resulting in energy savings. For example, the increased production of the animal feed co-products made during ethanol conversion results in a lower demand for other animal feeds, such as soybean meal. Using this method, ethanol is given an energy co-product credit based on the energy saved from producing less soybean meal. The replacement method allocates most of the energy to the primary product – 81% of the energy used is allocated to ethanol (Shapouri *et al.*, 2002). In this chapter, we used two approaches to allocate energy among co-products; the first approach used the process-based allocation method in the conversion subsystem and mass-based allocation was used in the feedstock and transportation subsystems; the second approach used the replacement method to allocate energy among the co-products.

Low Heat Versus High Heat Value

In energy balance calculations, some studies use higher heating values (HHVs) of energy products, while others use lower heating values (LHVs). The difference between HHV and LHV is the energy in vapor-generated during fuel combustion. HHV includes the energy in combustion vapor, while LHV does not. For fuel combustion in stationary facilities such as steam generation plants and power generation plants, some of the energy in combustion vapor can be recovered for use. In motor vehicles, where fuel ethanol is used, energy in combustion vapor cannot be recovered. Methodologically, it is incorrect to use HHV for some activities but LHV for other activities in a single study. This could create distorted results for the energy products in evaluation. As long as an NEB and NER analysis uses HHV or LHV consistently, either measurement will yield similar results, especially when the aim is to compare differences among fuels. Results presented in this chapter are based on LHV.

5.2.2 Description of GREET Fuel-Cycle Analysis

In 1996, with support from the US Department of Energy, the Argonne National Laboratory developed the Greenhouse Gases, Regulated Emissions, and Energy Use in Transportation (GREET) model for researchers and practitioners to evaluate energy use and air emissions of various transportation fuels and vehicle technologies. Since then, Argonne has continued to update and upgrade the GREET model. Transportation fuels in the GREET model include gasoline, diesel, compressed natural gas, liquefied petroleum gas, methanol, ethanol, biodiesel, electricity, and hydrogen. The GREET model, along with supporting documents, are available to the public at the Argonne's GREET Website (Argonne National Laboratory, 2004).

Figure 5.1 shows the stages and activities covered in GREET simulations of transportation fuel cycles. The stages of the fuel cycle are similar to the subsystems identified in the NEB analysis, including feedstock and fuel production stages, except that NEB calculations usually stop at the fuel pump at refuelling stations for ethanol. The energy values identified in the NEB inventory are entered into the GREET model, which in turn converts energy values into GHG emissions. In addition to the feedstock and fuel stages, GREET has a vehicle operation stage that estimates the fuel emissions associated with operating a vehicle. The feedstock stage and fuel stage together make up the well-to-pump (WTP) or upstream, stages (Figure 5.1). The vehicle operation stages make up the pump-to-wheel (PTW), or downstream, stages. The GREET model estimates fuel-cycle energy and emissions separately for each of these three stages.

Argonne and others have applied the GREET model to evaluate energy and emission effects of various transportation fuels including gasoline and ethanol. The GREET model utilizes a Microsoft Excel spreadsheet in a way that users can readily input their own data and assumptions into the model. For ethanol, GREET includes ethanol produced from both corn and cellulosic biomass. The model contains a large amount of data on growing corn for ethanol and the air emissions associated with using ethanol in gasoline engines. The USDA and Argonne National Laboratory worked together to design the feedstock and fuel production stages of the model. Results of corn ethanol from GREET simulations have been reviewed and used by US Department of Energy, the US Environmental Protection Agency, the auto industry, the petroleum industry, and other organizations. The USDA recently updated the corn farming and ethanol production stages of GREET to generate the results discussed in this chapter.

Estimating Greenhouse Gases

Each stage of the biofuel system consumes fossil fuels, thus generating greenhouse gas (GHG) emissions. GHG emissions for transportation fuels, including fuel ethanol, are

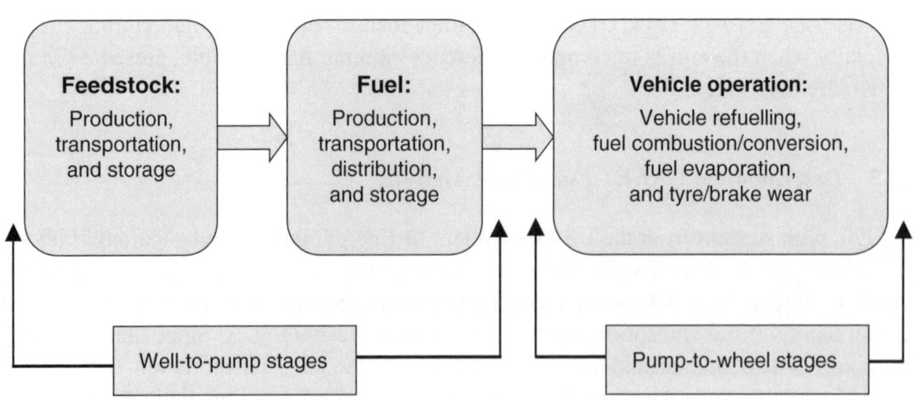

Figure 5.1 Stages covered in the GREET fuel-cycle analysis

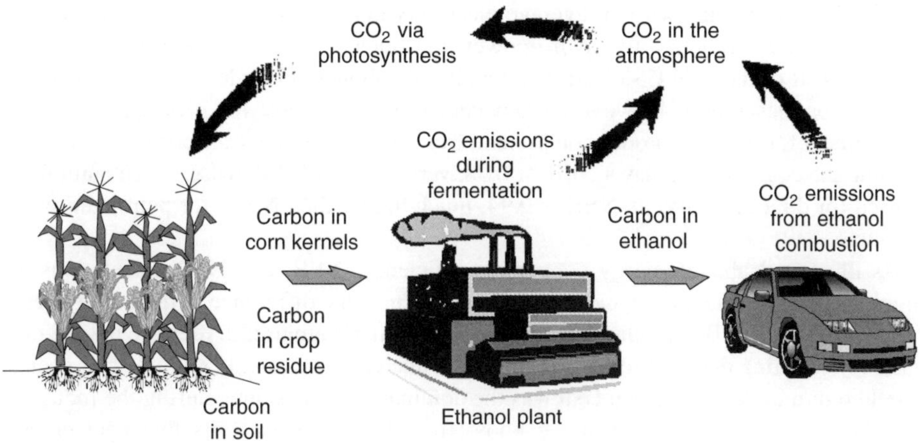

Figure 5.2 *Recycling of carbon dioxide in ethanol fuel cycle*

carbon dioxide (CO_2), methane (CH_4), and nitrous oxide (N_2O). These are three major GHGs identified by the Intergovernmental Panel on Climate Change for fuel combustion. Although GHG emissions, especially CO_2 emissions, are largely generated during fossil fuel combustion, there are significant non-combustion GHG emissions. In the case of corn ethanol, a large amount of N_2O emissions are generated in cornfields during the process of nitrification and denitrification of nitrogen fertilizer. Thus, fossil energy use identified in the NEB calculations may not account for all the GHG emissions generated over a fuel cycle. GHG emissions from non-combustion activities need to be identified and evaluated in order to complete the fuel-cycle analysis.

One obvious benefit of biofuels in terms of GHG emissions is that the carbon in biofuels is derived from the air during feedstock growth via the photosynthesis process. In a way, the carbon from biofuel combustion is recycled during feedstock growth (Figure 5.2). However, there are activities in the biofuel system that consume fossil fuels and produce GHG emissions that must be counted. The net GHG effects of biofuels versus conventional petroleum gasoline take into account both carbon recycling and GHG emissions of system activities.

5.3 Energy Balance of Fossil Fuel versus Biofuel

Although the NEBs of fossil fuels are negative, NEB analysis is still a useful tool for making energy value comparisons between fuels. For example, the production of gasoline requires energy for petroleum recovery, transporting the petroleum from oil fields to petroleum refineries, petroleum refining, and transporting the gasoline from petroleum refineries to refuelling stations. Taking all these activities into account, gasoline produced in the United States has a NER of about 0.8. On the other hand, refining petroleum into diesel requires less energy than refining it into gasoline. Consequently, diesel fuel from petroleum has a NER of about 0.85, a little higher than that for gasoline. In the USA, about 51% of electricity is generated from coal-fired power plants and 17% from natural gas-fired power plants, with an average electricity conversion efficiency of about 35%. The generation efficiency for hydro-power and nuclear power is slightly higher. The average NER of the total US electric system is only about 0.45 far less when compared to gasoline or diesel fuel. However, when coal is used directly for heating, the energy requirement for the total production system, from coal mining, to transporting the coal, to burning the coal for heat, has a NER of 0.98, very close to 1. Likewise, when natural gas is used for direct heating, its NER is 0.94, much higher than NERs for petroleum-based products.

As illustrated above, energy conversion efficiency, NEB, and NER calculations are highly dependent on end use and they fail to recognize the difference in quality of energy products. While NEB and other energy efficiency criteria are useful, they are not the only basis for selecting energy products. If this were the case, we might conclude that society should return to the days when coal was the dominant energy source during the Industrial Revolution, 150 years ago. But we know from historical accounts that coal created serious environmental and health problems in industrialized cities, leading to the adoption of other energy products. Society recognizes the quality of energy products that better serve our needs. Even though electricity and transportation fuels have lower NEBs compared to direct heating from burring coal or natural gas, they provide a specific kind of energy that accomplishes the consumer's needs at a particular time and place (Doering, 2004). Clearly, energy efficiency is not the only criterion for judging the social or economic value of a fuel.

Of all the renewable fuels, corn ethanol is perhaps the most studied because it is the only renewable liquid transportation fuel to achieve commercial success. Net energy balancing reports for corn ethanol began to surface in the mid-1970s. Early studies concluded that the NEB of corn ethanol was slightly negative, but as ethanol producers and farmers became more energy efficient, many researchers began to report positive NEBs for ethanol (Shapouri *et al.*, 2003). The example given below is based on a series of studies conducted by USDA and Argonne National Laboratory over the past 10 years. The accuracy and credibility of any net energy balance analysis depend greatly on the data used to build the production system inventory. In the case of renewable energy sources, such as corn, the best source of energy consumption information for farming is the USDA Agricultural Resource Management Survey (ARMS). This unique survey is conducted every five years to collect detailed information on the amount and type of energy used to produce corn, including gasoline, diesel fuel, natural gas, propane, and electricity. In addition, the amount of fertilizers and chemicals used by farmers to help increase corn yields is reported. The energy used for custom services performed by contractors, e.g., custom drying and purchased irrigation water are also included in the energy use estimates. The amount of fuel used to transport inputs to farms and haul corn to the initial point of sale is also included.

The most recent data on energy used for corn production is from the 2001 ARMS. The ARMS gathers data from 19 states, however, the example below focuses only on the major ethanol and corn-producing states: Illinois, Indiana, Iowa, Minnesota, Nebraska, Ohio, Michigan, South Dakota, and Wisconsin. In 2001, these nine States accounted for 79 and 92% of US corn and ethanol production, respectively. The survey reports energy use estimates on a per acre basis, e.g., gallons per acre, kilowatts per acre, and cubic feet per acre. The average corn yield per acre for each state is reported annually by USDA's crop production survey (National Agricultural Statistics Service, 2003). The GREET model was used to estimate the energy used to transport one kilogram of corn from local storage facilities or grain elevators to an ethanol plant.

The energy input data are converted from a per acre basis to a per kilogram basis with the three-year (2000–2002) average corn yield for each state. Three-year average corn yield is used to reduce the effect of weather variation, diseases, and other uncertainties that affect annual corn yield. State corn yields and energy use estimates are weighted by total production to derive a nine-State weighted energy use estimate for each energy input on a per kilogram of corn basis, e.g., liters of diesel fuel used to produce a kilogram of corn. Finally, energy inputs are converted to joules and aggregated to derive total energy use per kilogram of corn. All energy inputs were adjusted by energy efficiency factors for individual fuels, including transmission losses for electricity.

The inventory data for the energy conversion subsystem were collected by a phone survey of 28 ethanol plants in 2001. The questionnaire was designed to provide estimates of the amount of energy used to produce ethanol and ethanol co-products using both dry and wet mill processing. The survey revealed that most dry mill plants buy electricity and purchase natural gas to produce steam. Respondents were asked to report the amount of energy purchased on a per gallon of ethanol basis. The electricity and natural gas used to produce a gallon of ethanol by each dry mill plant was weighted by plant production and aggregated into average energy use per liter of ethanol. The estimates of natural gas and electricity were converted to joules and adjusted for energy and transmission losses during fuel production and transportation.

Unlike dry milling plants, wet mills are mostly energy self-sufficient and produce both thermal and electrical energy within the plants by burning coal or natural gas. Wet mill respondents provided the amount of cubic feet of natural gas and/or pounds of coal used to produce one gallon of ethanol. These estimates were adjusted if a wet milling plant sold excess electricity or purchased electricity from the suppliers. The amount of energy required to produce ethanol by each wet mill was adjusted by energy-efficiency factors and weighted by plant production. Energy use for each plant was converted to joules and aggregated to derive the average amount of joules required to produce one liter of ethanol in a wet milling plant. As described above, the GREET model is used to estimate the energy used in the product distribution subsystem. It includes a combination of fuels that are likely to be used to ship ethanol by various means of transportation from ethanol plants to refiners or blenders. Since ethanol is used primarily as a gasoline additive, the GREET model also provides energy use estimates for blending ethanol with gasoline and shipping the fuel from distribution centers to retailers. Finally, the GREET model adds the energy required to pump the fuel into a vehicle.

Since ethanol production results in multiple products, the energy used exclusively to produce the corn, transport the corn to the ethanol plant, and convert the corn to ethanol must be distinguished from the energy used to produce the other co-products. To demonstrate the difference that co-product allocation methods have on results, we used three approaches in this example. First, co-product allocation was ignored and the energy used for all the co-products was allocated to ethanol. Table 5.1 shows the input energy requirements for producing one liter of ethanol for each production subsystem, before energy is allocated among co-products. Energy estimates are provided for dry and wet milling, as well as a weighted industry average. The ethanol conversion subsystem required the most energy, followed by corn production subsystem. Total energy used, on a weighted average basis is 20.08 MJ per liter of ethanol, before energy is allocated among co-products. Using a low heating value of ethanol, which is 21.28 MJ per liter, results in an NEB of 1.19 MJ. Thus, corn ethanol has a positive energy balance, even before subtracting the energy used to produce the coproducts.

In the second approach, the process-based allocation method was used to separate the energy used for ethanol from the energy used for the other co-products. This was accomplished by modeling the energy use of ethanol and ethanol co-products for both wet and dry milling plants, using the Aspen Plus production simulator (Aspentech, 2004; McAloon *et al.*, 2004).

The Aspen model estimates the thermal and electrical energy used by each co-product in the ethanol conversion subsystem for both wet and dry milling. Energy used for corn production and transporting it from the farm was allocated among the coproducts using the mass weight of the corn-starch versus the entire corn kernel. In other words, the relative share of energy credited to ethanol was based on the weight of the starch content of the corn, which is 66% on average. The amount of energy allocated to the ethanol co-products was based on the weight of the non-starch materials in the corn kernel, such as fiber, oil, and protein. Co-products include distillers' dried grains with solubles from dry milling; and corn oil, corn gluten meal, and corn gluten feed from wet milling. Table 5.2

Table 5.1 *Ethanol results without co-product allocation*

Production subsystem	*Milling process*		*Weighted average*
	Dry	*Wet*	
	Million joules per liter		
Corn production	5.261	5.171	5.216
Corn transport	0.596	0.586	0.591
Ethanol conversion	13.133	14.591	13.862
Ethanol distribution	0.414	0.414	0.414
Total energy used	19.404	20.726	20.083
Net energy balance	1.871	0.513	1.192
Net energy ratio	1.10	1.02	1.06

Table 5.2 *Ethanol results with Aspen-process/starch-based allocation*

Production subsystem	*Milling process*		*Weighted average*
	Dry	*Wet*	
	Million joules per liter		
Corn production	3.472	3.413	3.442
Corn transport	0.393	0.387	0.390
Ethanol conversion	7.748	9.338	8.525
Ethanol distribution	0.414	0.414	0.414
Total energy used	12.023	13.547	12.766
Net energy balance	9.253	7.729	8.509
Net energy ratio	1.770	1.570	1.670

Table 5.3 Ethanol results with replacement allocation

Production subsystem	Milling process		Weighted average
	Dry	Wet	
	Million joules per liter		
Corn production	4.314	4.188	4.251
Corn transport	0.489	0.474	0.481
Ethanol conversion	10.769	11.819	11.294
Ethanol distribution	0.414	0.414	0.414
Total energy used	15.572	16.482	16.027
Net energy balance	5.704	4.794	5.249
Net energy ratio	1.370	1.290	1.330

presents the energy estimates after total energy use has been allocated among ethanol coproducts, using the Aspen-process/starch-based method.

The third approach utilizes the replacement method for allocating energy among co-products. It was assumed that the animal feed products produced during ethanol production would replace soybean meal on a protein equivalent basis. The energy that would have been required to produce the replaced soybean meal, including growing soybeans, transporting the soybeans to a crushing facility, and crushing the soybeans was deducted from the total energy used in the ethanol production system. A comparison of Tables 5.2 and 5.3 shows the difference that the choice of co-product energy allocation method can have on the final results. When the Aspen Plus simulator is used to isolate the energy used to produce only ethanol, the conversion energy requirement is 8.53 MJ per liter of ethanol when using the weighted average production for dry and wet mills (Table 5.2). When allocating energy among co-products on a starch-mass basis in the corn production subsystem, the energy required to produce one liter of ethanol is 3.44 MJ for weighted average production. An examination of Table 5.3 shows that using the replacement allocation method results in significantly higher energy use estimates. The total energy required to produce one liter of ethanol using the replacement method is 16.03 MJ compared to 12.77 MJ when using the Aspen-process/starch-based method in the weighted average case. The NEB for ethanol using the replacement allocation method is 5.25 MJ per liter compared to 8.51 MJ per liter when using the Aspen-process/starch-based method. The NER for ethanol is 1.33 when using the replacement method and using the Aspen-process/starch-based method results an NER of 1.67, i.e., there is a 67% energy gain from producing corn ethanol.

5.4 Greenhouse Gas Emissions from Corn Ethanol Production

Concern about potential global warming effects of major GHGs has led to recognition of the need to reduce anthropogenic GHG emissions worldwide. Using ethanol to fuel motor vehicles helps reduce GHG emissions. With updated data on corn farming and ethanol production from USDA, the GREET model was used to simulate fuel-cycle GHG emissions and energy impacts of corn-based ethanol. The corn-ethanol cycle includes fertilizer production, fertilizer transportation, corn farming, corn transportation, ethanol production, ethanol transportation, and ethanol use in motor vehicles in the form of E10 (10% ethanol and 90% gasoline by volume) and E85 (85% ethanol and 15% gasoline by volume).

To determine GHG emission reductions when replacing petroleum gasoline with corn ethanol, the GREET model was also used to estimate fuel-cycle GHG emissions of gasoline. This cycle includes crude oil recovery, crude oil transportation, petroleum refining, gasoline transportation, and gasoline use in motor vehicles.

Greenhouse gases in the evaluation include CO_2, CH_4, and N_2O. These three are combined together with their global warming potentials (1 for CO_2, 23 for CH_4, and 296 for N_2O) to derive CO_2-equivalent GHG emissions. In the corn-ethanol cycle, N_2O emissions during fertilizer production and from cornfields are also included. In the petroleum gasoline cycle, non-combustion CO_2 emissions from crude oil recovery in oil fields and petroleum refining are included. In the vehicle operation phase, GREET simulates the use of E10 and E85 in a conventional gasoline vehicle compared to the effects of using gasoline in the same vehicle.

Analogous to NEB co-product estimation, GHG emissions must also be allocated among co-products to isolate the emissions attributable to ethanol. In order to be consistent with our NEB estimation, we used the two allocation methods described earlier to derive the NEB results shown in Figures 5.2 and 5.3. First, the energy allocation for co-products in the fuel production stage was based on the starch weight of each product, as identified by the Aspen Plus model. The energy co-product value of ethanol in the feedstock stage was allocated on a mass basis, using the weight of the starch content of the corn. The second method used the energy replacement method to allocate energy co-product values in the feedstock and fuel production stages.

GREET simulations showed that the GHG emissions from the petroleum gasoline fuel cycle were significantly greater than those from ethanol blends. The use of one liter of gasoline generated 3012 grams of CO_2-equivalent GHG emissions. Replacing one liter of gasoline with ethanol on a one-to-one energy basis requires 1.5 liter of ethanol because ethanol has a lower heating value than gasoline. After adjusting for energy equivalency, replacing one liter of gasoline with ethanol, using the Aspen-process/starch-based allocation method, reduced GHG emissions by 40% when ethanol is produced in dry mills, 36% with wet mill production, and 38% using the weighted industry average. With the replacement allocation method, the substitution of one liter of gasoline with ethanol reduced GHG emissions by 26% with dry mill production, 24% using wet mill production, and 25% for the weighted industry average.

5.5 Summary

Net energy balancing has become the accepted practice for measuring and comparing the energy efficiency and sustainability of various energy sources, most notable ethanol versus gasoline. Biofuels are generally regarded to be more sustainable than fossil fuels because much of the energy content of biofuels is derived from renewable sources. However, energy efficiency does not follow a strict linear relationship with environmental burden. Thus, NEB and other energy efficiency measurements do not provide sufficient information to analyze the environmental sustainability of biofuels. To address environmental issues, a more comprehensive method should be used, such as fuel-cycle analysis. Together, net energy balancing and fuel-cycle analysis provide a powerful analytical tool for evaluating the sustainability of biofuels. The data requirements for these procedures are extensive and accurate results depend on the most current and reliable data available. The examples used in this chapter to demonstrate the application of NEB and fuel-cycle analysis are based on farm and ethanol production data from USDA. Argonne National Laboratory's GREET model was used to simulate GHG emissions for the energy fuel cycle of ethanol compared to the gasoline fuel cycle. Results showed the advantage ethanol has over gasoline for both energy and environmental sustainability. When using the Aspen-process/starch-based allocation method to estimate co-product energy values, ethanol achieved an average NER of 1.67, i.e., there is a 67% energy gain from producing ethanol on an industry-average basis. This compares to a 0.80 NER for petroleum gasoline. Using the replacement allocation method, which is a more conservative approach, resulted in an average NER of 1.33. Ethanol also produced fewer GHG emissions compared to gasoline. With the Aspen-process/starch-based allocation method, substituting gasoline with ethanol resulted in a 38% GHG emission reduction. When using the replacement allocation method, ethanol reduced GHG emissions by 25%.

References

Argonne National Laboratory (2004) *The Greenhouse Gases, Regulated Emissions, and Energy Use in Transportation (GREET) Model*, Transportation Technology R&D Center, US Department of Energy (accessed 2004). Available from *http://www.transportation.anl.gov/software/GREET/index.html.*

Aspentech (2004) Advanced System for Process Engineering, Aspen Technology Inc., Cambridge, MA. Available at: http://www.aspentech.com/index.cfm

Delucchi, M. A. (1991) *Emissions of Greenhouse Gasses from the Use of Transportation Fuels and Electricity*, ANL/ESD/TM-22. US Department of Energy, Argonne National Laboratory, Center for Transportation Research, Argonne, IL.

Delucchi, M. A. (2003) *A Life-Cycle Emission Model (LEM): Life-Cycle Emissions from Transportation Fuels, Motor Vehicles, Transportation Modes, Electricity Use, Heating and Cooling Fuels, and Materials*, UCD-ITS-RR-03-17, Main Report, University of California, Davis.

Doering, O. C., III (2004) Energy systems integration: fitting biomass energy from agriculture into U.S. energy systems, commissioned paper presented at the Agricultural as a Producer and Consumer of Energy Conference, sponsored by Farm Foundation and Office of Energy Policy and New Uses, USDA, June 24–25, Arlington, VA (accessed 2004). Available at: http://www.farmfoundation.org/

Energy Information Administration (2004) *Electric Power Annual 2003*, DOE/EIA-0348 (2003), Office of Coal, Nuclear, Electric and Alternative Fuels, Energy Information Administration, the US Department of Energy, Washington, DC.

Keeney, D. R. and DeLuca, T. H. (1992) Biomass as an energy source for the midwestern U.S., *American Journal of Alternative Agriculture*, **7**(3), 137–143.

McAloon, A., Yee, W. and F. Taylor (2004) Production of ethanol from corn by the dry grind process, US Department of Agriculture, Agricultural Research Service. Paper presented at Corn Utilization and Technology Conference, Indianapolis, Indiana, 7–9 June 2004.

National Agricultural Statistics Service (2003) *Crop Production, 2002 Summary*, CR PR 2-1 (03), US Department of Agriculture, National Agricultural Statistics Service. Available at: http://www.nass.usda.gov:81/ipedb/

Pimentel, D. (1991) Ethanol fuels: energy security, economics, and the environment, *Journal of Agricultural and Environmental Ethics*, **4**, 1–13.

Pimentel, D. (2003) Ethanol fuels: energy balance, economics, and environmental impacts are negative, *Natural Resources Research*, **12**(2), 127–134.

Shapouri, H., Duffield, J.A. and Graboski, M. (1995) *Estimating the Net Energy Balance of Corn Ethanol*, AER Report No. 721. US Department of Agriculture, Economic Research Service, Washington, DC.

Shapouri, H., Duffield, J.A. and Wang, M. (2002) *The Energy Balance of Corn Ethanol: An Update*, Agricultural Economic Report No. 814, US Department of Agriculture, Office of Energy Policy and New Uses, Washington, DC.

Shapouri, H., Duffield, J.A. and Wang, M. (2003) The energy balance corn ethanol revisited. *American Society of Agricultural Engineers*, **46**(4), 959–968.

Stout, B. A. (1990) *Handbook of Energy for World Agriculture*, Elsevier Applied Science. London.

Turhollow, A. F. and Perlack, R. D. (1991) Emissions of CO_2 from energy crop production, *Biomass and Energy*, **1**(3), 129–135.

US Department of Energy (1980) *Energy Balances in the Production and End-Use of Alcohols Derived from Biomass*, DOE/PE/70151-T5, US Department of Energy, US National Alcohol Fuels Commission and the Office of Advanced Technology R&D, Washington, DC.

Vigon, B. W., Tolle, D. A., Cornaby, B. W., Latham, H. C., Harrison, C. L., Boguski, T. L., Hunt, R. G. and Sellers, J. D. (1993) *Life-Cycle Assessment: Inventory Guidelines and Principles*, EPA/600/R-92/245, Environmental Protection Agency, Risk Reduction Engineering Laboratory, Office of Research and Development, Cincinnati, OH.

Wang, M., Saricks, C. and Santini, D. (1999) *Effects of Fuel Ethanol Use on Fuel-Cycle Energy and Greenhouse Gas Emissions*, ANL/ESD-38. Center for Transportation Research, Argonne National Laboratory, Argonne, IL.

6

Life Cycle Assessment as an Environmental Sustainability Tool

Adisa Azapagic

6.1 Introduction

Life cycle assessment (LCA) is an environmental management tool that enables identification and quantification of environmental impacts of a product, process or activity from 'cradle to grave', or from extraction of raw materials to final disposal of waste. Although LCA has been used in some industrial sectors (e.g. energy) for several decades, it has received wider attention and methodological development only since the beginning of the 1990s. Its widespread use has been promoted in particular by the incorporation of life cycle thinking within various environmental management standards and legislative acts, including the EU Eco-Management and Audit Schemes (EMAS) (EC, 1993), ISO 14000 Environmental Management Systems (EMS) (ISO, 1996) and the EC Directive on Integrated Pollution Prevention and Control (IPPC) (EC, 1996).

Today, LCA is a well-established tool and is used in a variety of applications in industry, research and policy-making. Some of the applications include identification of life cycle 'hot spots', identification of opportunities for improvements, system optimization and design for the environment. A review and examples of some of these applications of LCA to various products and processes can be found in Azapagic (1999, 2002). Here, the focus is on the use of LCA as an environmental sustainability tool with a particular application to the renewable-based systems, which is illustrated by two different case studies: hydro-electricity generation and production of alcoholic spirits. However, a brief overview of the state-of-the-art of the LCA methodology is presented first.

Renewables-Based Technology Edited by J. Dewulf and H. Van Langenhove

6.2 The LCA Methodology: A Brief Overview

The LCA methodology is standardized by the ISO 14040–14043 standards (ISO, 1997, 1998a, 1998b, 1998c). As defined by ISO 14040 (ISO, 1997), LCA is a compilation and evaluation of the inputs, outputs and the potential environmental impacts of a product throughout its life cycle, from acquisition of raw materials through production, use and waste disposal (i.e. from cradle to grave). Figure 6.1 shows the life cycle stages normally considered in the LCA of a product. Although the standard refers to products only, LCA can also be used to calculate the environmental impacts of processes and technologies (Azapagic, 1999, 2002), services or activities (ISO, 1997).

According to ISO 14040 (ISO, 1997), the LCA methodology comprises the following four phases:

1 goal and scope definition (defined by ISO 14041);
2 inventory analysis (defined by ISO 14041);
3 impact assessment (defined by ISO 14042); and
4 interpretation (defined by ISO 14043).

The main steps in each phase and their interactions are shown in Figure 6.2. The following sections give a brief overview of the methodology.

6.2.1 *Goal and Scope Definition*

The process of conducting an LCA as well as its outcomes are largely determined by the goal and scope of a study. For example, the goal of the study may be to identify the 'hot spots' in a manufacturing process and to use the results internally in a company to reduce the environmental impacts from the process. Alternatively, the company may wish to use the result externally, either to provide the LCA data to customers who use their product as a raw material, or perhaps to market their product on the basis of LCA results. In each

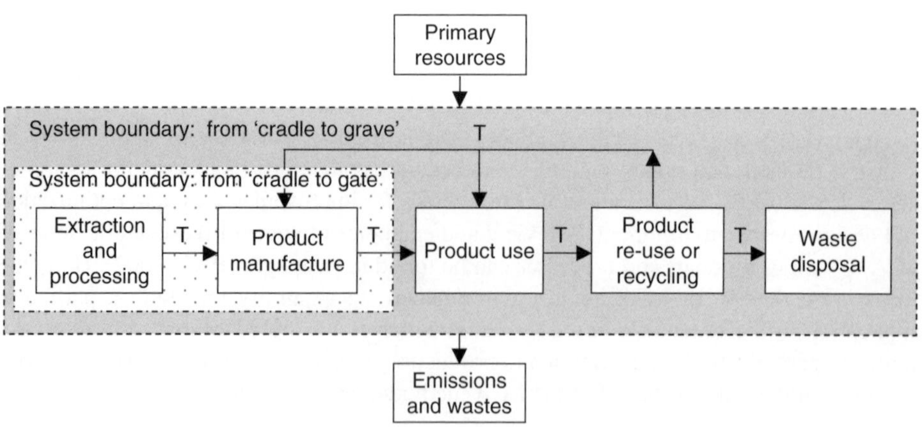

Figure 6.1 *Stages in the life cycle of a product from 'cradle to gate' and from 'cradle to grave'*
Source: Adapted from Azapagic (2004), reproduced with permission from John Wiley & Sons, Ltd.
Note: T = transport.

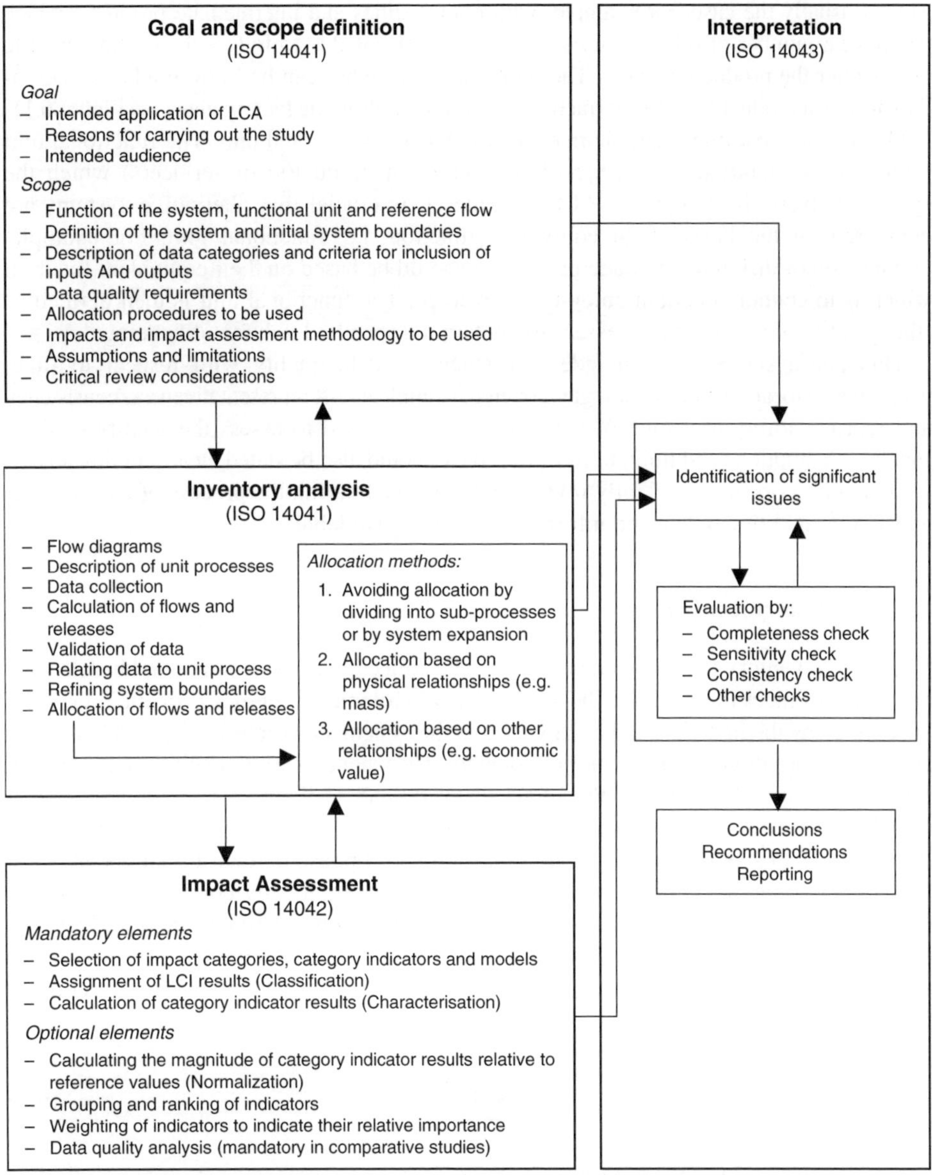

Figure 6.2 *The LCA methodological framework as defined by ISO 1040–14043*

case, the assumptions, data and system boundaries may be different so that it is important that these are defined in accordance with the goal of the study.

In full LCA studies, the system boundary is drawn to encompass all stages in the life cycle from extraction of raw materials to the final disposal, i.e. from cradle to grave. This is depicted in Figure 6.1. However, in some cases, the scope of the study will demand a different approach, where it is not appropriate or even possible to include all the stages in the life cycle.

This is usually the case, for example, with commodities and intermediate products, which can have a number of different uses so that it is not possible to follow their numerous life cycles after the production stage. The scope of such studies can be from 'cradle to gate' as they follow a product from the extraction of raw materials to the factory gate (see Figure 6.1).

One of the most important elements of LCA is the functional unit. The functional unit represents a quantitative measure of the output of product(s) or service(s) which the system delivers. In comparative LCA studies it is crucial that alternative systems are compared on the basis of an equivalent function, i.e. functional unit. For example, comparison of different beverage packaging should be based on their equivalent function which is to contain a certain amount of beverage. The functional unit is then defined as 'the quantity of packaging necessary to contain the specified volume of beverage'.

This phase should also include assessment of data quality with respect to time, geographical location and technologies covered. Completeness, representativeness, consistency and reproducibility are some of the criteria that are used to assess the quality of data. Finally, assumptions and limitations of the study should also be stated clearly in this phase.

Goal and scope are constantly reviewed and refined during the process of carrying out an LCA, as additional data and information become available.

6.2.2 *Inventory Analysis*

Life cycle inventory (LCI) analysis involves the collection of environmental burdens data necessary to meet the goals of the study. The environmental burdens (or interventions) are defined by the materials and energy used in the system, emissions to air, liquid effluents and solid wastes discharged into the environment. The main LCI steps are given in Figure 6.2; further details can be found in ISO 14041 (ISO, 1998a).

Following a preliminary system definition in the goal and scope definition phase, detailed system specification is carried out in the LCI phase to identify data needs. The system is defined as a collection of materially and energetically connected operations (for example, a manufacturing process) that performs some defined function. Detailed system characterization involves its disaggregation into a number of inter-linked subsystems. Environmental burdens are then quantified for each subsystem according to the equation:

$$B_j = \sum_{i=1}^{1} b_{j,i} x_i \tag{6.1}$$

where B_j is the total burden from the system, $b_{j,i}$ is burden (or intervention) j from process or subsystem i and x_i is a mass or energy flow associated with that subsystem.

If the system under study produces more than one functional output, then the environmental burdens from the system must be allocated among these outputs. This is the case, for example, with co-product, re-use and recycling systems; in LCA, such systems are known as multiple-function systems. Allocation is the process of assigning to each function of a multiple-function system only those environmental burdens which that function generates. ISO 14041 (1998a) recommends three methods for dealing with allocation:

- if possible, allocation should be avoided by disaggregating the given process into different sub-processes or by system expansion;
- if it is not possible to avoid allocation, then the allocation problem must be solved by using system modelling which reflects the underlying physical relationships among the functional units;
- where physical relationships cannot be established, other relationships, including economic value of the functional outputs, can be used.

The allocation method used will usually influence the results of LCA study so that the identification of an appropriate allocation method is crucial. Sensitivity analysis should be carried out in cases where the use of different allocation methods is possible to determine the influence of the allocation method on the results. For further discussion on allocation, see, for example, Azapagic and Clift (1999).

6.2.3 Impact Assessment

Life cycle impact assessment (LCIA) is the third LCA phase and its main purpose is to translate the environmental burdens quantified in LCI into the related potential environmental impacts (or category indicators). As shown in Figure 6.2, this is carried out within the following three mandatory steps (ISO, 1998b):

1 selection of impact categories, category indicators and LCIA models;
2 classification;
3 characterization.

The selection of impact categories, category indicators and LCIA models must be consistent with the goal and scope of the LCA study and must reflect the environmental issues of the system under study. Classification involves aggregation of environmental burdens into a smaller number of environmental impact categories to indicate their impacts on human and ecological health and the extent of resource depletion. The identification of impacts of interest is then followed by their quantification in the next, characterization step, as follows:

$$E_k = \sum_{j=1}^{j} e_{k,j} B_j \tag{6.2}$$

where $e_{k,j}$ represents characterization factor k for burden B_j showing its relative contribution to impact E_k. The characterization factors are calculated using appropriate LCIA models (see Section 6.3 for more detail on LCIA models).

A further three optional steps are also included within this phase:

1 normalization;
2 grouping;
3 weighting of impacts.

The impacts can be normalized with respect to the total emissions or extractions in a certain area and over a given period of time. This can help assess the extent to which an activity contributes to the regional or global environmental impacts. However, normalization results should be interpreted with care because of the lack of reliable data for many impacts at both the regional and global scales.

Grouping involves qualitative or semi-quantitative sorting and/or ranking of impacts and it may result in a broad ranking or hierarchy of impact categories with respect to their importance. For example, categories could be grouped in terms of high importance, moderate importance and low priority issues. Some methods that include grouping are the verbal-argumentative approach and the ranking method (Pennington *et al.*, 2004).

The final stage within LCIA is weighting of impacts, often referred to as valuation. It involves assigning weights of importance to the impacts to indicate their relative importance. As a result, all impact categories are aggregated into a single environmental impact function EI as follows:

$$\mathrm{EI} = \sum_{k=1}^{K} w_k E_k \tag{6.3}$$

where w_k is the relative importance of impact E_k.

Weighting is probably the most controversial step of the methodology mainly because it involves social, political and ethical value choices (Finnveden, 1997). At present, there is no consensus on how to aggregate the environmental impacts into a single environmental impact function, nor even on whether such aggregation is conceptually and philosophically valid.

6.2.4 *Interpretation*

The main objectives of this phase are to analyse results, reach conclusions, explain limitations and provide recommendations based on the findings of LCI and/or LCIA (ISO, 1998c). The main stages in this phase are listed in Figure 6.2.

Quantification of environmental impacts carried out in LCI and LCIA enables identification of the most significant issues and life cycle stages that contribute to these issues. This information can then be used to target these 'hot spots' for system improvements or innovation.

Sensitivity analysis should be carried out before the final conclusions and recommendations of the study are made. Data availability and reliability are some of the main issues in LCA since the results and conclusions of an LCA study will be determined by the data used. Sensitivity analysis can help identify the effects that data variability, uncertainties and data gaps have on the final results of the study and indicate the level of reliability of the final results of the study.

Finally, the findings and conclusions of the study are reported in accordance with the intended use of the study. The report should give a complete, transparent and unbiased account of the study as detailed in ISO 14040 (ISO, 1997). If the study is used externally, critical review by an independent agent should be carried out. Further details on the LCA methodology can be found in the ISO 14040–14043 standard series (ISO, 1997, 1998a–c).

6.3 LCIA Impact Categories as Indicators of Environmental Sustainability

Because of its ability to quantify environmental interventions and the related impacts, LCA lends itself naturally as a tool for assessing environmental sustainability. The use of LCA for these purposes can be twofold: identification of relevant or significant environmental indicators for a particular system (ISO, 1997) and assessment of its overall environmental sustainability, either for system improvements or for comparison with alternative systems.

Depending on the goal and scope of the study, environmental sustainability indicators could be identified either at the inventory or the impact assessment level. The former type of indicators are defined by the environmental burdens, i.e. the use of biotic (renewable) and abiotic (non-renewable) resources, air emissions, water discharges and solid waste. Examples of the indicators defined at the LCIA level include depletion of biotic and abiotic resources, global warming potential, ozone depletion and human toxicity. The use of environmental burdens as sustainability indicators is fairly straightforward as they are calculated by carrying out mass and energy balances for the system under study (Azapagic and Perdan, 2000). However, the use of appropriate impact categories as the indicators of environmental sustainability is much more complex and the results are more difficult to interpret. Therefore, the rest of this section focuses on the use of various LCIA methods to identify the appropriate sustainability indicators for products, processes and services.

A number of LCIA methods exist, but they are divided in two general groups:

1 problem-oriented approaches;
2 damage-oriented methods.

In the problem-oriented methods the environmental burdens are aggregated according to their relative contribution to the environmental effects that they might cause. The impacts most commonly considered in the problem-oriented approach include resource depletion, global warming, ozone depletion, acidification, eutrophication, photochemical oxidant formation, human toxicity and ecotoxicity. Typical examples of the problem-oriented approaches are the CML method (Heijungs *et al.*, 1992; Guinée *et al.*, 2001) and EDIP (Hauschild and Wenzel, 1998) method. Problem-oriented approaches are often referred to as ‘midpoint’ approaches because they link the environmental interventions from LCI somewhere at the intermediate position between the point of intervention and the ultimate damage caused by that intervention (see Figure 6.3). Damage-oriented methods, on the other hand, model the ‘endpoint’ damage caused by environmental interventions to ‘areas of protection’, which include human health, natural and human-made environment (Udo de Haes and Lindeijer, 2002). Typical examples of damage-oriented methods are EPS 2000 (Steen, 1999) and Eco-Indicator 99 (Goedkoop and Spriensma, 2001). The newly proposed IMPACT 2002+ method (Jolliet *et al.*, 2003) attempts to link the problem- and damage-oriented approaches in a common framework.

The following sections give a brief overview of the three most widely used LCIA methods: CML 2 Baseline, EPS 2000 and Eco-indicator 99. A simple example is used to illustrate how they can be used to identify relevant (or significant) indicators and to assess environmental sustainability of renewable-based energy systems in comparison with energy generated from non-renewable resources.

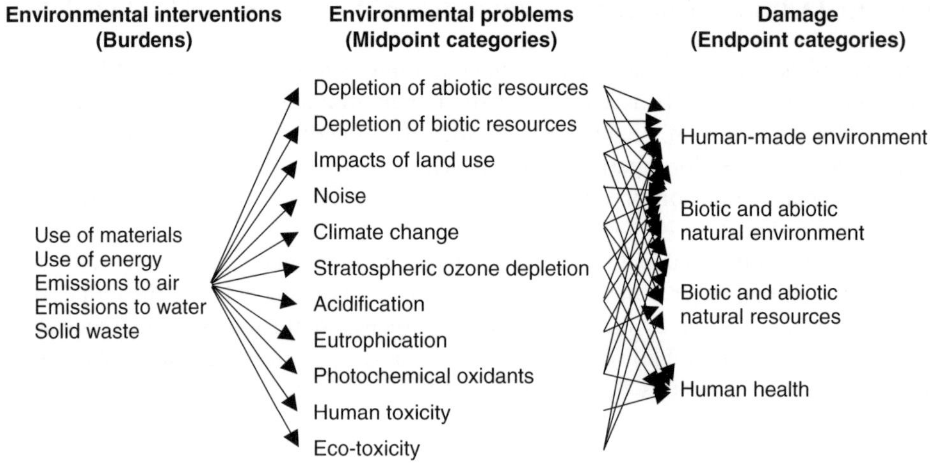

Figure 6.3 *The link between environmental interventions, problems (midpoint categories) and damage (endpoint categories) to the environment and human health*

6.3.1 CML 2 Baseline Method

In this method, described in detail by Guinée *et al.* (2001), environmental burdens are aggregated according to their relative contributions to the environmental problem or impact that they can potentially cause. The following impacts, defined as midpoint categories, are included (see Figure 6.3):

- depletion of abiotic resources;
- impacts of land use;
- climate change (global warming);
- stratospheric ozone depletion;
- acidification;
- eutrophication;
- photochemical oxidant formation (photochemical or summer smog);
- human toxicity;
- eco-toxicity (freshwater, marine and terrestrial).

Their definitions are given in Box 6.1. As shown in Box 6.1, the impacts are calculated relative to the characterization factor of a reference substance. For example, CO_2 is a reference gas for determining global warming potentials of other related gases, such as CH_4 and other VOCs; therefore, its characterization factor is 1 kg $CO_{2\ eq}$/kg CO_2 while that of CH_4 is 21 kg $CO_{2\ eq}$/kg CH_4. Note that the equations for calculating the impacts given in Box 6.1 are based on the general equation (6.2) for characterization of environmental impacts, defined in Section 6.2.3.

The impacts calculated by this method are categorized as potential rather than actual as they are quantified at the intermediate position between the point of environmental intervention and the damage caused, rather than at the endpoint.

Box 6.1 CML 2 Baseline Method: Definition of Environmental Impact Categories

Abiotic resource depletion potential includes depletion of fossil fuels, metals and minerals. The total impact is calculated as:

$$ADP = \sum_{j=1}^{J} ADP_j B_j \qquad \text{(kg Sb eq.)}$$

where B_j is the quantity of abiotic resource *j* used and ADP_j represents the abiotic depletion potential of that resource. This impact category is expressed in kg of antimony used, which is taken as the reference substance for this impact category. Alternatively, kg oil eq. can be used instead.

Impacts of land use are calculated by multiplying the area of land used (A) by its occupation time (t):

$$ILU = A \times t \qquad (m^2/yr)$$

Climate change is calculated as global warming potential (GWP), which equals to the sum of emissions of greenhouse gases multiplied by their respective GWP factors, GWP_j:

$$GWP = \sum_{j=1}^{J} GWP_j B_j \qquad \text{(kg } CO_2 \text{ eq.)}$$

where B_j represents the emission of greenhouse gas *j*. GWP factors for different greenhouse gases are expressed relative to the global warming potential of CO_2, which is therefore unity. The values of GWP depend on the time horizon over which the global warming effect is assessed. GWP factors for shorter times (20 and 50 years) provide an indication of the short-term effects of greenhouse gases on the climate, while GWP for longer periods (100 and 500 years) are used to predict the cumulative effects of these gases on the global climate.

Stratospheric ozone depletion potential (ODP) indicates the potential of emissions of chlorofluorohydrocarbons (CFCs) and other halogenated hydrocarbons to deplete the ozone layer and is expressed as:

$$ODP = \sum_{j=1}^{J} ODP_j B_j \qquad \text{(kg CFC-11 eq.)}$$

where B_j is the emission of ozone depleting gas *j*. The ODP factors are expressed relative to the ozone depletion potential of CFC-11.

Human toxicity potential (HTP) is calculated by taking into account releases toxic to humans to three different media, i.e. air, water and soil:

$$HTP = \sum_{j=1}^{J} HTP_{jA} B_{jA} + \sum_{j=1}^{J} HTP_{jW} B_{jW} + \sum_{j=1}^{J} HTP_{jS} B_{jS} \quad \text{(kg 1,4 - DB eq.)}$$

Box 6.1 (Continued)

where HTP_{jA}, HTP_{jW}, and HTP_{js} are toxicological classification factors for substances emitted to air, water and soil, respectively, and B_{jA}, B_{jW} and B_{jS} represent the respective emissions of different toxic substances into the three environmental media. The reference substance for this impact category is 1,4-dichlorobenzene.

Eco-toxicity potential (ETP) is also calculated for all three environmental media and comprises five indicators ETP_n:

$$ETP_n = \sum_{j}^{J} \sum_{i=1}^{I} ETP_{i,j} B_{i,j} \qquad \text{(kg 1,4-DB eq.)}$$

where n (n = 1–5) represents freshwater and marine aquatic toxicity; freshwater and marine sediment toxicity and terrestrial ecotoxicity, respectively. $ETP_{i,j}$ represents the eco-toxicity classification factor for toxic substance j in the compartment i (air, water, soil) and $B_{i,j}$ is the emission of substance j to compartment i. ETP is based on the maximum tolerable concentrations of different toxic substances in the environment by different organisms. The reference substance for this impact category is also 1,4-dichlorobenzene.

Photochemical oxidants creation potential (POCP) is related to the potential of VOCs and NOx to generate photochemical or summer smog. It is usually expressed relative to the POCP classification factors of ethylene and can calculated as:

$$POCP = \sum_{j=1}^{J} POCP_j B_j \qquad \text{(kg ethylene eq.)}$$

where B_j is the emission of species j participating in the formation of summer smog and $POCP_j$ is its classification factor for photochemical oxidation formation.

Acidification potential (AP) is based on the contribution of SO_2, NO_x and NH_3 to the potential acid deposition. AP is calculated according to the equation:

$$AP = \sum_{j=1}^{J} AP_j B_j \qquad \text{(kg } SO_2 \text{ eq.)}$$

where AP_j represents the acidification potential of gas j expressed relative to the AP of SO_2 and B_j is its emission in kg.

Eutrophication potential (EP) is defined as the potential of nutrients to cause over-fertilisation of water and soil, which can result in increased growth of biomass. It is calculated as:

$$EP = \sum_{j=1}^{J} EP_j B_j \qquad \text{(kg } PO_4^{3-} \text{ eq.)}$$

where B_j is an emission of species such as N, NO_x, NH_4^+, PO_4^{3-}, P, and COD; EP_j represent their respective eutrophication potentials. EP is expressed relative to PO_4^{3-}.

See Guinée *et al.* (2001) for a full description of the methodology.

Although this method could be used to measure sustainability of both non-renewable and renewable-based systems, the only impact category directly related to the use of renewables is land use. Other impacts relevant to these types of system, such as biodiversity and depletion of biotic resources, are not considered, which is an obvious shortcoming in this LCIA method. Nevertheless, a simple case study below illustrates how the method can be used to identify relevant sustainability indicators for renewable-based systems as well as those based on non-renewable resources.

Figure 6.4 shows the life cycle impacts of the generation of 1000 MJ of coal, oil and hydro-electricity. It is apparent that for all three electricity generation systems all the impacts shown in Figure 6.4 could potentially be relevant for assessing their environmental sustainability. However, judging by the absolute values of the impacts alone, abiotic resource depletion and global warming potential appear to be more 'significant' as they are one to two orders of magnitude higher than the other impacts. Using the same 'value' system, ozone depletion appears to be the least significant impact in this example. Other, more sophisticated, value systems can be used in the valuation stage to reveal stakeholder preferences and identify the most important impacts.

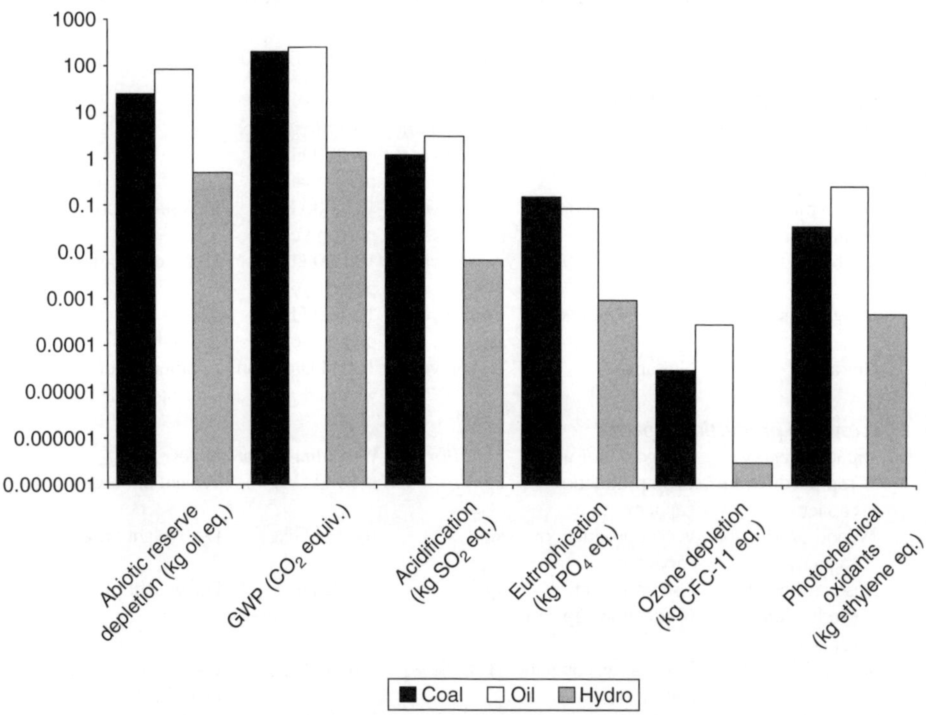

Figure 6.4 Using the CML problem-oriented method to identify relevant sustainability indicators and to compare environmental sustainability of electricity generation from non-renewable (coal and oil) and renewable (hydro) resources; only selected impact categories shown

Note: Functional unit: 1000 MJ.

6.3.2 Environmental Priority Strategies (EPS) 2000

The EPS 2000 method has been developed mainly as a tool for product development; however, it can be used externally for environmental declarations, purchasing decisions or for environmental accounting (Steen, 1999). The following 'areas of protection' are considered in this method:

- human health;
- ecosystem production capacity;
- abiotic stock resources;
- biodiversity;
- cultural and recreational values.

The indicators included under each area of protection are given in Box 6.2. Compared to the CML method, EPS 2000 includes more impact categories directly relevant to sustainability assessment of renewable-based systems, particularly the ecosystem production capacity and biodiversity indicators.

The pathway-specific characterization factors included in this method are determined by using three types of models: empirical, equivalency and mechanistic (Steen, 1999). In

Box 6.2 EPS 2000: Definition of Environmental Impact Categories

The following impact categories are included as default categories in the EPS 2000 method:

1 ***Human health impact categories***

Impact category	*Category indicator*	*Indicator unit*	*Weighting factor*	*Notes*
Life expectancy	Years of lost life (YOLL)	person years	85 000 ELU/ person years	
Severe morbidity and suffering	Severe morbidity	person years	100 000 ELU/ person years	Incl. starvation
Morbidity	Morbidity	person years	10 000 ELU/ person years	Like a cold or flu
Severe nuisance	Severe nuisance	person years	10 000 ELU/ person years	Causes a reaction to avoid the nuisance
Nuisance	Nuisance	person years	100 ELU/person years	Irritating, but not causing any action

2 ***Ecosystem production capacity***

Impact category	*Category indicator*	*Indicator unit*	*Weighting factor*	*Notes*
*Crop production capacity	Crop production capacity	kg	0.15 ELU/kg	Weight at harvest
*Wood production capacity	Wood production capacity	kg	0.04 ELU/kg	Dry weight basis
*Fish and meat production capacity	Fish and meat production capacity	kg	1 ELU/kg	Full weight of animals
Soil acidification	Base cation capacity soil	H^+ mole eq.	0.01 ELU/kg	Used if other indicators unavailable
*Production capacity for water	Production capacity for irrigation water	kg	0.003 ELU/kg	Must be acceptable for irrigation
*Production capacity for water	Production capacity for drinking water	kg	0.03 ELU/kg	Fulfilling WHO criteria on water (1997)

* *Expressed as decreased production capacity*

3 *Abiotic stock resources*

Impact category	*Category indicator*	*Indicator unit*	*Weighting factor*	
Depletion of element reserves	Reserves of an element	kg	Different values for each element	
Depletion of fossil reserves	Natural gas reserves	kg	1.1 ELU/kg	
Depletion of fossil reserves	Oil reserves	kg	0.506 ELU/kg	
Depletion of fossil reserves	Coal reserves	kg	0.0498 ELU/kg	
Depletion of mineral reserves	Reserves of a mineral	kg	Different values for each mineral	

4 *Biodiversity*

Impact category	*Category indicator*	*Indicator unit*	*Weighting factor*	*Notes*
Extinction of species	Normalized extinction of species	–	1.10E+11	Normalization is made with respect to the species extinct during 1990

5 *Cultural and recreational value*

These indicators are difficult to describe as they are highly specific and qualitative in nature; they are therefore defined on a case-by-case basis.

the empirical method, the characterization factor for a substance is determined by dividing the value of an indicator with an emission of the substance. The equivalency method uses equivalency factors to calculate the characterization factors. For example, to calculate the impacts of CO_2 on severe morbidity via global warming, the characterization factor for CO_2 with respect to severe morbidity is used and multiplied by the GWP for CO_2. The mechanistic method typically estimates the portion of an emitted amount of a substance that will reach a sensitive target and use dose-response information to calculate the response per mass of the emitted substance.

The environmental impacts are aggregated into a single impact function in the valuation stage as defined by equation 6.3. The weights w_k are obtained using various contingent valuation methods, e.g. 'willingness to pay' (WTP) or 'willingness to accept' (WTA). In WTP the weights are derived by asking people how much they would be prepared to pay to protect an environmental asset while the WTA method is based on willingness to accept loss of that asset.

The default weights in EPS 2000 are based on the willingness to pay to restore impacts on the areas of protection, as measured among OECD inhabitants (Steen, 1999). The weights are expressed in environmental load units (ELU) per unit of the indicator, with 1 ELU being equivalent to 1 Euro. Some of the weighting factors used in this method are listed in Box 6.2. In addition to WTP, other contingent valuation methods used include market and hedonistic pricing. Different uncertainty factors are incorporated into the EPS method to reflect the uncertainties inherent in the contingent valuation methods.

Figure 6.5 shows how the EPS 2000 method can be used to assess and compare environmental sustainability of non-renewable and renewable-based systems for the example of coal, oil and hydro-electricity. Figure 6.5 shows that the generation of 1000 MJ of electricity from oil costs around 60 ELU; the same amount of hydro-electricity costs only a fraction of this (0.25 ELU).

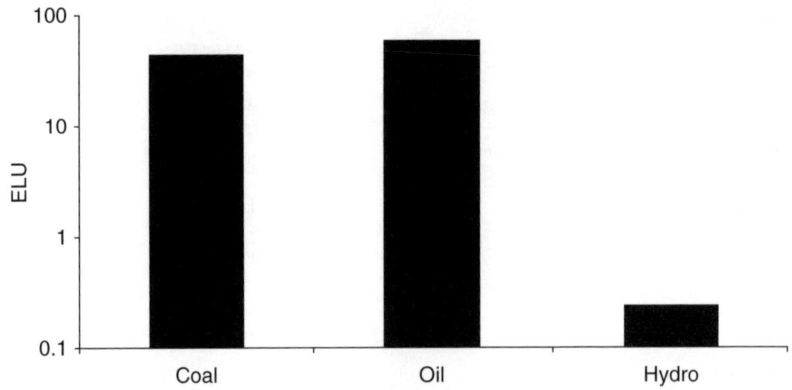

Figure 6.5 An example of the use of the EPS 2000 method to assess and compare environmental sustainability of electricity generation from non-renewable (coal and oil) and renewable (hydro) resources
Note: Functional unit: 1000 MJ.

6.3.3 Eco-Indicator 99

This method links the midpoint and endpoint impact categories, as shown in Figure 6.6. Similar to EPS 2000, its main intended use is as a tool for product design. Three areas of protection or types of damage are considered within Eco-Indicator 99:

- damage to human health, expressed in Disability Adjusted Life Years (DALYs);
- damage to ecosystem quality, expressed in terms of Potentially Disappeared Fraction (PDF; expressed as percentage) of plant species in a certain area over certain time;
- damage to mineral and fossil resources, expressed as additional energy requirement in MJ to extract future lower-grade ores.

The damage categories are defined briefly in Box 6.3 a full description of the methodology can be found in Goedkoop and Spriensma (2001).

As illustrated in Figure 6.6, following the LCI phase, the indicators quantifying these three damage categories are calculated in three steps:

1 resource and land use and fate analysis;
2 exposure and effect analysis;
3 damage analysis.

Various empirical fate and exposure models are used within Eco-Indicator 99 (see Goedkoop and Spriensma, 2001). Damage analysis involves calculation of the damage categories by multiplying the environmental interventions by the appropriate characterization factors (see equation 6.2). The damage categories can then be normalized using the normalization factors for Europe. Normalized results are often divided by the number of people in Europe, to show damage caused by one person per year.

To indicate the importance or contribution of different environmental impacts to each of the three damage categories, different weighting factors can be applied. The weights are derived from the cultural theory (Thompson *et al.*, 1990), taking into account three

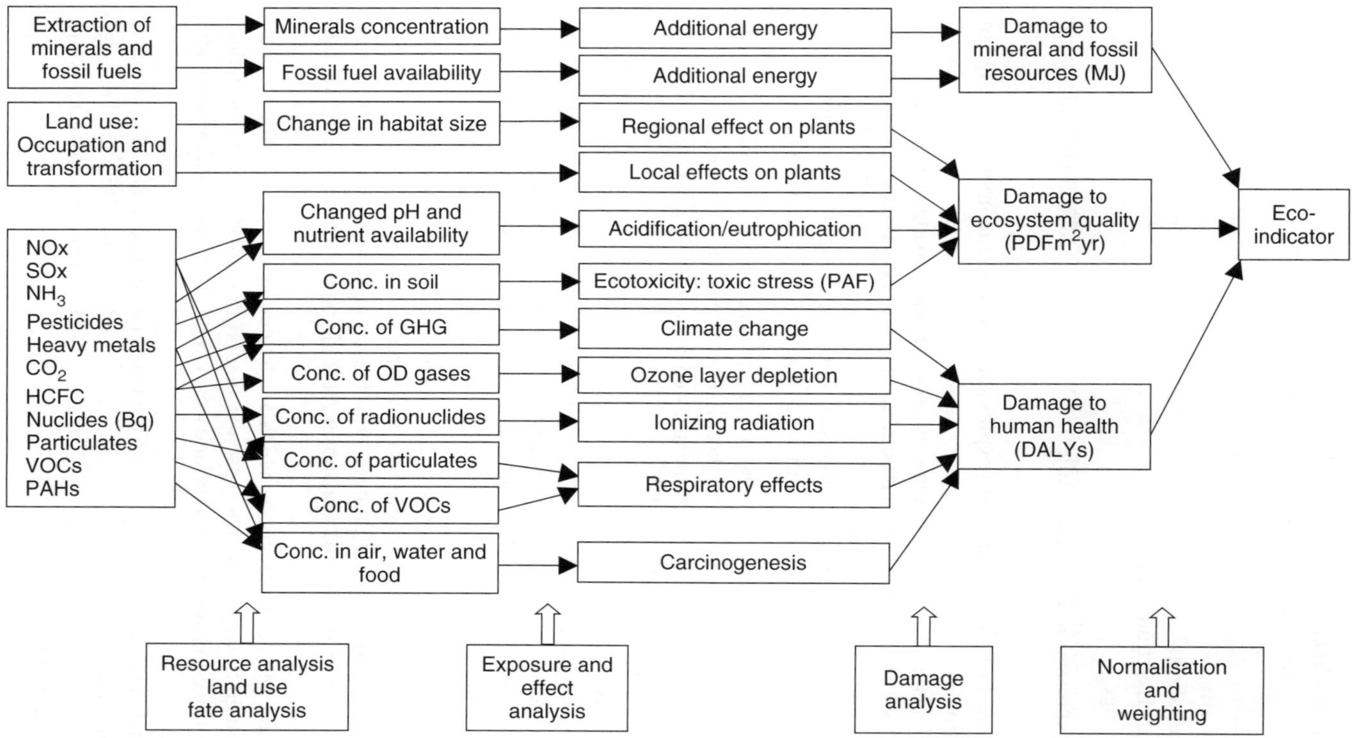

Figure 6.6 *Summary of the Eco-indicator 99 method*

Source: Based on Goedkoop and Spriensma (2001).

Notes: VOCs – volatile organic compounds; PAHs – polynuclear aromatic hydrocarbons; GHG – greenhouse gases; OD – ozone depletion; PAF – potentially affected fraction.

Box 6.3 Eco-Indicator 99: Definition of the Damage (Endpoint) Categories

1 ***Damage to human health***

This damage category comprises the following indicators:

- carcinogenesis;
- respiratory effects;
- ionizing radiation;
- ozone layer depletion; and
- climate change.

They are all expressed in Disability Adjusted Life Years (DALYs) and calculated by carrying out:

1 fate analysis, to link an emission (expressed in kg) to a temporary change in concentration;
2 exposure analysis, to link the temporary concentration change to a dose;
3 effect analysis, to link the dose to a number of health effects, such as occurrence and type of cancers; and finally,
4 damage analysis, to link health effects to DALYs, using the estimates of the number of Years Lived Disabled (YLD) and Years of Life Lost (YLL).

For example, if a cancer causes a ten-year premature death, this is counted as 10 YLL and expressed as 10 DALYs. Similarly, hospital treatment due to air pollution has a value of 0.392 DALYs/yr; if the treatment lasted 3 days (or 0.008 years), then the health damage is equal to 0.003 DALYs.

2 ***Damage to ecosystem quality***

The indicators within this damage category are expressed in terms of Potentially Disappeared Fraction of Plant Species (PDF) due to the environmental load in a certain area over certain time. Therefore, damage to ecosystem quality is expressed as PDF m^2 yr. The following indicators are considered:

- ecotoxicity, which is expressed as the percentage of all species present in the environment living under toxic stress (Potentially Affected Fraction or PAF). As this is not an observable damage, a rather crude conversion factor is used to translate toxic stress into real observable damage, i.e. convert PAF into PDF;
- acidification and eutrophication, which are treated as one single impact category. Damage to target species (vascular plants) in natural areas is modelled. The model used is for the Netherlands only, and it is not suitable to model phosphates;
- land use and land transformation, which are based on empirical data of occurrence of vascular plants as a function of land use types and area size. Both local damage in the area occupied or transformed and regional damage to ecosystems are taken into account.

For ecosystem quality, two different approaches are used:

(a) Toxic, acid and the emissions of nutrients go through the following three steps:

1. fate analysis, linking the emissions to concentrations;
2. effect analysis, linking concentrations to toxic stress or increased nutrient or acidity levels; and
3. damage analysis, to link these effects with the PDF of plant species.

(b) Land use and transformation are modelled on the basis of empirical data on the quality of ecosystems, as a function of the type of land use and area size.

3 *Damage to resources*

Two indicators are included here: depletion of minerals and fossil fuels. They are expressed as additional energy in MJ that will be needed for extraction in the future due to a decreasing amount of minerals and fuels. Geostatical models are used to relate availability of a mineral resource to its remaining amount or concentration. For fossil fuels, the additional energy is based on the future use of oil shale and tar sands.

Resource extraction is modelled in two steps:

1. resource analysis, which is similar to fate analysis, as it links an extraction of a resource to a decrease in its concentrations (through geostatical models);
2. damage analysis, linking decreased concentrations of resources to the increased effort for their extraction in the future.

See Goedkoop and Spriensma (2001) for a full description of the methodology.

different general types of people and their perspectives on life: hierarchist, egalitarian and individualist. For example, in the human health category, the hierarchist would assign the highest importance to the respiratory effects from inorganic substances, a relatively low importance to carcinogenic effects and almost no importance to radiation effects.

Finally, the damage categories are aggregated into a single environmental impact function – an eco-indicator – by using a further sets of weights, again applying the hierarchist, egalitarian and individualist perspectives. In the hierarchist approach, the contribution of human health and ecosystem quality to the total eco-indicator is 40% each. If the egalitarian perspective is applied, ecosystem quality contributes 50% to the eco-indicator, while human health and damage to resources contribute 30% and 20%, respectively. In the individualist perspective, human health is by far the most important category, contributing 55% to the total impact. For all three perspectives, land use, respiratory effects from inorganic substances and fossil fuels are the most important impact categories while respiratory effects from organic substances (summer smog), ionizing radiation and ozone depletion play an insignificant role.

Therefore, this method involves a two-stage weighting process: first, the impact categories are weighted to aggregate them into the three damage categories, followed by the weighting of the damage categories to derive the eco-indicator. The individual weights for the impact categories for different perspectives can be found in Goedkoop and Spriensma (2001).

Figure 6.7 illustrates the type of results obtained by the Eco-Indicator 99 method for the example of coal, oil and hydro-electricity. The damage analysis, carried out for the

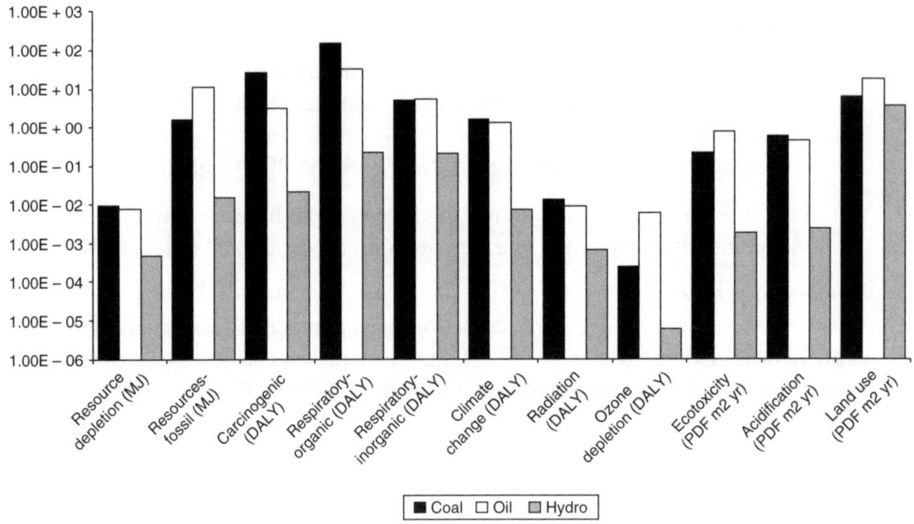

Figure 6.7 *Using Eco-indicator 99 (H, A) to identify relevant sustainability indicators and to compare environmental sustainability of hydro- and coal-based electricity*
Notes: Functional unit: 1000 MJ, H – hierarchist perspective, A – average weighting set: Human Health = 40%; Ecosystem Quality = 40%, Resources = 20%.

hierarchist perspective and using the average weights (see Figure 6.7 for details), shows that the most significant (or relevant) impacts are respiratory effects from organic substances, carcinogenic effects, and depletion of fossil resources and land use. Similar to the results obtained by the CML method, ozone depletion appears to be the least significant for all three systems shown. It should be noted that this method also uses limited number of categories directly relevant to renewable-based systems, e.g. land use and transformation.

6.3.4 Choosing the LCIA Method and Indicators

The advantages and disadvantages of using different LCIA methods and indicators have been extensively discussed (see e.g. Bare *et al.*, 2002). Some practitioners prefer midpoint indicators that describe the impacts earlier in the cause–effect mechanism because of the additional uncertainties and forecasting considered necessary when modelling indicators closer to an endpoint (Pennington *et al.*, 2004). These methods are also more transparent and easier to understand by non-specialists as they do not incorporate *a priori* weighting factors but allow decision-makers to derive their own weights by using various socio-economic techniques (Bare *et al.*, 2002; Finnveden *et al.*, 2002; Udo de Haes *et al.*, 2002). A typical example of such methods would be various multi-criteria decision analysis approaches, such as analytic hierarchy process, multi-attribute utility theory, and so on (see, e.g. Azapagic and Perdan, 2005a).

Others, on the other hand, argue that choosing indicators later in the cause–effect mechanism (closer to an endpoint) facilitates more structured and explicit weighting across impact categories (Bare *et al.*, 2002; Pennington *et al.*, 2004; Udo de Haes *et al.*, 2002). However, the main problem with most of these methods is that they assume that the weights derived in fairly specific and controlled conditions are applicable universally in all decision situations (Azapagic and Perdan, 2005b). Furthermore, the weighted results are difficult to interpret and could easily be misunderstood or misused.

In principle, any of the above LCIA methods could be used for assessing environmental sustainability of products, technologies or services. Although both the EPS and Eco-Indicator methods have been developed with product design in mind, there is little precluding their use in any other applications. Therefore, the choice of the LCIA method is open to the practitioner and stakeholders. However, the choice of the 'right' method may depend on the goal and scope of the LCA study, the type of system to be assessed (e.g. non-renewable or renewable-based) and data available (ISO, 1997). Sometimes it may even be necessary to use several methods, particularly in comparative studies, as part of sensitivity analysis to find out if different methods favour different alternatives being compared.

This is illustrated in Figures 6.4, 6.5 and 6.7. In this simple example, all three LCIA methods used show that hydro-electricity is clearly preferred environmentally over electricity generated from coal and oil, regardless of the fact that the EPS 2000 method includes more indicators directly relevant to the renewable-based systems. So, if the goal of the study is to identify the most sustainable system of the alternatives compared, in this particular case there is no difference in the final outcome, whichever LCIA method used. However, if the goal is identification of the relevant or most significant impacts, the use of different methods may give different answers. Thus, the choice of the LCIA method, should be made carefully and the results interpreted with care.

The following section gives further, more complex examples of the use of LCA for the assessment of environmental sustainability of renewable-based systems.

6.4 Using LCA to Assess Environmental Sustainability

This section illustrates the use of LCA to assess and compare the environmental sustainability of two renewable-based systems producing alcoholic spirits: whisky produced from cereal grain and neutral spirit (e.g. vodka) produced from whey, a waste by-product from cheese production. Although these two alcoholic spirits have a different taste, they deliver the same (or at least similar) 'social' function, so that a comparison of their environmental sustainability is legitimate as, for instance, is comparison of any alternative food products. The idea here is to find out if using a waste material from one industry (i.e. applying the industrial ecology principles) is environmentally more sustainable than using (renewable) agricultural feedstock.

Whey is a traditional by-product from cheese manufacturing and is often discharged into the environment as waste. As it contains lactose, it can be utilized as a viable feedstock for neutral spirit production by fermentation and distillation to produce almost pure ethanol. If desired, the neutral spirit can be flavoured by the addition of different plant extracts (e.g. juniper or coriander) to produce flavoured spirits, such as gin. One of the obvious advantages of using waste materials such as whey to manufacture neutral alcoholic spirit is that pre-processing requirements are minimized due to sugars already being present in the waste material. However, potential disadvantages may be the dilute nature of whey which contains low level of ethanol, thus increasing demands for processing energy to produce the spirit.

The whisky production system is shown in Figure 6.8. The system boundaries include all activities from extraction of raw materials, through crop farming and the manufacturing process to the product leaving the factory gate. The whisky manufacturing sub-system includes the malting plant, grain distillery and maturation warehouse. Cereal grain, either wheat or maize, is used as the main raw material for the process. The malting plant provides malted barley used in the distillery plant as a source of enzymes. The whisky produced in the distillery is then matured for a minimum of three years and usually up to eight or 12 years, during which time a considerable amount of the product is lost through evaporation (2% v/v per year). In this example, a three-year-old whisky is considered. The use and product disposal stages are outside the scope of the study and are therefore not included here, making this in effect a 'cradle-to-gate' study. Further detail on this system can be found in Azapagic (2002).

The cradle-to-gate LCA process of the whey alcohol system is shown in Figure 6.9. The system comprises the whey alcohol plant and fuel, water and energy supply sub-systems. As whey is considered a waste material in this study, no environmental burdens have been allocated to its production. The functional unit for both systems is defined as the 'production of one litre of pure alcohol'.

Comparison of the LCA results for the two systems is given in Figure 6.10. Due to space limitations, only the results obtained by the (modified) CML problem-oriented method are presented and discussed; however, either of the two damage-oriented problems discussed in the preceding section could be used for the same purposes. As shown in

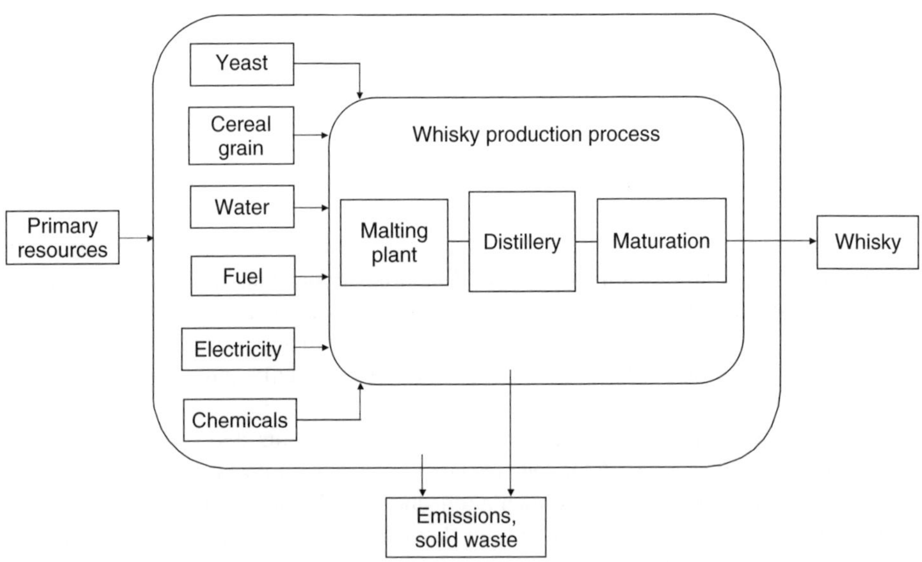

Figure 6.8 The life cycle of the whisky production system from 'cradle to gate'

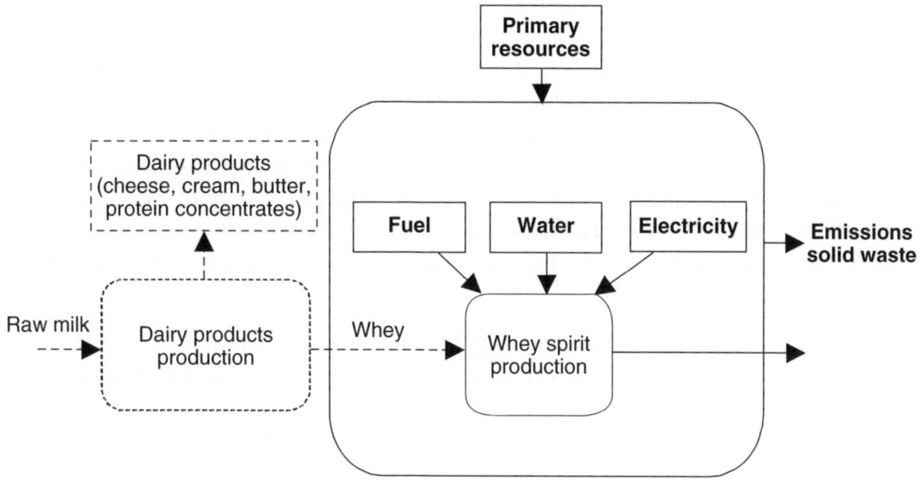

Figure 6.9 The life cycle of the whey spririt production system from 'cradle to gate'

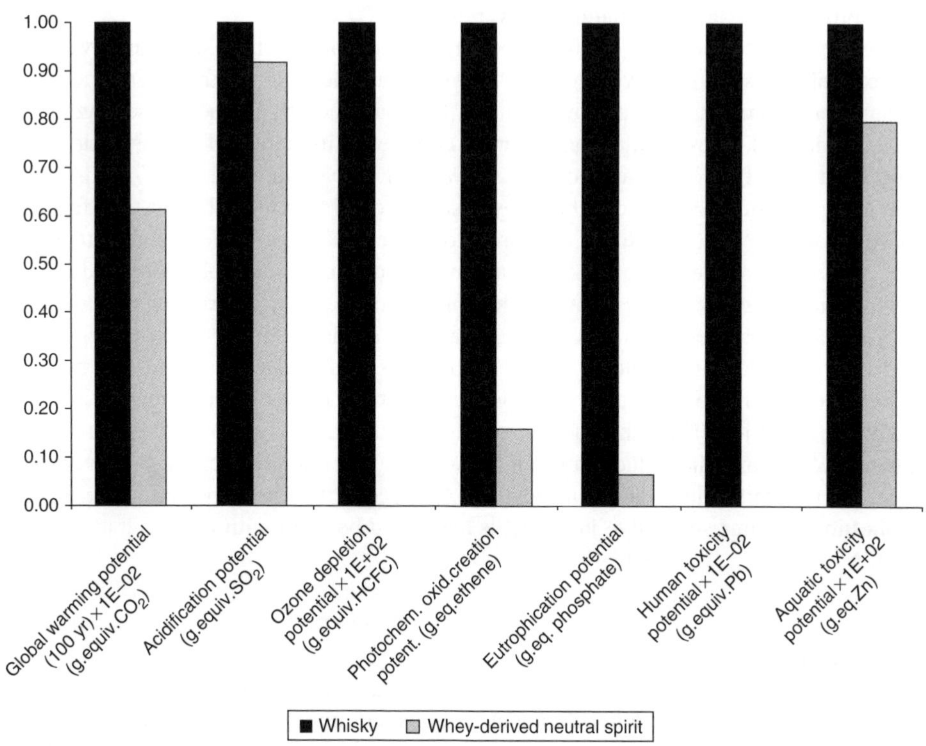

Figure 6.10 Comparison of the whisky and whey-derived spirit systems
Note: Functional unit: 1 litre of pure alcohol.

Figure 6.10, per litre of pure alcohol, all environmental impacts from the whey-alcohol system are lower than those from the three-year-old whisky. For human toxicity, eutrophication and ozone depletion, the difference is almost 100% in favour of the whey-derived spirit.

Thus, on the basis of environmental criteria alone, the choice seems obvious: the whey-derived alcoholic spirit is more sustainable than whisky. However, as the whisky production system is based on a traditional, 100-year-old manufacturing process, it would be interesting to examine if various improvements to the system could make it more environmentally sustainable than the whey-based alcohol. This is discussed next.

Three 'hot spots' are responsible for almost all environmental impacts from the whisky system:

- manufacturing of fertilizers for application in the farming subsystem;
- farming activities to produce cereal grain for use in the whisky manufacturing process;
- the whisky manufacturing process.

Acidification, eutrophication, human and aquatic toxicity are mainly attributable to the farming activities and the manufacture of fertilizers. Global warming is contributed to equally by farming, manufacture of fertilizers, whisky manufacturing and maturation process. Ozone depletion and photochemical oxidants arise mainly from the manufacturing process and the spirit maturation warehouse.

Therefore, these three stages should be targeted for maximum system improvements. For example, feasible improvement options would include sourcing the cereal grain from organic farms and various improvements in the manufacturing process (see Azapagic, 2002, for details). These could lead to almost complete elimination of human toxicity and significant reduction of aquatic toxicity (80%) and global warming (45%). This would then make aquatic toxicity from the whey system around five times higher than from the improved whisky system; global warming would be 20% higher. However, at the same time, these improvements would lead to a two-fold increase in eutrophication and acidification from the whisky system, due to the application of organic fertilizer (manure) and the associated emissions of nitrates to water as well as the lower productivity per area of land in organic farming (Audsley, 1997). Therefore, whey-alcohol would still be more sustainable with respect to eutrophication, acidification, photochemical oxidant creation and ozone depletion. Thus, the question is then whether a nearly complete elimination of aquatic toxicity and the reduction of global warming are more important than a twofold increase in eutrophication and acidification. As discussed earlier, answering these types of question requires subjective judgements and could be dealt within the valuation stage of LCA.

6.5 Summary

A move towards sustainable development requires a more holistic life cycle approach to understanding and managing human interactions with the environment. LCA is a tool that enables and supports such an approach as it embodies life cycle thinking and can provide a full assessment of the environmental consequences of various human activities. This chapter has attempted to summarize and illustrate the use of LCA as a tool for assessing environmental sustainability of some of these activities, with a particular focus on the renewable-based systems.

The LCA methodology is now at the stage where LCA as a tool can be applied consistently and in a standardized form. However, some further methodological developments are still needed for the LCIA models and methods. The three most widely used LCIA methods have been reviewed in this chapter: CML 2 Baseline, EPS 2000 and Eco-Indicator 99. Their application has been demonstrated in several examples to illustrate how LCA can be used as a tool for the identification of environmental sustainability indicators, for identification of the most dominant stages or 'hot spots' and opportunities for system improvements. The use of LCA for a sustainability comparison of alternative systems has also been discussed.

LCA is a powerful tool; however, it must be applied in a rigorous manner and its results interpreted carefully if it is to play a more important role as a sustainability decision-support tool. Although LCA has come a long way since the 1990s when the methodology started to be developed, it still has a long way to go before it is accepted fully and used without scepticism. For this to happen, further development of LCA has to be directed towards its meeting the needs and expectations of stakeholders and decision-makers. It is only then that we can expect LCA to become integrated fully into corporate and public policy decision-making.

References

Audsley, E. (1997) *Harmonisation of Environmental Life Cycle Assessment for Agriculture*. Final Report. Concerted Action AIR3-CT94-2028. Silsoe, UK.

Azapagic, A. (1999) Life cycle assessment and its application to process selection, design and optimisation, *Chemical Engineering Journal*, **73**, 1–21.

Azapagic, A. (2002) Life cycle assessment: a tool for identification of more sustainable products and processes, in Clark, J. and Macquarrie, D. (eds), *Handbook of Green Chemistry and Technology*, Blackwell Science, Oxford, pp. 62–85.

Azapagic, A. (2004) Life cycle thinking and life cycle assessment (LCA), in Azapagic, A., Perdan, S. and Clift, R. (eds), *Theory and Practice of Sustainable Development: Case Studies for Engineers and Scientists,* John Wiley & Sons, Chichester, pp. 426–437.

Azapagic, A. and Clift, R. (1999) Allocation of environmental burdens in multiple-function systems, *Journal of Cleaner Production*, **7(2)**, 101–119.

Azapagic, A. and Perdan, S. (2000) Indicators of sustainable development for industry: a general framework. *IChemE Transactions B*, **78(B4)**, 243–261.

Azapagic, A. and Perdan, S. (2005a) An integrated sustainability decision-support framework: problem structuring, Part I. *International Journal of Sustainable Development & World Ecology*, **12**, 112–131.

Azapagic, A. and Perdan, S. (2005b) An integrated sustainability decision-support framework: methods and tools for problem analysis, Part II, *International Journal of Sustainable Development and World Ecology*, **12**, 98–111.

Bare J. C., Udo de Haes, H. A. and Pennington, D. W. (2002) An international workshop on life-cycle impact assessment sophistication, EPA/600/R-00/023, Environmental Protection Agency, Cincinnati, OH. Available at: www.epa.gov/ORD/WebPubs/EPA600r-00023.htm

EC (1993) Council Regulation No. 1836/93 of 29 June 1993 Allowing voluntary participation by companies in the industrial sector in a community eco-management and audit scheme (1993), *Official Journal of the European Communities*, L168, 10 July, pp. 1–18, HMSO, London.

EC (1996) Council Directive 91/61/EC Concerning Integrated Pollution Prevention and Control, *Official Journal of the European Communities*, No. L257, 10 October, HMSO, London.

Finnveden, G. (1997) Valuation methods within LCA: where are the values? *International Journal of LCA*, **2**, 163–169.
Finnveden, G., Hofstetter, P., Bare, J., Basson, L., Ciroth, A. and Mettier, T. (2002) Normalization, grouping and weighting in life cycle impact assessment, in Udo de Haes, H., Jolliet, O., Finnveden, G., Goedkoop, M., Hauschild, M. and Hertwich, E. (eds), *Life-cycle Impact Assessment: Striving Towards Best Practice*, SETAC Press, Pensacola, FL.
Goedkoop, M. and Spriensma, R. (2001) *The Eco-indicator 99: A Damage Oriented Method for Life Cycle Assessment, Methodology Report*, 3rd edn, Pré Consultants, Amersfoort, The Netherlands.
Guinée, J. B., Gorrée, M., Heijungs, R., Huppes, G., Kleijn, R., van Oers, L., Wegener Sleeswijk, A., Suh, S., Udo de Haes, H. A., de Bruijn, H., van Duin, R. and Huijbregts, M. A. J. (2001) *Life Cycle Assessment: An Operational Guide to the ISO Standards*, Parts 1, 2a and 2b, Kluwer Academic Publishers, Dordrecht.
Hauschild, M. and Wenzel, H. (1998) *Environmental Assessment of Products*, vol. 2: *Scientific Background*, Chapman and Hall, London.
Heijungs, R., Guinée, J. B., Huppes, G., Lankreijer, R. M., Udo de Haes, H. A., Wegener Sleeswijk, A., Ansems, A. M. M., Eggels, P. G., van Duin, R. and Goede H. P. (1992) *Environmental Life Cycle Assessment of Products: Guide*, Report 9266, Centre of Environmental Science (CML), Leiden, The Netherlands.
Jolliet, O., Margni, M., Charles, R., Humbert, S., Payet, J., Rebitzer, G. and Rosenbaum, R. (2003) IMPACT 2002+: a new life cycle impact assessment methodology, *International Journal of LCA*, **8(6)**, 324–330.
ISO (1996) *ISO 14001: Environmental Management Systems – Specification with Guidance for Use*, ISO, Geneva.
ISO (1997) *ISO/DIS 14040: Environmental Management – Life Cycle Assessment – Principles and Framework*, ISO, Geneva.
ISO (1998a) *ISO/DIS 14041: Environmental Management – Life Cycle Assessment – Goal and Scope Definition and Life Cycle Inventory Analysis*, ISO, Geneva.
ISO (1998b) *ISO/CD 14042: Environmental Management – Life Cycle Assessment – Life Cycle Impact Assessment*, ISO, Geneva.
ISO (1998c) *ISO/CD 14043: Environmental Management – Life Cycle Assessment – Life Cycle Interpretation*, ISO, Geneva.
Pennington, D. W., Potting, J., Finnveden, G., Lindeijer, E., Jolliet, O., Rydberg, T. and Rebitzer G. (2004) Life cycle assessment, Part 2: current impact assessment practice. *Environment International*, **30**, 721–739.
Steen, B. (1999) *A Systematic Approach to Environmental Priority Strategies in Product Development (EPS), 2000-General System Characteristics & 2000-Models and Data*, Chalmers, CPM report 1999: 4 and 5. Available at: http://eps.esa.chalmers.se/download.htm
Thompson, M., Ellis, R. and Wildavsky, A. (1990) *Cultural Theory*, Westview Press, Boulder, CO.
Udo de Haes, H. A. and Lindeijer, E. (2002) The conceptual structure of life-cycle impact assessment, in Udo de Haes, H., Jolliet, O., Finnveden, G., Goedkoop, M., Hauschild, M. and Hertwich, E. (eds), *Life-cycle Impact Assessment: Striving Towards Best Practice*, SETAC Press, Pensacola, FL.
Udo de Haes, H., Jolliet, O., Finnveden, G., Goedkoop, M., Hauschild, M. and Hertwich, E. (eds) (2002) *Life-cycle Impact Assessment: Striving Towards Best Practice*, SETAC Press, Pensacola, FL.

7

Exergy

Jo Dewulf and Herman Van Langenhove

7.1 Introduction

The term 'sustainability' has numerous definitions. Currently, the definition by Brundlandt of 1987 is widely accepted: 'Sustainable development is a development that meets the needs of the present generation without compromising the needs of the future.' According to this definition, there is no doubt that sustainable development is a topic in engineering and technology. Indeed, technology plays a central and essential role in sustainable development: it delivers the goods to fulfil the needs. At the same time, it threatens sustainable development due to its effects on the environment. There are several concepts today providing guidelines on how to organize the environmental component of technology in a more sustainable way: Clean Technology (Clift and Longley, 1995), Green Engineering (Anastas and Zimmerman, 2003) and Industrial Ecology (Graedel and Allenby, 1996; Lowe *et al.*, 1997). All the principles covered in these concepts are implicitly based on the consideration of exchanges, conversions and effects of mass and energy that are related to the life cycle of a product or a process chain. In these concepts, it is obvious that renewable resources can play a major role in achieving environmental sustainability.

Basically, technology is embedded in the physical chemical realm, as exemplified in Figure 7.1. Technology is situated in the industrial ecosystem that consists of a network of production and consumption of goods, and a reservoir of waste materials that can be considered as resource material. Optimizing the network within the industrial ecosystem is typically studied in industrial ecology. However, technology also interacts with the natural environment. Natural resources pass the natural/industrial ecosystem boundary and are taken into the production chain. The resources can be both of renewable and

Renewables-Based Technology Edited by J. Dewulf and H. Van Langenhove

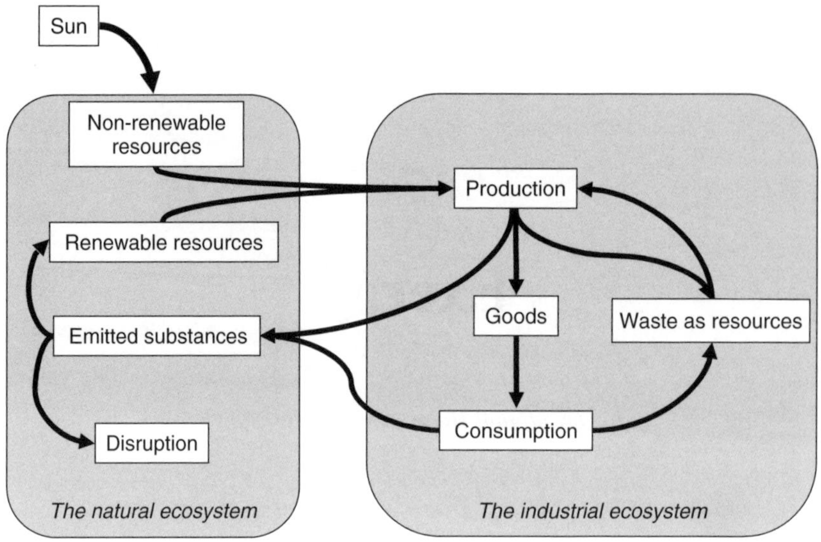

Figure 7.1 *Physical chemical realm. Energy and substance flows and processes directly and indirectly induced by technology development*

non-renewable origin, where the latter does not allow closing material cycles. In the other direction, emissions are taken into the natural ecosystem due to emissions generated at the production stage and later on by consumption and disposal of the generated products. These emissions contribute to two opposite phenomena in the natural environment. First, they are known to disrupt the natural system, ending up in global warming, habitat destruction, tropospheric ozone creation, stratospheric ozone depletion, eutrophication... Second, if they originate in renewable resources, e.g. wood and crops, they bring elements such as carbon back into the natural ecosystem, allowing the material cycle to be closed back through the production of renewable resources by photosynthesis.

From the substance and energy flow diagram, a scheme that is idealized in terms of sustainability can be drawn (Figure 7.2). The industrial ecosystem there consists of production that results only in utilizable goods and in consumer products that can be either re-used as a resource or end up in emissions into the natural ecosystem, closing the material cycle without disruptions. At the same time, the production only makes use of renewable resources and recovered waste materials. This system is driven by solar energy. Non-renewable resources are no longer used and no 'waste' is produced that threatens the ecosystem and human health. By accomplishing these conditions, the method of material exchange does not endanger the possibilities of the future generations to fulfil their needs. However, the current industrial ecosystem represented in Figure 7.1 is not solely driven by solar energy embedded in renewable resources, but also by non-renewable inputs. A number of emissions are not closing the materials cycle, e.g. gaseous carbon dioxide emissions from fossil resource combustion.

Apart from compatibility with the natural environment through the appropriate nature of resources taken in and emissions generated, technology needs to be integrated as much as possible within the industrial network. Good integration results in an indirect environmental

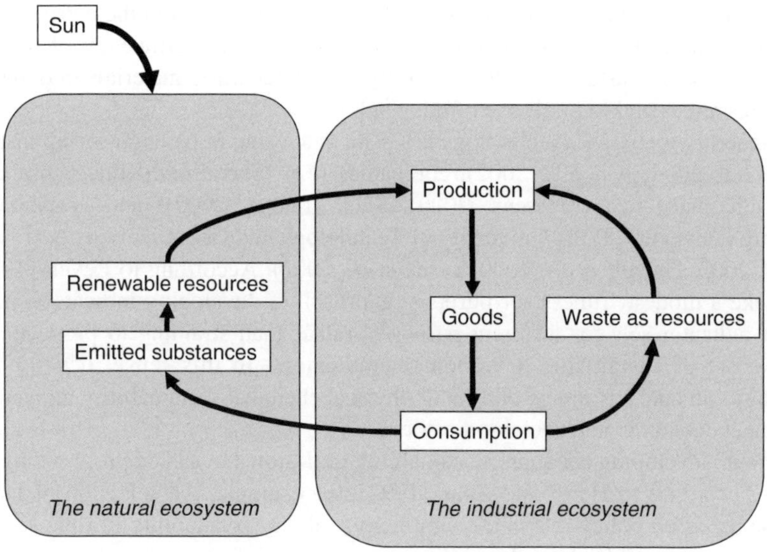

***Figure** 7.2 Physical chemical realm idealized in terms of sustainability. Energy and substance flows and processes directly and indirectly induced by a sustainable technology development*

benefit. Indeed, design of technology that uses waste materials as a resource saves virgin resource intake. At the same time, if the designed technology ends up in waste products that have resource potential, it not only reduces the disruption by emissions because it avoids them, but simultaneously it allows the industrial ecosystem to save virgin resources intake.

Finally, technological sustainability benefits from high efficiency because of both economic and ecological reasons. Due to an efficient resource-to-product flow, it saves resources, whether they are renewable, non-renewable or waste. At the same time, high efficiency means a high share of products in the outgoing energy and substance flow, implicitly indicating limited waste emission. As a result, it can be stated that efficiency is the third concern with respect to the design of sustainable technology, next to environmental and industrial ecology concerns.

7.2 Assessment of Sustainability of Technology: Developing Metrics

If one aims at creating a global framework to assess the sustainability of technology, it should deal with – apart from the economic and social realms – the physical chemical realm with its energy and substance flows illustrated in Figures 7.1 and 7.2. Thus, there is a need to develop an assessment methodology that is able to cope with it in a consistent and quantifiable way. Stating that new technologies are more sustainable has to be argued by using metrics (Darton, 2002; Dewulf *et al.*, 2000; Lems *et al.*, 2002; Winterton, 2001). Metrics are substantial in the sustainability debate, especially because the term sustainability has a very broad sense, making it subject to both use and misuse.

Metrics that aim at quantifying the physical chemical sustainability of technology should pay attention to effects on the natural ecosystem, the optimization of the industrial ecosystem and efficiency. Renewability of resources extracted from the ecosystem is one of the key issues, next to the disrupting effect of emissions, efficiency of the process chain, the re-use of waste materials, the ability to transfer waste materials into the pool of waste as resources and to make use of this pool.

Sustainability metrics for technology assessment are sought by engineering institutions (e.g. CWRT, 2002; IChemE, 2002), companies (e.g. GlaxoSmithKline (Smith, 2001), Janssen Pharmaceutica (Liessens, 2000), Shell (Lange, 2002)) and academics (e.g. Oldenburg University, Delft University of Technology and Ghent University (Eissen and Metzger, 2002, Dewulf *et al.*, 2000; Lems *et al.*, 2002). According to Levett (1998), we should take a modest 'fitness-for-purpose' approach in developing indicators, i.e. using different indicator sets for different purposes, rather than straining to produce a single definitive set of sustainable development indicators. In this sense, development of technology can take advantage of a set of physical chemical sustainability indicators, that complement economic and social indicators.

Criteria in developing adequate sustainability indicators have been proposed by several authors. According to Hardi and Zdan (1997), for example, the selection of indicators should be based on policy relevance, simplicity, validity, availability of time-series data, good quality, affordable data, the ability to aggregate information, sensitivity to small changes and reliability. Specifically in the field of environmental indicators, the OECD (1998) has established a set of three main criteria. The first criterion deals with policy relevance and utility for all users, i.e. being representative, simple, easy to interpret, able to show trends over time, responsive to changes, providing a basis for international comparisons, being either national in scope or applicable to regional environmental issues of national significance, and having a threshold or reference value against which to compare it. Second, the indicator should be analytically sound: theoretically well founded in technical and scientific terms, based on international standards and with international consensus about its validity and lending itself to being linked to economic models, forecasting and information systems. The third criterion is measurability so that the indicators are readily available or made available at a reasonable cost–benefit ratio, adequately documented, of known quality and updated at regular intervals in accordance with reliable procedures.

Another issue in developing indicators is the number of indicators, typically exemplified by the so-called information pyramid (Singh and Moldan, 2002). Starting from numerous raw data, information is aggregated into a limited number of indicators or even into one index. It is evident that higher aggregation levels make the information more comprehensive and surveyable, but simultaneously make it less transparent and more able to be lost.

In developing sustainability indicators, the above considerations should be taken into account, although the experience shows that it is already hard to find indicators that comply with a subgroup of the criteria (Levett, 1998).

7.3 A Thermodynamic Basis for Developing Sustainability Assessment Metrics: Exergy

Both the natural and the industrial ecosystems (Figure 7.1) are subject to the basic laws of thermodynamics. Thermodynamics is a discipline in natural sciences that studies the transformation of energy and materials through processes. The basic laws of thermodynamics are rather abstract and were for a long time hard to apply in practice. In the late twentieth century, applications in industrial engineering were discovered, exemplified by the handbooks by Çengel and Boles (1994) and Bejan (1997). Also applications in the study of natural ecosystems were reported (Jorgensen and Müller, 2000). Nowadays thermodynamics is one of the basic courses in numerous curricula, e.g. in physics, biology, chemistry, engineering etc.

Two laws, the so-called first and second laws, are the basis of thermodynamics. The first law states that, in general terms, energy cannot be created or disappear, but can only be transformed. All industrial and natural processes only transform energy and materials, without loss of energy if one makes a balance between inputs into and outputs out of the considered processes. Inputs and outputs are all kinds of energy and material: kinetic energy (e.g. wind energy), radiation energy (e.g. solar radiation), potential energy (e.g. hydropower), renewable resources (biomass), non-renewable resources (fossil resources), waste materials, etc. The second law of thermodynamics states that all processes generate entropy, reflecting a quality loss of the input energy. This second law is attracting increasing attention (Minkel, 2002). Due to entropy generation, the energy that can be made available from the outputs (exergy embodied in outputs) is less than the energy that can be made available from the inputs (exergy embodied in inputs), although the total energy of the outputs equals the total energy of the inputs. This quality degradation takes place in all physical chemical processes, whether they take place in the natural ecosystem (production of biomass, disruption processes...) or in the industrial ecosystem (production, consumption...) and it is quantifiable by the loss of exergy. This degradation is illustrated in Figure 7.3. Starting from inputs with relatively high exergy content, a process degrades the exergy content. The exergy approach provides a quantitative base with the same unit (Joule) for all types of energy carriers and materials. The exergy

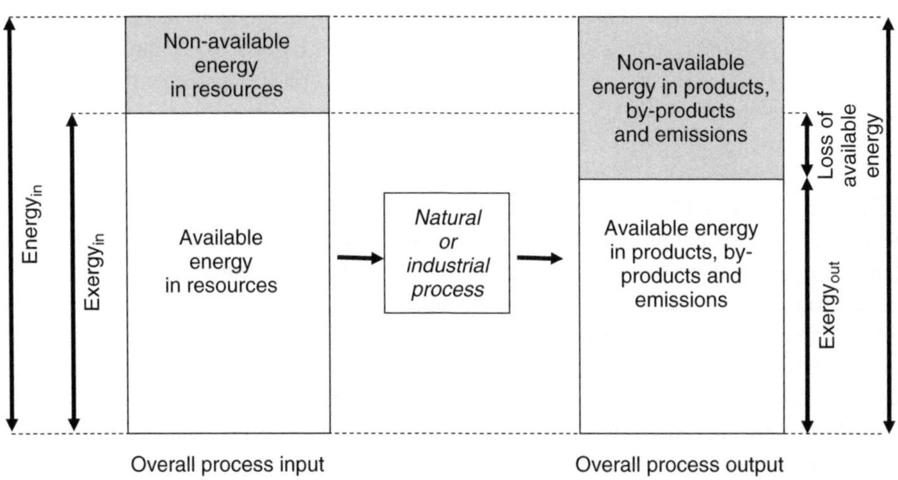

Figure 7.3 *Analysis of a process with the basic laws of thermodynamics. The first law states that all energy going into the process is equal to the energy leaving the process (energy$_{in}$ = energy$_{out}$); the second law states that the available energy or exergy embodied in products, by-products and emissions is lower (exergy$_{in}$ > exergy$_{out}$) than the exergy brought into action*

concept has the ability to serve as a basis to assess the overall physical chemical realm induced by technology quantitatively.

7.4 Technology Assessment by Exergy Analysis

The first applications of exergy analysis in the 1980s focused on production process analysis in terms of efficiency, see, for example, the handbooks *The Exergy Method of Thermal Plant Analysis* by Kotas (1985) and *Exergy Analysis of Thermal, Chemical and Metallurgical Processes* by Szargut *et al.* (1988). The outcome of these analyses is twofold. For single processes, they show to what extent exergy embodied in resources being brought into action is transferred in exergy included in products, by-products, waste and irreversibilities. Second, whole production chain analysis, i.e. starting from natural resources to the final product (mining, transport, manufacturing...), results in the so-called cumulative exergy consumption (CExC) indicator. The cumulative exergy consumption is the total amount of useful energy or exergy that has to be extracted from the natural ecosystem in order to deliver the desired product. For example, to produce 1 kg of aluminum, polyethylene plastics and paper, 250.2, 86.0 and 59.9 MJ have to be extracted. The exergy embodied in the respective products amounts up to 33.0, 46.5 and 16.5 MJ (Szargut *et al.*, 1988; Dewulf *et al.*, 2001). This exergy analysis shows the overall efficiency of the production chain through the ratio of exergy embodied in the product over the cumulative exergy consumption. It clearly shows the depletion of environmental resources induced by product generation, an issue that cannot be overlooked if one is aiming at development of sustainable technology.

Later on, exergy analyses of whole industrial societies are reported. Frequently studied metabolisms are national economies with natural resource intake, on one hand, and products delivered, on the other. This is illustrated by the example of Japan in 1985, showing a resource conversion system with a total input of about 18 EJ or 150 GJ/capita and a net output 3.8 EJ or 31 GJ/capita (Wall, 2002). Ayres *et al.* (2003) presented a historical overview of the nature of the US resource intake for the twentieth century, illustrating the growing share of fossil fuels as exergy to drive the US economy (Figure 7.4). Recently, Ukidwe and Bakshi (2004) developed a thermodynamic accounting of the ecosystem contribution, illustrated by the 1992 data for 91 industry sectors.

Since the development of the industrial ecology theory and the growing role of waste as a resource, exergy analysis has had to cope with this fact: the cumulative exergy required to produce products is no longer 100% virgin material. Dewulf and Van Langenhove (2002a) developed an exergy-based industrial metabolism assessment. Different solid products were studied: cardboard, newspaper and plastics. Both production and waste disposal options were considered, taking into account the consequences they generate for other products in the industrial metabolism. Indeed, if, for example, solid waste products are incinerated with heat and electricity production, they save virgin resource intake for the energy production. Simultaneously, the incineration option implicitly demands a continued virgin resource intake for the manufacturing of the solid products, e.g. plastics incineration that results in heat and electric power does not deliver plastics and therefore requires continued production of plastics from virgin fossil resources. Recycling, however, results in savings on this latter virgin resource intake but results in continued virgin resource requirements for heat and electricity production.

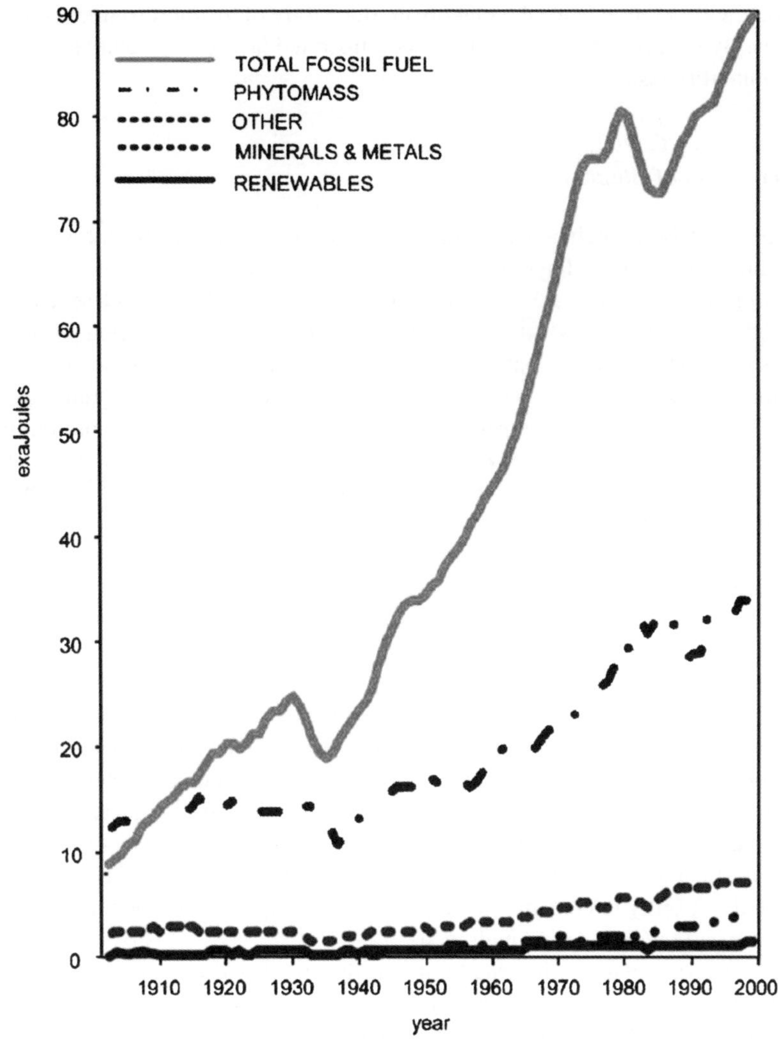

Figure 7.4 *Breakdown of exergy input to the US economy; 1900–1998*
Source: Reprinted from Energy, 28, Ayres *et al.*, Copyright 2003, with permission from Elsevier.

Exergy analysis of the overall metabolism of polyethylene shows that recycling is more efficient than incineration with energy production and landfilling with methane gas recovery for energy purposes. The quantitative information on exergetic efficiencies of industrial metabolisms with internal loops, rather than of single processes and linear production chains, is useful if one aims at integration of industrial ecology principles in technology assessment.

7.5 Exergy-Based Indicators: How to Assess the Role of Renewables

When one aims at the assessment of the sustainability of technology, one should not only pay attention to process chain efficiency and the embedding of the designed technology within the industrial metabolism: the nature of virgin resources is one of the key issues. By means of two cases, it will be illustrated how exergy analysis copes with this sustainability issue.

7.5.1 The Case of Ethanol

Ethanol is a widely used chemical produced by industry as well as from fossil resources via cracking to ethylene and hydration to ethanol, as by a combination of agriculture and fermentation. Dewulf *et al.* (2000) evaluated the exergetic efficiency of several technological pathways for the production of ethanol. First, they examined ethanol production from fossil resources and the pathway via agriculture (wheat) and fermentation. Next, they examined the production via hydrogen by use of solar energy captured by photovoltaic (PV) cells. The electric current generated is used to split water into oxygen and hydrogen. Ethanol is then synthesized from this hydrogen and carbon dioxide, the latter being captured from flue gases from power plants. The basic idea was to enhance the content of renewable resources (solar energy) in the final product and to close the material cycle as much as possible. As in agriculture, the basic resources are water and carbon dioxide, converted by means of solar energy into high quality products. These products are degraded in nature into the initial low quality products, CO_2 and H_2O, closing the material cycle.

The agriculture/fermentation pathway to ethanol is mainly based on solar energy. However, the pathway also requires mineral resources and fossil fuels for the production of nutrients and pesticides, and power for the use of machinery in agriculture and fermentation. Furthermore, this method of production delivers also wheat straw (from agriculture), gluten and cake (both from fermentation), all useful products. The resources required in the agricultural step must be attributed in a proportional way to ethanol and to the other useful products. Also the PV-based route requires substantial non-renewable inputs mainly in order to construct the PV cells and to operate the chemical synthesis of ethanol from carbon dioxide and hydrogen.

From the exergy analysis, it turned out that the fossil resources-based route was the most efficient, with a cumulative exergy requirement of 60.13 MJ to deliver 1 kg ethanol. After an exergy-based allocation of the by-products in the bio-route, it was found that a non-solar input of 7.6 MJ is necessary to produce 1 kg ethanol. This value is almost negligible compared to the required solar irradiation for the photosynthesis process (4240 MJ). The route through PV-generated electricity and electrolysis of water to produce hydrogen for the synthesis of 1 kg ethanol requires only 338 MJ of solar irradiation but 26.4 MJ non-solar resources are required.

This study shows that cumulative exergy consumption as such cannot be the sole indicator used to analyse the sustainability of the three options. In other words, not only efficiency should be taken into account, but the nature of the resources as well. The study mentioned above delivered two exergy-based parameters to assess the sustainability: efficiency and renewability of resources. With respect to efficiency, fossil resources-based technology proved to be better than the PV-route and the bio-route (49 versus 7 and 0.7 %, respectively). On the other hand, the different nature of resources showed a share of 0.02, 91.1, 99.8 % of renewables in the input, respectively. This illustration shows that the PV route is largely based on renewables, but is far more efficient than the bio-route, the latter being limited by the efficiency of the photosynthesis process. In conclusion, this example shows that exergy analysis is able to take into account quantitatively the role of renewables in technology development based on a scientifically sound basis.

7.5.2 Bio-Fuels

Currently, a number of 'bio-fuels', i.e. alcohols, esters, ethers and other chemicals made from biomass, are the focus of attention. These bio-fuels are intended to drive our industrial society towards a more renewables-based society. Two major types of bio-fuels from agricultural crops can be distinguished. The first type is a fermentation-based route starting with crops with high carbohydrate content. The product ethanol can be obtained from a large number of crops: corn, sugar beet, sugar cane, wheat, etc. Second, from crops with a substantial fraction of triglycerides such as rapeseed and soy, methyl esters are produced through chemical transesterification. Here, extraction of the triglycerides from the seeds is followed by a chemical reaction with methanol in order to deliver the methyl ester bio-fuel and the by-product glycerol. These fuels such as soybean methyl ester (SME) and rapeseed methyl ester (RME) have properties that are comparable to diesel. Bio-fuels have a number of characteristics allowing them to be called 'sustainable fuels'. Indeed, they are based on the ultimate renewable energy resource: solar irradiation. At the same time, they close the carbon cycle, making them CO_2 neutral.

With these advantages in mind, one might think that all bio-fuels contribute to sustainability to the same extent. However, this is not the case. First, a number of non-renewable resources are required both for the agricultural and industrial production stages, so that the renewability degree cannot be 100%. This was illustrated for wheat-based ethanol in the previous section. Next, bio-fuels require agricultural land, which will no longer be available for other needs, such as food or feed production. This means that efficiency continues to play an eminent role in the agricultural-industrial production chain: the less efficient, the higher the land requirements for the same fuel production. In third place, the assessment of bio-fuels is complicated by so-called joint production. Indeed, a number of crops not only result in bio-fuels after an industrial post-harvest conversion, but also are turned into marketable chemicals, fertilizers of fodders. An accurate sustainability assessment of bio-fuels must take into consideration an accurate procedure to allocate agricultural and industrial input materials. Exergy analysis allows one to do so, as shown in the study by Dewulf *et al.* (2005).

To do this type of analysis, a detailed scheme of the production routes is necessary. In the case of rapeseed methyl ester (RME), the considered agricultural inputs for the rapeseed production were seeds, solar energy, burnt lime, nitrogen-phosphorous-potassium and calcium ammonium nitrate fertilizers, pesticides (glyphosate, deltamethrin, metazachlor, surface active Lissapol and Manganese micronutrient Mantrac), and fuel consumption for agricultural machinery and transportation. Outputs are rapeseeds and straw. At the industrial production with four sub-stages, i.e. pre-treatment (drying, cleaning and storage), extraction, refining and esterification, considered input materials are rapeseeds, power consumption (electricity, steam, fuels) and chemicals (phosphoric acid, hexane, bleaching material, sodium hydroxide, hydrogen chloride, methanol, sodium methylate). Products are meal, RME and glycerol. Quantities of products are

3000 kg rapeseeds and 1190 kg RME.ha^{-1}.yr^{-1}. In case of soybean methyl ester (SME), the production process is similar to the RME production with similar agricultural inputs and outputs. Product data are 2554 kg soybeans and 418.5 kg SME per ha and year. Corn-based ethanol analysis should take into account the agricultural production through the use of seeds, pesticides, fertilizers, fuels and solar energy, being similar to rapeseed and soy production, delivering kernel and stover. The industrial stage, however, is different since the main process is a biological fermentation preceded by milling and hydrolysis and followed by liquid/solid separation and distillation. Next to kernel, it requires energy and chemicals (enzymes, yeast, sodium chloride, calcium oxide, ammonia, denaturants). Products are ethanol and distiller's dried grains with solubles (DGGS). The production of 7600 kg kernel and 2330 kg EtOH.ha^{-1}.yr^{-1} were taken into account.

The results of the exergy analysis of the agricultural production stages for rapeseed, soy bean and corn indicate that solar radiation is the major input source accounting for more than 99.9%. This is illustrated for the RME case in Figure 7.5 of an input of 32100 GJ solar exergy and 5.38 non-renewables, 1 hectare of agricultural land delivers 86 GJ rapeseeds and 62 GJ of straw, being a total of 148 GJ. In this sense, it can be stated that the agricultural products are largely renewable. Based on the huge solar energy input, the production, however, is again not efficient. Overall output to input ratios are 0.46, 0.17 and 0.47% for the rapeseed, soy and corn production, respectively, due to the limited ability of the photosynthesis process to convert solar exergy into chemical exergy embodied in vegetable products.

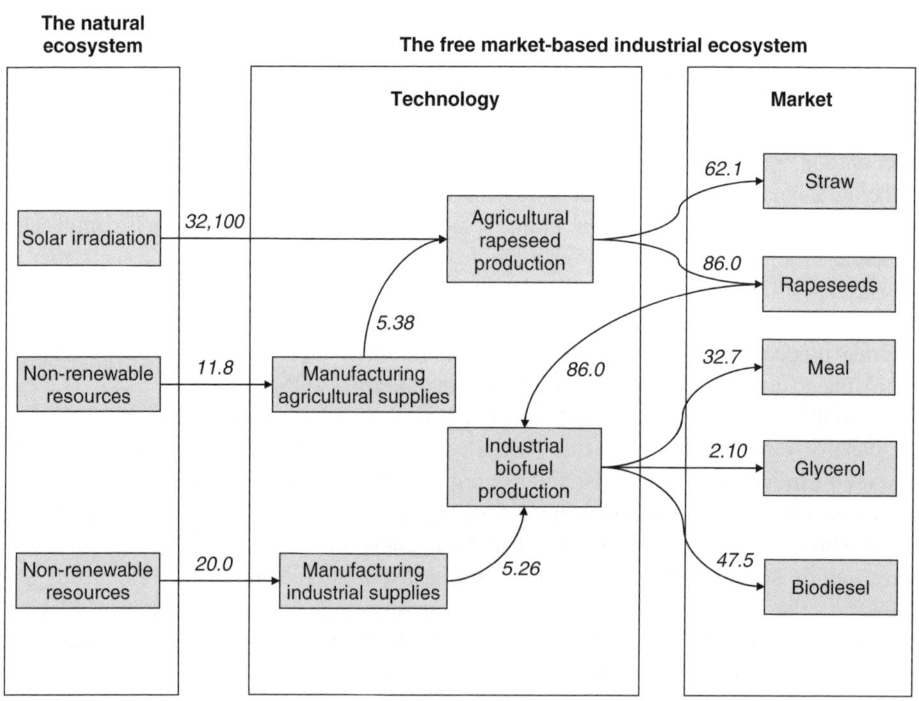

Figure 7.5 *Overall RME biodiesel production in exergy (GJ.ha^{-1}.yr^{-1})*
Source: Dewulf *et al.* (2005).

When the three agricultural production routes are compared, it turns out that corn production delivers the largest amount of exergy, 220 $GJ.ha^{-1}.yr^{-1}$, better than rapeseed (148) and soy (92). It has also to be noticed that the by-products straw and stover have an important share in the output, being very similar in the three cases: 42.0, 41.6 and 39.0 % for rapeseed, soy and corn respectively. These by-products can be either used as fodder or can stay in the field as fertilizer for the next crop.

The sustainability analysis of the bio-fuel production not only requires analysis of the agriculture stage, but also of the industrial conversion of crops into bio-fuels. To convert the crops rapeseeds, soy beans and corn kernel into bio-fuels from 1 ha, with exergy contents of 86.0, 53.8 and 134 GJ respectively, non-renewable inputs (fuels, electricity, steam and chemicals) of 5.26, 5.72 and 5.39 GJ are required. Next to bio-fuels, the conversion results in a significant amount of by-products that can be marketed. The shares of the bio-fuels in the global output are limited to 57.7 and 29.8% for RME and SME respectively, due to the high exergy content of the by-products glycerol and particularly meal. Ethanol shows the highest share in industrial output: 61.4%. With overall outputs of 82, 55 and 112 GJ for rapeseeds, soy beans and kernel conversion, efficiencies of 90.1, 92.4 and 80.6% are obtained, respectively. It should be noted that the auxiliary inputs also require their own industrial production starting from non-renewables. For example, to supply the required 5.26 GJ to process rapeseeds into RME, extraction of 20.0 GJ non-renewables from the environment have to be taken into account. The overall chain analysis is illustrated in Figure 7.5. This illustration allows allocation of renewable and particularly non-renewable resources to the delivered bio-fuels. Shares of rapeseeds, soy beans and corn kernel in the agricultural output are 58.0, 58.4 and 61.0% respectively. Shares of RME, SME and EtOH in the industrial output are 57.7%, 29.8% and 61.4%, respectively. This means that the agricultural inputs to be allocated to the final bio-fuels are limited to 33.5, 17.5 and 37.4%. It turns out that allocated agricultural and industrial resource inputs for 47.5 GJ of RME are 4.84 GJ (0.335×5.38 GJ $+ 0.577 \times 5.26$).

These analyses can be used in an exergy breeding assessment. The exergy Breeding Factor (BF_{ex}) can be defined in a same way as the energy breeding factor (BF_{en}) (Johansson, 1992) or the net energy (NEnV) (Shapouri *et al.*, 2002). This approach aims at the quantification of the amount of so-called renewables that can be harvested from the investment of non-renewable resources. To do so, renewable inputs have to be subtracted from the total input. With a non-renewable input of 4.84 GJ and a bio-fuel production of 47.5 GJ RME, a BF_{ex} of 9.8 is obtained. If one considers the whole industrial metabolism for this bio-fuel production, i.e. not only the agricultural-industrial bio-fuel production chain but also the agricultural and industrial resource supply chains, one ends up with an allocated non-renewable resource input of 15.5 GJ ($0.335 \times 11.8 + 0.577 \times 20.0$), or an overall BF_{ex} value of 3.1. In other words, it can be stated that the renewable fraction of RME is limited to 67.6%. Corn-based ethanol proves to be the most suitable bio-fuel in terms of efficiency and renewability. Indeed it shows an overall industrial metabolism exergy breeding factor (overall BF_{ex}) of 4.17, resulting in the lowest non-renewable fraction being about one-quarter (24.0%). RME and SME show overall BF_{ex} values of about 3 and a non-renewable fraction of about one-third.

7.6 Exergy-based Indicators: Integrating the Role of Renewables in an Overall Physical Chemical Sustainability Assessment

The previous sections have shown that exergy-based technology is able to adequately describe the renewability of the virgin resource intake and the efficiency of a production chain. Based on this, two sustainability indicators, being α for renewability and η for efficiency have been proposed by Dewulf *et al.* (2000). Both parameters are scaled between 0 and 1, where values of α and η of zero means a zero fraction of renewables in the resource intake and an efficiency of zero respectively; values of 1 mean 100% renewables-based and efficient processes. These parameters do not reflect the two other important physical chemical processes illustrated in Figure 7.1: the disruption potential of emitted waste and the integration of the studied technology within the industrial ecosystem. If one aims at a complete sustainability assessment of the physical chemical realm associated with a technology, indicators additional to α and η need to be brought forward.

Exergy-based approaches have been used to quantify the disruption potential of emissions. First, Cornelissen (1997) and Dewulf *et al.* (2000) considered the exergy that is required to abate the emissions through end-of-pipe technologies such as waste water treatment plants and Denox installations for waste gas treatment. By preventing the emissions through these technical solutions, disruption processes in the environment are anticipated and substituted by industrial end-of-pipe processes that can be considered a part of the technology chain. The exergy to construct and operate these processes results in a decrease of the parameter η because they cost exergy without levelling up the set of marketable products. This approach meant that the overall efficiency of fossil resources-based ethanol production was reduced from 49 to 36.5% due to the high CO_2 abatement cost. The abatement costs for the solar irradiation-based ethanol products hardly affected the efficiency parameter η, since these routes closed the carbon cycle.

Emission limits in legislation are not necessarily equal to non-disruption levels, making this approach controversial. Therefore, Dewulf and Van Langenhove (2002b) considered the emissions as they are released in practice and presented a thermodynamic analysis of the disruption process. The deterioration process, i.e. exposure of the ecosystem and human population to these emissions, ending up in deteriorated environmental systems and lost human lives, is also subject to the second law of thermodynamics. By doing so, technology results in two types of exergy consumption: the cumulative exergy consumption embodied in the extracted virgin resources (see above) and the cumulative exergy degradation in the environment that is induced by the emissions.

As well as efficiency, the renewability of virgin resources, the ability of emissions to disrupt natural processes, physical chemical sustainability assessment should include industrial ecology concerns. It should handle the waste-to-resource principle quantitatively. Recently, Dewulf and Van Langenhove (2005) proposed a set of five universal environmental sustainability indicators for the assessment of products and production pathways, taking into account the three interacting sections presented in Figure 7.1: the production and consumption chain (middle), processes within the natural ecosystem (left) and processes in the industrial metabolism (right). For the production chain, the efficiency parameter η was taken, whereas for the natural ecosystem the renewability parameter α (see above) and the (non)toxicity τ of emissions, based on the cumulative degree of degradation (see above) were proposed. Two parameters reflect the industrial

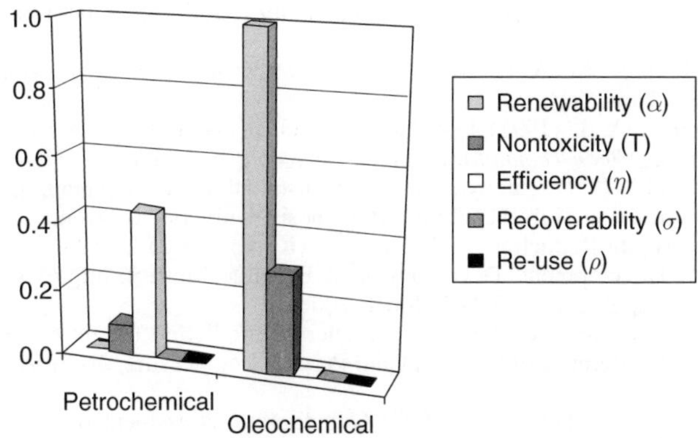

Figure 7.6 *Scores of petrochemical and oleochemical based alcohols on the developed set of 5 physical-chemical sustainability indicators. Scores are scaled between 0 (non-sustainable) and 1 (full sustainable)*
Source: Reprinted from *Resources, Conservation and Recycling*, 43, Dewulf and Van Langenhove, 2005: 419–432. Copyright 2005, with permission from Elsevier.

ecosystem integration. First, ρ is the re-use indicator, representing the fraction of waste used as a resource in the overall package of resources brought into action. Second, σ is the recoverability indicator, showing the fraction of the generated product that can be recovered later on. The set of five indicators, all scaled between 0 and 1, was illustrated for different production pathways of alcohols for the manufacture of surfactants. The results are illustrated in Figure 7.6. They show that renewability is one of the five key issues in assessing the physical chemical sustainability of technology.

7.7 Summary

This chapter has illustrated that thermodynamics can serve as a strong basis to assess the sustainability of technology, including the role of renewable resources in this perspective. Based on a scientifically sound basis, the exergy concept provides quantitative sustainability metrics if one starts from a detailed process analysis. The concept also allows one to cope with the allocation issue, being frequently observed in joint production of marketable renewables-based products. Finally, applications of the exergy concept have brought forward a complementary set of well-scaled indicators where renewability is one of the key issues. The set of indicators covers the physical chemical realm; integration with indicators that handle economic and social issues puts the overall sustainability assessment in prospect.

References

Anastas, P. T. and Zimmerman, J. B. (2003) Design through the 12 principles of Green Engineering, *Environmental Science and Technology*, **37**, 94A–101A.

Ayres, R. U., Ayres L. W. and Warr, B. (2003) Exergy, power and work in the US economy, 1900–1998, *Energy*, **28**, 219–273.

Bejan, A. (1997) *Advanced Engineering Thermodynamics*, 2nd edn, John Wiley and Sons, Ltd, Toronto.

Çengel, Y. A. and Boles, M. A. (1994) *Thermodynamics: An Engineering Approach*, 2nd edn, McGraw-Hill Inc., New York.

Clift, R. and Longley, A. J. (1995) Introduction to clean technology, In Kirkwood, R. C. and Longley, A. J. (eds), *Clean Technology and the Environment*, Blackie, London, pp. 174–198.

Cornelissen, R. L. (1997) Thermodynamics and sustainable development, the use of exergy analysis and the reduction of irreversibility, PhD thesis, Universiteit Twente.

CWRT (Center for Waste Reduction Technologies, AIChE) (2002) Collaborative Projects: Focus Area Sustainable Development: Development of Baseline Metrics, http://www.aiche.org/cwrt/pdf/BaselineMetrics.pdf (accessed 27 October, 2004).

Darton, R. (2002) Sustainable development and energy: predicting the future, paper presented at 15th International Conference of Chemical and Process Engineering, 25–29 August 2002, Praha, Czech Republic.

Dewulf, J. and Van Langenhove, H. (2002a) Quantitative assessment of solid waste treatment systems in the industrial ecology perspective by exergy analysis, *Environmental Science and Technology*, **36**, 1130–1135.

Dewulf, J. and Van Langenhove, H. (2002b) Assessment of the sustainability of technology by means of a thermodynamically based life cycle analysis, *Environmental Science and Pollution Research*, **9**, 267–273.

Dewulf, J. and Van Langenhove, H. (2005) Integrating industrial ecology principles into a set of environmental sustainability indicators for technology assessment, *Resources, Conservation and Recycling*, **43**, 419–432.

Dewulf, J., Van Langenhove, H. and Dirckx, J. (2001) Exergy analysis in the assessment of the sustainability of waste gas treatment systems, *The Science of the Total Environment*, **273**, 41–52.

Dewulf, J., Van Langenhove, H., Mulder, J., van den Berg, M. M. D., van der Kooi, H. J. and de Swaan Arons J. (2000) Illustrations towards quantifying the sustainability of technology, *Green Chemistry*, **2**, 108–114.

Dewulf, J., Van Langenhove, H. and Van De Velde, B. (2005) Exergy-based efficiency and renewability assessment of biofuel production, *Environmental Science and Technology*, **39**, 3878–3882.

Eissen, M. and Metzger, J. O. (2002) Environmental performance metrics for daily use in synthetic chemistry, *Chemical European Journal*, **8**, 3581–3585.

Graedel, T. E. and Allenby, B. R. (1996) *Design for Environment*, Prentice Hall, Englewood Cliffs, NJ.

Hardi, P. and Zdan, T. J. (1997) *Assessing Sustainable Development: Principles in Practice*, International Institute for Sustainable Development, Winnipeg, Manitoba.

IChemE (Institute of Chemical Engineers) (2002) The sustainability metrics, sustainable development progress metrics recommended for use in the process industries. Available at: http://www.getf.org/file/toolmanager/O16F26202.pdf (accessed 27 October, 2004).

Johansson, A. (1992) *Clean Technology*, Lewis Publishers, Boca Raton, FL.

Jorgensen, S. E. and Müller, F. (2000) *Handbook of Ecosystem Theories and Management*, CRC Press-Lewis Publishers, Boca Raton, FL.

Kotas, T. J. (1985) *The Exergy Method of Thermal Plant Analysis*, Butterwoods, London.

Lange, J.-P. (2002) Sustainable development: efficiency and recycling in chemicals manufacturing, *Green Chemistry*, **4**, 546–550.

Lems, S., van der Kooi, H. J. and de Swaan Arons, J. (2002) The sustainability of resource utilization, *Green Chemistry*, **4**, 308–313.

Levett, R. (1998) Sustainability indicators – integrating quality of life and environmental protection, *Journal of the Royal Statistical Society Series A – Statistics in Society*, **161**, 291–302.

Liessens, J. (2000) Sustainability performance assessment: a sustainable development pilot project at Janssen Pharmaceutica, paper presented at Euro Environment Conference, 18–20 October, 2000, Aalborg, Denmark.

Lowe, E. A., Warren, J. L. and Moran, S. R. (1997) *Discovering Industrial Ecology: An Executive Briefing and Sourcebook*, Battelle Press, Columbus, OH.

Minkel, J. R. (2002) The meaning of life, *New Scientist*, **176 (2363)**, 30–33.

OECD (1998) *Towards Sustainable Development: Environmental Indicators*, Organization for Economic Co-operation and Development, Paris.

Shapouri, H., Duffield, J. A. and Wang, M. (2002) *The Energy Balance of Corn Ethanol: An Update*, Report No. 813, US Department of Agriculture, Washington, DC.
Singh, A. and Moldan, B. (2002) Environmental indicators and indices, discussion paper for UNEP-SCOPE workshop on 'Making science more policy relevant': a new generation of assessments, 13–14 June 2002, Prague.
Smith, P. (2001) How green is my process? A practical guide to green metrics, paper presented at conference on Green Chemistry: Sustainable Products and Processes, University of Wales, Swansea, April 3–6, 2001.
Szargut, J, Morris D. R. and Steward F. R. (1988) *Exergy Analysis of Thermal, Chemical and Metallurgical Processes*, Hemisphere Publ. Corp, New York/Springer Verlag, Berlin.
Ukidwe, N. and Bakshi, B. (2004) Thermodynamic accounting of ecosystem contribution to economic sectors with application to 1992 U.S. economy, *Environmental Science and Technology*, **38**, 4810–4827.
Wall, G. (2002) Conditions and tools in the design of energy conversion and management systems of a sustainable society, *Energy Conversion and Management*, **43**, 1235–1248.
Winterton, N. (2001) Twelve more green chemistry principles, *Green Chemistry*, **3**, G73–G75.

8

Material Flow Analysis and the Use of Renewables from a Systems Perspective

Stefan Bringezu

8.1 Introduction

Material flow analysis (MFA) examines the throughput of technical and natural processes. In general, the focus is on processes which are controlled by humans, starting with the extraction or harvest of primary materials, through physico-chemical transformation, manufacturing, consumption, recycling, down to the disposal of as well as loss of materials. MFA is based on accounts in physical units (usually in terms of Mg, i.e. tonnes) quantifying the inputs and outputs of those processes. The subjects of the accounting are chemically defined substances (e.g. carbon or carbon dioxide), on the one hand, and natural or technical compounds or 'bulk' materials (e.g. coal, wood), on the other.

MFA is rooted in various scientific disciplines (Fischer-Kowalski, 1999). Recently MFA application has undergone a change of focus in the course of development of environmental policy. Whereas in the 1980s the analysis of critical substances such as heavy metals and nitrogen were commissioned in the context of pollution control, the analysis of the physical basis of economies in terms of total material throughput attracted increasing attention in the 1990s. During this time, supported by a growing research field of MFA, policy-makers became increasingly aware of the economy as a physical system where the outflows to the environment are determined by the inputs taken from nature. In order to strengthen the policy relevance of research and to foster an exchange between scientists and those who demand MFA results, the ConAccount network was established.

Renewables-Based Technology Edited by J. Dewulf and H. Van Langenhove

A comprehensive research agenda was defined (Bringezu *et al.*, 1998), and regular meetings have been organized. The development of MFA contributed to the formation of new institutions such as the International Society for Industrial Ecology where the topic has constituted an important element from its inauguration (Erkman, 1997).

The idea of an industrial or societal metabolism to a large extent imprints the conceptual background of MFA, i.e. the notion that, analogously to an organism, the functioning of industry, society and economy depends on the flow of substances, materials and energy (Ayres and Ayres, 2002). The organization of the internal flows and the resulting physical exchange with the environment through resource extraction and final waste disposal and emissions critically determine the environmental performance. It is assumed that sustainable development of society will require certain conditions for that exchange and the physical growth of economies which allow the coexistence of man and nature (Bringezu, 2002). MFA is an instrument to better understand how that metabolism works, and to search for options how it can be further developed to fulfil those requirements, while meeting the various needs for material use.

With regard to policy relevance, MFA can be and has been used to plan and control the implementation and effectiveness of environmental policies (Bringezu, 2004). On the one hand, MFA can be used to track and control the flow of hazardous or critical substances, thus contributing to the detoxification of the societal metabolism. On the other hand, economy-wide MFA has been developed to derive indicators of the volume and structure of the whole material throughput and materials accumulation of national or regional economies. This research approach is designed to analyze the dynamics of the metabolism with regard to decoupling material use and economic growth, and to support the search for options for the dematerialization of the economy and its consequences for the environment and the economy. Based on indicators derived from MFA, the German and Japanese governments have decided to increase materials productivity by certain factors. At the European and international level, MFA has contributed to official reporting (Eurostat, 2001; EEA, 2003; OECD, 2004), and to the development of the knowledge-gathering function of the thematic strategy on sustainable use of resources (Bringezu, 2002; Moll *et al.*, 2003; EU Commission, 2003). In future, one can expect that at least certain elements of MFA may be used also for regular impact assessment of policies.

This chapter will provide a short overview of the MFA methodology, and then describe major types of MFA which have been conducted to analyze the implications of the use of renewables, i.e. biomass, from harvest to final disposal. It will become clear that it is not biomass *per se* which exerts various impacts to the environment, but the substance and materials flows linked to its production and consumption.

8.2 Overview of the Methodology

With regard to the primary interest, two basic types of material flow-related analyses can be distinguished, although in practice a continuum of different approaches exists (Table 8.1). On the one hand, the concern about environmental impacts of certain substances (e.g. heavy metals), bulk materials (e.g. timber), or products (e.g. beverage containers), and therefore flows of substances and materials linked to these entities are studied (Type I). On the other hand, interest lies with the (environmental) performance of firms, sectors

Table 8.1 Material flow-related types of analysis

Type I			Type II		
a	b	c	a	b	c
Specific environmental problems related to certain impacts per unit of flow of			Problems of environmental concern related to the throughput of		
Substances	*Materials*	*Products*	*Firms*	*Sectors*	*Regions*
e.g. Cd, Cl, Pb, Zn, Hg, N, P, C, CO_2, CFC	e.g. wooden products, energy carriers, excavation, biomass, plastics	e.g. diapers, batteries, cars	e.g. single plants, medium and big companies	e.g. production sectors, chemical industry, construction	e.g. total or main throughput, mass flow balance, total material requirement
within certain firms, sectors, regions			associated with substances, materials, products		

Source: Bringezu and Moriguchi (2002).

(e.g. construction), or whole regions or national economies, and therefore the throughput of substances and/or materials of these entities is analyzed (Type II). Whereas Type I analyses are often performed from a natural science or technical engineering perspective, Type II analyses are more directed towards the analysis of socio-economic relationships. More details and examples are given in Bringezu and Moriguchi (2002). Here we will provide selected examples in relation to the use of renewables.

This chapter will start with examples of Type Ia, Ib studies. Type Ic is usually addressed as a product LCA. Type IIa comprises accounting, bookkeeping and analysis of materials in companies, and is dealt with elsewhere. Furthermore, this chapter will also provide examples of Type IIb and IIc studies.

Although there are a variety of methodological approaches to material flow accounting and analysis, MFA studies have some essential features in common. One basic feature is the perspective of a systems analysis. The interconnection and physical linkages between various processes are analyzed with different levels of detail. Aggregated information, e.g. on the resource consumption of economies, is required to get an overview on the overall performance. Detailed information, e.g. on the processes consuming most resources, is necessary to find effective levers for improvement measures.

An MFA process usually comprises four steps: (1) goal and systems definition; (2) accounting and balancing; (3) modelling; and (4) evaluation (Bringezu and Moriguchi, 2002; Brunner and Rechberger, 2003; van der Voet, 2002).

The systems definition comprises the formulation of the target questions, the definition of scope and systems boundary. Target questions are defined according to the primary objectives. In all types of analysis, it has to be determined which flow categories will have to be accounted for in order to determine the path and quantify the magnitude of the flows, and to discover those flows which are most relevant and crucial for the problems of primary interest, and those factors most responsible for these flows. The scope defines the spatial, temporal and sometimes functional extent of the studied objects. The categorized flows are studied along the path that is related to spatially defined compartments or regions or to functionally defined industrial sectors. The flows are always accounted for based on a temporally defined period. The scope may be similar for Type I and Type II. The system boundary defines the start and the end of the material flows which are accounted for. It is – at least – partly determined by the scope but may comprise additional functional elements, e.g. the border between the environment and the economic sectors of a region. Scope and system boundary are not necessarily identical, especially when regionally oriented accounts are combined with product chain-oriented accounts. For instance, if the transnational material requirements of a national economy is determined (e.g. as part of TMR, see below), the scope remains national while the system boundary is defined functionally on a larger scale.

At a certain level of detail, the process chain and flow analysis defines the processes for which the inputs and outputs are to be determined quantitatively by accounting and balancing. Here the fundamental principle of mass conservation is used to balance inputs and outputs of processes and (sub)systems. The balancing is used to check accuracy of empirical data, to improve consistency and to 'fill in' missing data. This is usually performed on the basis of stochiometric or technical coefficients and may be assisted by computer simulation based on mathematical modelling.

The accounting leads to an inventory and quantitative description of flows. This can be used for 'bookkeeping' and monitoring, and represents the basis for static and dynamic modelling. The latter may be used to analyze scenario-based alternatives for the status quo and ongoing dynamic of the material flow system studied. The evaluation of the results is related to the primary interest and basic assumptions. Often it is performed on the basis of environmental pressure indicators. These indicators may represent either generic pressures associated with the amount of primary resources input (e.g. primary energy, primary materials, water), or specific pressure indicators (e.g. GWP, ODP) (Bringezu *et al.*, 2003; van der Voet *et al.*, 1999).

8.3 Examples of MFA Studies in the Context of Renewables

8.3.1 Type Ia Studies: SFA – Agriculture, Nitrogen and Heavy Metals

Substance flow analysis (SFA) has been used to determine the main entrance routes to the environment, the processes associated with these emissions, the stocks and flows within the industrial system as well as the trans-media flows, chemical, physical, biological transformations and resulting concentrations in the environment. Spatio-temporal distribution is of high concern in SFA. Results from these analyses are often used as inputs for further analyses to quantitatively assess risks to substance-specific endpoints.

For instance, studies on the flow of nitrogen reveal that agriculture has the highest impact on distorting the natural nitrogen cycle. Losses of mineral fertilizer to groundwater and disproportionate application of manure in the open field through intensive cattle production represent the largest outflows to the environment (e.g. van der Voet, 1996; Bringezu *et al.*, 2001). Thus, SFA can be used to determine target processes for policy interventions, e.g. to reduce eutrophication of water bodies. It can also evaluate the effectiveness of such measures, for example, of adjusting cattle density to soil absorption capacities and of measures to increase efficiency of fertilizer use.

SFA may also reveal the relevance of different sources of heavy metal contamination of agricultural soil such as mineral or organic fertilizer, air deposition, etc. Those studies may also compare different farming systems with regard to the risk that certain thresholds for soil accumulation, plant up-take and leaching to groundwater will be exceeded (Moolenaar, 2002). Thus, the side effects on substance flows through management schemes of biomass production may also be evaluated by SFA.

8.3.2 *Type Ib Studies: Analysis of Selected Bulk Materials – Timber Products*

Selected bulk material flows have been studied for various reasons. At a global level, the flow of biomass from human production has been studied to relate it to biomass production in natural ecosystems in order to evaluate the pressure to species diversity and the domination of nature by human beings (Haberl, 1997; Vitoussek *et al.*, 1997).

At a regional level, analysis of forest growth and timber harvest flows may be used to determine unused potentials for use of forestry products. This has been done, for instance, for a German region north-west of Berlin, Prignitz-Ruppin (Thrän, 2003), and a Swiss region (Müller *et al.*, 2004), based on the analysis of wood flows (Figure 8.1). In the German region only one-third of the 'sustainable cut' was used, and the rest added to the forestry stock. The analysis provided the basis to determine regional targets for timber product management, e.g. for total timber production, energy wood production, in relation to processing capacities.

Note that long-term sustainable resource management will also need to consider to what extent the nutrients extracted from the forest through wood products are lost during use and final disposal and what this implies for recycling requirements.

8.3.3 *Type IIb Studies: Analysis of Sectors – Construction and Energy Supply*

Although there is no single definition, a sector usually comprises a variety of process chains which are managed to pursue a main purpose. On the one hand, econonomic statistics

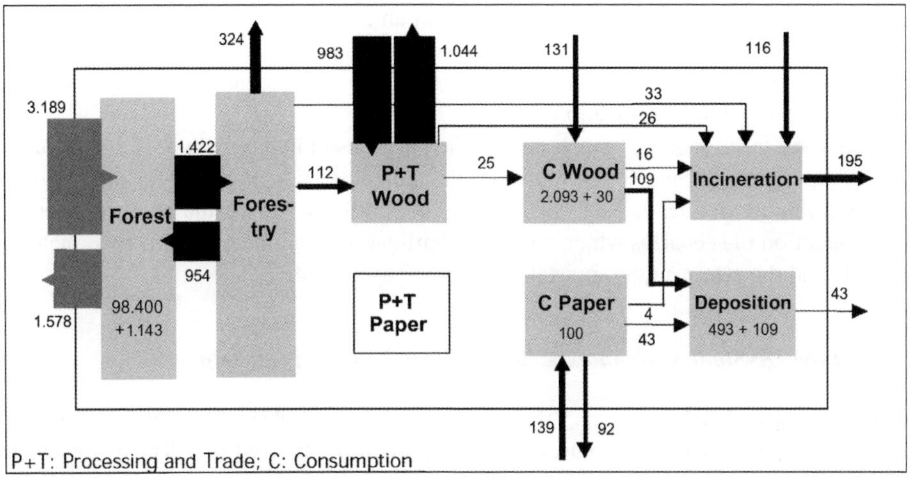

Figure 8.1 *Analysis of wood flows in a German region was used to determine unused potentials of renewables*
Note: Material flows and stock changes in [kg DW/ (cap*a)] for 1995, (DW = dry weight).
Source: After Thrän (2003). Reproduced with permission of Daniela Thrän.

may refer to the classification of branches and product groups of the NACE system which define e.g. construction or vehicle production. These branches or sectors are part of the consistent system of national accounts which is used for econometric analysis and modelling. This type of definition is appropriate especially for inter-sectoral comparative studies. On the other hand, a more functional definition may orientate towards the fields of final demand, such as housing, mobility, nutrition, etc. This type of definition may be appropriate for the analysis of intra-sectoral flows and options for improvement.

The increased use of renewables in the form of forestry products for housing and dwelling purposes has been studied in scenario-based research for Germany in order to elucidate the main material flows, priority processes and options for policy measures to increase sustainability of the housing sector (Buchert *et al.*, 2003). Starting from a small proportion of single and semi-detached wooden houses of about 5% in 2000, a theoretical increase to 50% of all new build private houses until 2025 would thus contribute to a certain reduction of the material intensity of construction, although the use of mineral resources would only be reduced by 33% (excluding the hidden flows, see below).

The potentials to use biomass for energy have been studied by a comprehensive, scenario-based analysis for Germany. Real and potential flows of biomass from agriculture, forestry and nature conservation practices were studied. Results showed that biomass from harvest residuals and from set-aside agricultural land (which is no longer used for the production of food) can contribute a significant share to fuel and heat/electricity supply (Fritsche *et al.*, 2004). These scenarios, however, did not consider that the biomass from set-aside land could primarily be used for materials purposes, e.g. plastics made from starch or plant oil.

The use of biomaterials seems promising. Based on a review of comparative LCA of biomass versus fossils' use for heat/electricity, fuels, and materials, Weiss *et al.* (2004) showed that in general higher environmental benefits are associated with biomass being used as source for generation of energy and material goods instead of utilizing it for the production of bio-fuels for full substitution of fossil fuels, especially diesel. Considerable environmental burden from biomass utilization is the result of agricultural cultivation practices and the conventional use of mineral fertilizers. The introduction of less polluting farming practices in combination with an intelligent selection of crops for the various purposes of utilization, the efficient processing of biomass, and the use of agricultural residues are expected to significantly reduce the negative environmental effects associated with large-scale biomass utilization.

In the long term, it seems much more efficient to use biomass which is available for non-food purposes primarily for biomaterials and 'only' subsequently for energy, thus gaining a double dividend. Analysis of these possibilities will be the challenge for the next generation of scenarios which will consider the use of biomass for non-food material products and the effect on the societal metabolism as a whole.

8.3.4 Type IIc Studies: Economy-wide MFA and Derived Indicators

A major field of MFA concerns the analysis of the metabolism of cities, regions, and national or supranational economies. Accounting can be directed to selected substances and materials or to total material throughput. In recent years economy-wide MFA and derived indicators have been developed as a growing field of research.

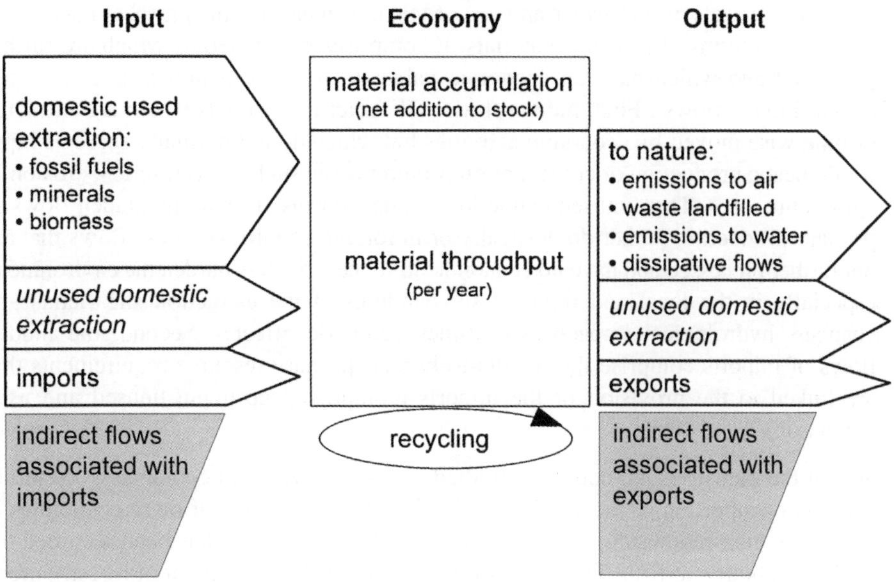

Figure 8.2 *Overview scheme of economy-wide MFA*
Source: Adapted from Eurostat (2001).

Economy-wide MFA accounts for material exchange between national or regional economies and (1) the environment (via resource extraction on the input side and waste deposition, and releases to air and water on the output side); and (2) other economies (via trade) through measuring flows in physical units (Mg) (Eurostat, 2001). Economy-wide MFA are supposed to form a physical complement to the monetary System of National Accounts (Figure 8.2). The mass differences between material inputs and outputs relate to the physical stock changes within the national economy. At the overview level, economy-wide MFA does not account for the internal material flows within the economy (e.g. between production units). Those are shown in more detail, e.g. by Physical Input–Output Tables (PIOT).

Economy-wide MFA constitutes the basis from which a variety of material flow-based indicators can be derived (Table 8.2).

- *DMI* (*Direct Material Input*) is defined as measuring the input of materials into the domestic economy which are of economic value and which are processed and used in production and consumption activities. DMI comprises domestic extraction used such as fossil fuels (coal, oil), minerals (metal ores, construction minerals, industrial minerals), biomass (timber, cereals), plus the imports.
- *DMC* (*Domestic Material Consumption*) accounts for the input of materials used for final consumption in the economy. DMC is defined as DMI minus exports. It indicates the amount of materials for short life or long life products which finally end up as waste or emission to the environment within the country.
- *TMR* (*Total Material Requirement*) is defined as accounting for the domestic resource extraction and the resource extraction associated with the supply of the imports (all

primary materials except water and air). TMR thus measures the physical basis of an economy in terms of primary materials. It comprises raw materials which are further processed and which have an economic value (= 'used extraction'), as well as so-called 'hidden flows'. First, hidden flows (HF) refer to materials which are extracted or otherwise moved by economic activities but which do not normally serve as input for domestic production or consumption (mining waste such as overburden, erosion in agriculture, etc.). This 'unused extraction' relates, for instance, to the hidden flows of primary production (either domestically or in foreign countries). These flows that are not further processed and have no economic utility nevertheless burden the environment, especially in the local and regional surroundings of the extraction site (landscape changes, hydrological impacts, sometimes eco-toxic effects). Second, the hidden flows of imports comprise the 'cradle-to-border' primary resource requirements that are linked to the provision of the imports (comprising upstream unused and used extraction).

In most of the industrialized countries studied so far, renewables, i.e. biomass, constitute only a minor proportion of the TMR (Table 8.3). The metabolism of these economies is 90% based on non-renewables. The highest amount of renewables has been recorded for Finland with about one-fifth. This is due to the exceptionally large input of forestry products to the Finnish industry which is also used to a large extent for exports, e.g. of pulp and paper products. As a consequence of the use of forest and processing residuals for

Table 8.2 *Indicators derived from economy-wide MFA*

Indicator classes	*Indicators or aggregates*		*Accounting rules*
	Acronym	*Full name*	
Input			
	DMI	Direct material input	DMI = Domestic used extraction + imports
	TMR	Total material requirement	TMR = DMI + HF (or IF)
Output			
	DPO	Domestic processed output	DPO = emissions + waste
	DMO	Direct material output	DMO = DPO + exports
Consumption			
	DMC	Domestic material consumption	DMC = DMI – exports
	TMC	Total material consumption	TMC = TMR – exports – hidden or indirect material flows of exports
Balance			
	NAS	Net additions to stock	NAS = DMI – DPO – exports
	PTB	Physical trade balance	PTB = imports – exports
Efficiency			
	GDP/input or output indicator	Material productivity of GDP	GDP divided by indicators values
	Unused/used	Resource-intensity of materials extraction	Ratio of unused (hidden or indirect) to used (DMI) materials

Note: HF: hidden flows; IF: indirect flows
Source: After Bringezu and Schütz (2001a).

Table 8.3 *Composition of the TMR in selected countries and economic regions*

Year	*EU-15*	*D*	*J*	*USA*	*NL*	*SF*	*UK*	*PL*	*China*
	1997	*1999*	*1994*	*1994*	*1993*	*1999*	*1999*	*1997*	*1999*
DMI (t/capita)	17	22	16	26	28	47	16	14	3
DMC (t/capita)	16	19	15	24*	17	40	13	12	3
TMR (t/capita)	49	71	45	86	67	100	44	32	38
TMR shares (%)									
Non-renewables	91	90	94	93	90	75	85	90	87
Foreign resources	40	38	56	7	72	44	45	23	4
Domestic unused resources	31	39	22	67	10	15	25	38	89
TMR shares (t/capita)									
Fossil energy carriers	15	29	13	32	15	10	14	13	5
Metals	10	14	9	10	3	26	11	3	2
Minerals	12	13	10	11	7	27	8	7	2
Excavation	3	3	9	13	7	6	4	2	20
Biomass	4	7	3	6	6	21	6	3	5
a) domestic	4	3	1	6	3	17	4	3	4
b) foreign	0.4	3	2	1	3	4	2	0.2	0.3
Erosion	4	4	1	13	17	3	1	3	4
a) domestic	3	1	0.1	13	0.1	1	0.2	3	4
b) foreign	2	2	1	0.0	17	3	1	1	0.0
Other	0.3	1	1	1	11	5	0.4	0.1	0.0

Sources: EU-15 (European Union): Bringezu and Schütz (2001a) and updated for DMI by Wuppertal Institute database.
D (Germany): Schütz (2003).
J (Japan): Adriaanse *et al.* (1998), Moriguchi (2000).
USA (United Stated of America): Adriaanse *et al.* (1998).
* calculated based on relation of DMC/DMI from 1991.
NL (Netherlands): Adriaanse *et al.* (1998), Kleijn (2002).
SF (Finland): Mäenpää *et al.* (2002).
UK (United Kingdom): Bringezu and Schütz (2001b).
Pl (Poland): Schütz *et al.* (2002).
China: Liu and Wang (2005); Liu (2004).
After Bringezu (2004).

energy supply, the requirements of fossil fuel resources in Finland are comparatively low. In most of the countries domestic biomass production is in the order of magnitude of about 3 to 4 t/cap. Higher values are usually associated with production for export. Lower values of domestic production may be partially compensated for by imports or may be due to cultural specifics of nutrition like in Japan where protein-rich fish constitutes a major contribution. In some countries, such as The Netherlands, imports of biomass play a dominant role. In this case a significant portion of the input is going to food industries and trade for export, mainly to Western Europe.

A major hidden flow linked to biomass production is erosion on agriculture land. The rationale of the TMR indicator comprises all primary materials (besides water and air) which are extracted or moved within the natural environment by means of technology. Although erosion may be regarded as an indirect effect of human action, it is considered as a stand-in for the human-induced loss of top soil which is comparable to the loss of top soil due to direct interventions, e.g. for infrastructure building (which is considered under 'excavation' in Table 8.3); see also Bringezu *et al.* (2003). Although empirical data on erosion are often lacking, the available data indicate that erosion represents a material flow which for a variety of countries lies in the order of magnitude of the biomass harvested itself. The aggregated data for the EU conceal that there are significant differences between regions, e.g. between the Scandinavian and Mediterranean countries. Imports to the EU are also associated with a significant amount of erosion in other regions. This is mainly caused by imported products such as cacao and coffee.

Analysis of the dynamics of the societal metabolism revealed that DMI and TMR tend to increase with economic growth, although the relationship is nonlinear. There is also a high variation of material use and total material requirement between rich countries which shows that prosperity is possible at different levels of resource use. At the same time the overall trend indicates that with increasing economic activity (measured as GDP), material use per capita seems to stabilize. This means that there is a relative decoupling of economic growth and material and resource use. Currently, however, there is no evidence of an automatic absolute decoupling in the course of economic development. Single examples of absolute decoupling have been limited to cases of either direct or indirect political intervention (Bringezu *et al.*, 2004).

Furthermore, the analysis of past and ongoing trends showed that there is an increasing shift of resource-related environmental pressure of industrial countries to other regions. Whereas the domestic resource extraction stabilizes or even declines in industrial countries, a growing part of resource requirements is supplied via imports, which increases the environmental burden in other regions (Bringezu, 2002; Schütz *et al.*, 2004). For the European Union this development is mainly caused by the use patterns of metal resources where there are hardly any significant mining activities for basic metals within Europe. Those resources are to a growing extent delivered from developing countries.

For biomass, trade is also going to play an increasing role, although domestic production and trade within Europe are dominant. In the 1990s the domestic consumption of biomass was fluctuating between 4.0 and 4.3 tonnes per capita in EU-15, and between 3.2 and 3.6 tonnes per capita in AC-13. One might expect that in the course of technological and economic convergence of the enlarged EU the differences in consumption level will be reduced, and this will be facilitated by increasing trade between those regions. Indeed, while domestic production remained rather constant, from 1990 to 2000, in the EU-15 the imported biomass grew by 7%, while the exported biomass increased by 20%. In the AC-13, between 1992 and 2000 the imports of biomass exploded by 102%, while exports grew by 39%.

The effects of an increased use of renewables may also be evaluated by means of input–output (IO) analysis, using standard monetary IO tables in combination with NAMEA matrices showing the sectoral direct use of resources through domestic extraction or via import, and the impact potentials of certain direct emissions (e.g. GWP, acidification potential). NAMEA (National Accounting Matrix including Environmental Accounts) is an environmental accounting framework developed by Statistics Netherlands at the end of the 1980s. It consists of a conventional national accounting matrix extended with environmental accounts in physical units. Moll *et al.* (2003) analyzed the direct and indirect environmental pressures associated with the product groups delivered to final demand in

Germany in the late 1990s. Products from construction, food and beverages production, motor vehicles and trailers production, energy supply, basic metals, and products from agriculture turned out to contribute most to total material requirement as well as to specific environmental pressures such as global warming (measured as GWP). Food products and beverages induced 6% of primary energy supply, 9% of TMR, 9% of GWP, 25% of acidification effect, and 12% of waste (without bulky waste). Based on those calculations, a hypothetical growth of 50% in final demand for those products would *ceteris paribus* lead to an increase of the national pressure indicators by 3% of primary energy supply, 4% of TMR, 4.5% of GWP, 13% of acidification, and 6% of waste (without bulky waste).

Assessment of any increase in biomass production and use will have to consider other analytical instruments and information besides MFA. Limiting factors such as available land, and technological options and constraints will have to be taken into account. A very preliminary estimation may be used to exemplify a major problem in this context. In Germany, in 1999 about 8 Mio t of carbon were used as chemical feedstock for refinery purposes in order to produce plastics etc.; if the same amount of carbon were to be provided on the basis of rapeseed oil (which has a comparable energy content compared to naphtha), this would require about 8 Mio hectare of additional crop land based on current cultivation practice, i.e. about half of the German agriculture area. One has to consider that about half of Germany's land is used for agriculture, and that absolute global requirements for domestic consumption of agricultural goods are even 20% higher (Schütz *et al.*, 2003; Steger, 2005). In addition, productivity per hectare cannot be infinitely increased without jeopardizing a healthy environment (due to the linkage of biomass production to a certain minimum loss of inputs to production, even if efficiency potentials were used).

Thus, it becomes clear, that starting from the status quo of resource requirements – see the structure of TMR – a substitution of renewables (biomass) for the bigger part of non-renewables (minerals) will not be feasible nor sensible. Instead any shift towards renewables which is expected to contribute to a more sustainable supply will have to be combined with a more efficient use of all resources, either renewable or non-renewable. This increase in resource efficiency will have to be implemented across different economic sectors. The general aim will be to increase the relation of services and products provided per primary material input. This will also be a task for agricultural practices. Extended crop rotation, optimized fertilization, and whole plant harvest (for materials and energy use) are examples of options which may contribute to this end. Significant increases in resource efficiency will only be gained with an approach along the whole production–consumption chain. This includes improvements within food manufacturing (e.g. energy use) and integration of waste and wastewater management with agricultural processes in order to optimize recycling and cascading use of materials and energy.

Nevertheless, one might expect that biomass will play the dominant role for the metabolism of the future. Figure 8.3 shows the structure and volume of the EU metabolism at the end of the last century as well as target values for the long-term development of the main components under conditions of sustainability.

Currently, the EU's metabolism is still in a phase of physical growth. The net addition to stock (NAS) amounts to about 10 t/cap. It represents the mass of additional buildings and infrastructure, and thus, the physical expansion of the materials stock of the technosphere.

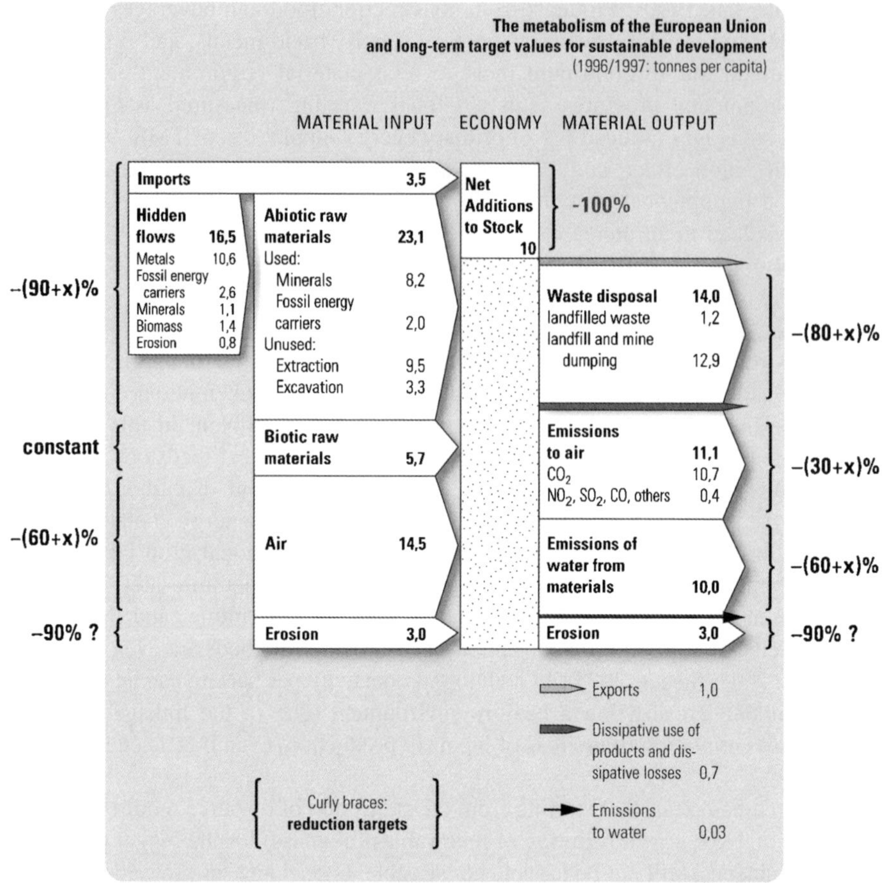

Figure 8.3 *The metabolism of the European Union and long-term target values for sustainable development*
Source: Bringezu (2004).

Naturally, this expansion cannot continue infinitely, instead the growth phase will be superseded by a 'maturity phase', characterized by a dynamic steady-state flow equilibrium between input and output. Consequently, NAS will approach values around zero. Still, however, it remains an open question, when and at which absolute level of materials stock this will happen.

Approaching a sustainable structure of the societal metabolism will probably include the phase-out of fossil fuel use, also in order to mitigate climate change. Both the abolition of fossil fuels and the reduction of net growth which is fed by construction minerals would be associated with a significant decrease in primary minerals input. In addition, one could argue that all other uses of non-renewables will not be compatible with a sustainable supply and regeneration of resources. Consequently, the use of all primary minerals should be reduced to a minimum. Starting from the status quo, the long-term target for the next 50 to 100 years would be a reduction of more than 90%. This is consistent with demands for a factor of 10 for the next 50 years (Schmidt-Bleek, 1994) and a factor of 4 for the next 30 years (von Weizsäcker *et al.*, 1995), although both factors originally have not been specified for certain components of the societal metabolism. In order to minimize the shift from domestic to foreign resource supply, this target may provide orientation for intra-European resource extraction as well as imports and their hidden flows.

Concerning the use of renewables, it is uncertain what volume of biomass can be used within Europe in the long term. Considering that potentials for harvest especially in forestry have not been exploited to full extent, that biomass cultivation under conditions such as organic farming could be applied on a large scale while increasing hectare productivity to a certain extent seems possible and the efficiency of biomass use could be improved significantly, one could assume that the order of magnitude of biomass harvest in agriculture and forestry could at least be maintained also under conditions of sustainable cultivation. This is a tentative hypothesis regarding the future sustainable use of biomass on a European scale which will have to be tested and further developed by detailed and comprehensive studies. Future research will have to clarify the quantity and quality of biomass production which can be guarantued while minimizing the use of primary minerals and controlling also other conditions of sustainability (e.g. concerning species diversity).

Another condition of sustainable supply with renewables will be the reduction of erosion on agriculture and forestry land down to levels which do not exceed the rate of regeneration. Erosion depends on many factors which vary considerably between locations and regions. The available data do not allow specification of an average rate of erosion which could be deemed tolerable in the long term at a European scale. Preliminary data for Germany indicated that the regeneration rate of arable soil is one order of magnitude below the average rate of erosion (Loske *et al.*, 1996). Further research is necessary to provide adequate data for European soils and to find ways of aggregating this information into the overall picture.

The target values for the other input and output flows shown in Figure 8.3 have been calculated by stoechiometric relations, based on the three pillars of ultimate orientation: (1) reduction of NAS by 100%; (2) decreased use of primary minerals by 90 + x%; and (3) more or less constant use of biomass. Under these conditions, biomass will be the dominant carrier of the EU's metabolism. The input volume will be equalled by the output of residuals stemming from respiration and combustion.

Naturally, this outline of a target material flow account of the EU is of hypothetical value. It may be used as possible reference for evaluation, and as a basis for further discussion and research. Future studies are deemed necessary to analyze the effect of increased substitution of renewables for non-renewables with regard to the whole societal metabolism in order to detect and possibly avoid problem shifting between material flows, between environmental media or between regions.

8.4 Summary

Material flow analysis (MFA) examines the flow of substances and (bulk) materials through technical and natural processes. It is used to describe, understand and develop the societal or industrial metabolism at various levels (e.g. local, regional, (supra-)national). MFA can be used to assess the environmental and socio-economic implications of resource use. The use of (natural) renewables, i.e. biomass, is linked to various material flows which exert pressure to the environment. MFA can be used to determine the relevant processes and to design adequate policy measures to control the flow of critical substances. MFA may also quantify unused potentials of biomass use, e.g. in forestry. Scenario-based analyses can, for instance, outline potential use for housing and energy supply. Economy-wide MFA and derived indicators are used to quantitatively describe the performance, past and possible future dynamics of the physical basis of economies. Currently, industrial economies are dominated by non-renewable resources. Extended use of renewables will require a further significant increase in resource efficiency, both of non-renewables and renewables, across all sectors and along the whole life cycle if the societal metabolism is to be developed towards a more sustainable condition.

Acknowledgements

The author would like to thank Helmut Schütz for provision and quality control of the economy-wide MFA data, Stephan Moll for input output analysis, Philipp Schepelmann for reading the article, the editors for valuable comments and Fabian Hanneforth for technical assistance.

References

Adriaanse, A., Bringezu, S., Hammond, A., Moriguchi, Y., Rodenburg, E., Rogich, D. and Schütz, H. (1998) *Resource Flows: The Material Basis of Industrial Economies*, World Resources Institute, Wuppertal Institute, Netherlands Ministry of Housing, Spatial Planning, and Environment, Japan's National Institute for Environmental Studies, Washington, DC: World Resources Institute Report.

Ayres, R. U. and Ayres, L. (eds) (2002) *Handbook of Industrial Ecology*, Edward Elgar Publishers, Cheltenham.

Bringezu, S. (2002) *Towards Sustainable Resource Management in the European Union*, Wuppertal Papers No. 121, Wuppertal Institute, Wuppertal. Available at: http://www.wupperinst.org/Publikationen/WP/WP121.pdf

Bringezu, S. (2004) *Erdlandung: Navigation zu den Ressourcen der Zukunft*, Hirzel Verlag: Stuttgart.

Bringezu, S., Doedens, H., Faulstich, M., Gallenkemper, B., Koch, M., Reich, J., Van Wickeren, P., Weber, G. and Werner, P. (2001) *Stickstoffbilanz in Deutschland. Landwirtschaft – Futtermittel- und Ernährungsindustrie – Abwasser- und Abfallentsorgung – Bezugsjahre 1995/1998*, ATV-DVWK, Deutsche Vereinigung für Wasserwirtschaft, Abwasser und Abfall e.V., ATV-DVWK-Arbeitsbericht, Hennef.

Bringezu, S., Fischer-Kowalski, M., Kleijn, R. and Palm, V. (eds) (1998) *The ConAccount Agenda: The Concerted Action on Material Flow Analysis and its Research and Development, Agenda*, Wuppertal Special 8. Wuppertal Institute, Wuppertal.

Bringezu S. and Moriguchi, Y. (2002) Material flow analysis, in Ayres, R. U. and Ayres, L. (eds) *Handbook of Industrial Ecology*, Edward Elgar Publishing, Cheltenham, pp. 79–90.

Bringezu, S. and Schütz, H. (2001a) Material use indicators for the European Union, 1980–1997: economy-wide material flow accounts and balances and derived indicators of resource use, Eurostat Working Papers 2/2001/B/2, 29 June 2001.

Bringezu, S. and Schütz, H. (2001b) *Total Material Resource Flows of the United Kingdom*, A study of the Wuppertal Institute for the Department for the Environment, Transport and the Regions, Final report Project DETR EPG 1/8/62, May 2001, Wuppertal, Institute, Wuppertal/DETR, London.

Bringezu, S., Schütz, H. and Moll, S. (2003) Rationale for and interpretation of economy-wide material flow analysis and derived indicators, *Journal of Industrial Ecology*, **7**(2), 43–64.

Bringezu, S., Schütz, H., Steger, S. and Baudisch, J. (2004) International comparison of resource use and its relation to economic growth: the development of total material requirement, direct material inputs and hidden flows and the structure of TMR, *Ecological Economics*, **51**, 97–124.

Brunner, P. and Rechberger, H. (2003) *Practical Handbook of Material Flow Analysis.* (Advanced Methods in Resource and Waste Management Series 1), Lewis Publishers, Boca Raton, FL.

Buchert, M., Fritsche U., Jenseit, W., Rausch, L., Deilmann, C., Schiller, G., Siedentop, S. and Lipkow, A. (2003) *Stoffflussbezogene Bausteine für ein nationales Konzept der nachhaltigen Entwicklung – Verknüpfung des Bereichs 'Bauen und Wohnen' mit dem komplementären Bereich 'öffentliche Infrastruktur'*, Umweltbundesamt, Berlin. Short English version available at: http://www.umweltbundesamt.de/uba-info-medien/dateien/2600.htm.

EEA, European Environment Agency (2003) *Europe's Environment: The Third Assessment*, Environmental Assessment Report No. 10, EEA, Copenhagen.

Erkman, S. (1997) Industrial ecology: an historical view, *Journal of Cleaner Production*, **5**, 1–10.

EU Commission (2003) *Towards a Thematic Strategy on Sustainable Use of Natural Resources. Communication from the Commission to the Council and the European Parliament*, COM (2003) 572 final, EU Commission, Brussels.

EUROSTAT, Statistical Office of the European Communities (ed.) (2001) *Economy-wide Material Flow Accounts and Derived Indicators: A Methodological Guide*, Edition 2000. European Communities, Luxembourg.

Fischer-Kowalski, M. (1999) Society's metabolism: the intellectual history of materials flow analysis, Part II, 1970–1998, *Journal of Industrial Ecology*, **2**(4), 107–136.

Fritsche, U. R., Dehoust, G., Jenseit, W., Hüneke, K., Rausch, L., Schüler, D., Wiegemann, K., Heinz, A., Hiebel, M., Ising, M., Kabasci, S., Unger, C., Thrän, D., Fröhlich, N., Scholwin, F., Reinhardt, G., Gärtner, S., Patyk, A., Baur, F., Bemmann, U., Groß, U., Heib, M., Ziegler, C., Flanke, M., Schmehl, M. and Simon, S. (2004) *Stoffstromanalyse zur nachhaltigen energetischen Nutzung von Biomasse*, Öko-Institut e.V., Darmstadt.

Haberl, H. (1997) Biomass flows in Austria: integrating the concepts of societal metabolism and colonisation of nature, in Bringezu, S., Fischer-Kowalski, M., Kleijn, R. and Palm, V. (eds) *Proceedings of the ConAccount workshop, 21–23 January, 1997*, Wuppertal Special 4, Wuppertal Institute, Wuppertal, pp. 102–107.

Kleijn, R. (2002) personal communication.

Liu, J. (2004) personal communication.

Liu, J. and Wang, Q. (2005) Direct material input and dematerialization analysis of the Chinese economy, *Resources Science*, in press.

Loske, R. *et al.* (1996) *Zukunftsfähiges Deutschland: Ein Beitrag zu einer global nachhaltigen Entwicklung*, Birkhäuser, Berlin.

Mäenpää, I., Mänty, E. and Härmä, T. (2002) Statistical compilation system of the time series of the Finnish natural resource use. TMRFIN version 1.3. Update 21.8.2002. Thule Institute, University of Oulu; Environmental Cluster Research Program of Ministry of the Environment Eco-efficient Finland project (1998–2000).

Moll, S., Bringezu, S. and Schütz, H. (2003) *Resource Use in European Countries: An Estimate of Materials and Waste Streams in the Community, Including Imports and Exports Using the Instrument of Material Flow Analysis (Zero Study)*, European Topic Centre on Waste and Material Flows, Copenhagen.

Moolenaar, S. W. (2002) Heavy metals in agrosystems, in Ayres, R. U. and Ayres, L. (eds) *Handbook of Industrial Ecology*, Edwar Elgar Publishers, Cheltenham, pp. 421–431.

Moriguchi, Y. (2000) personal communication.

Müller, D., Bader, H.-P. and Baccini, P. (2004) Physical characterization of regional timber management for a long-term scale, *Journal of Industrial Ecology*, **8**, 65–88.

OECD (2004) Recommendation of the Council on Material Flows and Resource Productivity. Adopted by the OECD Council on 21 April 2004, OECD, Paris.

Schmidt-Bleek, F. (1994) *Wieviel Umwelt braucht der Mensch? MIPS: Das Maß für ökologisches Wirtschaften*, Basel: Birkhäuser, Berlin.

Schütz, H. (2003) *Economy-Wide Material Flow Accounts, Land Use Accounts and Derived Indicators for Germany*, Report to the Commission of the European Communities, DG Eurostat/ B1, Wuppertal Institute, Wuppertal.

Schütz, H., Moll, S. and Bringezu, S. (2004) *Globalisation and the Shifting Environmental Burden: Material Trade Flows of the European Union*, Wuppertal Papers 134e, Wuppertal Institute, Wuppertal.

Schütz, H., Welfens, M. J. and Sleszynski, J. (2002) Sustainable development through dematerialization: the case of Poland, *Osteuropa-Wirtschaft*, **47**, 173–199.

Steger, S. (2005) *Der Flächenrucksack des Europäischen Außenhandels mit Agrarprodukten*, Wuppertal Papers 152, Wuppertal Institute, Wuppertal.

Thrän, D. (2003) Measuring sustainable development of rural regions on the basis of MFA: a bottom-up wood flow analysis, paper presented at workshop: 'Quo vadis MFA?' Material Flow Analysis – where do we go? Issues, trends and perspectives of research for sustainable resource use, 8–10 October 2003, Wuppertal.

Van der Voet, E. (1996) *Substances from Cradle to Grave (Development of a Methodology for the Analysis of Substance Flows Through the Economy and the Environment of a Region – with Case Studies on Cadmium and Nitrogen Compounds)*, Optima Druk, Molenaarsgraaf, The Netherlands.

Van der Voet, E. (2002) Substance flow analysis methodology, in Ayres, R. U. and Ayres, L. (eds) *Handbook of Industrial Ecology*, Edward Elgar Publishers, Cheltenham, pp. 91–101.

Van der Voet, E., van Oers, L., Guinée, J. B., Udo de Haes, H. A. (1999) Using SFA indicators to support environmental policy, *Environmental Science and Pollution Research*, **6**, 49–58.

Vitoussek, P. M., Mooney H. A., Lubchenco, J. and Melillo, J. M. (1997) Human domination of Earth's ecosystems, *Science*, **277**, 494–499.

Von Weizsäcker, E. U., Lovins, A. B. and Lovins, L. H. (1995) *Factor Four: Doubling Wealth Halving Resource Use,* Earthscan, London.

Weiss, M., Bringezu, S. and Heilmeier, H. (2004) Energie, Kraftstoffe und Gebrauchsgüter aus Biomasse: Ein flächenbezogener Vergleich von Umweltbelastungen durch Produkte aus nachwachsenden und fossilen Rohstoffen, *Zeitschrift für Angewandte Umweltforschung*, Jg. **15/16**, 361–378.

9

Ecological Footprints and Biocapacity

Essential Elements in Sustainability Assessment

William E. Rees

9.1 Introduction

It is no small irony that in the age of 'technological man' people actually play a greater role in ecosystems than ever. For example, *H. sapiens* has long been the most successful terrestrial carnivore ever to have walked the earth and, during the twentieth century, humans became the most voracious predator in the world's oceans. Remarkably, considering our unchallenged status as top carnivore, we are also the dominant herbivore in grasslands and forests all over the planet, particularly if we consider the demands of 'industrial metabolism' (Rees, 2003a; Fowler and Hobbs, 2003). Human impacts even transcend ecology. Earth scientists assert that economic activity has become the most significant geological force altering the face of the planet and climatologists agree that we are now actually beginning to affect global climate.

Despite such findings, modern humans – particularly city dwellers – are so psychologically alienated from nature that they rarely think of themselves as biological entities let alone as dependent components of the world's ecosystems (Rees, 2003b). Indeed, many economists and technological optimists argue that the human enterprise is 'decoupling' from the ecosphere. As the late Julian Simon hyperbolically proclaimed, 'Technology exists now to produce in virtually inexhaustible quantities just about all the products made by nature' (Simon, cited in Bartlett, 1996).

The overall aim of this chapter, therefore, is to help reground the sustainability debate in biophysical reality. More specifically, I hope to make the case that ecological footprint analysis is a sound basis from which to assess ecological sustainability and to illustrate

Renewables-Based Technology Edited by J. Dewulf and H. Van Langenhove

how the method stimulates questions about sustainable development that are invisible to conventional analyses.

9.2 Eco-Footprint Analysis

Despite the cascade of empirical evidence that even the present-scale human economic activity threatens to undermine the integrity of the ecosphere, there is little evidence in the international policy arena that mainstream institutions are seriously willing to consider abandoning the perpetual growth machine. Indeed, policy-makers generally believe that the Malthusian dilemma and concerns about 'limits to growth' have long been put to rest (Rees, 2002a). According to some authors, 'Because of increases in knowledge, the earth's "carrying capacity" has been increasing throughout the decades and centuries and millennia to such an extent that the term "carrying capacity" has by now no useful meaning' (Simon and Kahn, 1984). Growth optimists also insist that, by trading with neighbours, any region or country can eliminate any remaining constraints on growth imposed by limited domestic resource endowments.

Ecological footprint analysis (EFA) was introduced explicitly to reopen the debate on human carrying capacity (Rees, 1992, 1996; Rees and Wackernagel, 1994; Wackernagel and Rees, 1996). Indeed, the method gains much of its analytic strength by inverting the standard carrying capacity ratio. If carrying capacity asks, 'How large a population can a particular area support?' (a question that can be rendered seemingly irrelevant by trade), EFA asks, 'How large an area is required to support a particular population?' (a question that includes those areas that are effectively 'imported' through trade).

Answering this second question enables any population to compare its total biophysical demand on Earth to the biocapacity of its domestic land base, thus revealing the extent to which that population is living beyond its local ecological means. EFA also allows the population to assess the proposition that its consumption patterns are 'decoupling' from nature, i.e., a sequential time series of EFAs will reveal whether the population's lifestyles are really becoming less material-intensive and more ecologically benign.

EFA starts from a series of simple premises:

- Human beings are integral components of the ecosystems that sustain us. We can therefore best assess ecological sustainability using biophysical data.
- Most human impacts on ecosystems are associated with energy and material extraction and consumption.
- These energy and material flows can be converted to corresponding productive or assimilative ecosystems areas.
- There is a measurable, finite area of productive land and water ecosystems on Earth.

Every human population imposes an 'ecological footprint' on the Earth, equivalent to the amount of the planet's productive capacity required to supply that population with resources and waste assimilation. We therefore formally define the ecological footprint of a specified population as:

> the area of land and water ecosystems required on a continuous basis to produce the resources that the population consumes, and to assimilate (some of) the wastes that the population produces, wherever on Earth the relevant land/water may be located.
>
> (Rees, 2001)

A complete eco-footprint analysis would thus quantify the total ecosystem area that the population effectively 'appropriates' to meet its final demand for economic goods and services, including the area it needs to provide its share of certain life-support functions such as carbon sequestering. In practice, of course, we cannot consider separately every one of the tens of thousands of consumption items available today. However, most of these are included in major consumption categories such as 'meat products', 'legumes and pulses', 'plastics', 'organic chemicals' and similar categories that are compiled by national statistical agencies.

We are also effectively restricted to those categories of waste such as carbon dioxide and organic nutrients (e.g., nitrates and phosphates) that can readily be represented by an exclusive assimilation area. Indeed, some products such as minerals and concrete are represented in the eco-footprint largely through their carbon sink areas (because of the fossil fuels used to produce them) although relevant mine sites and related degraded landscapes are also included in the analysis. Sometimes we can ignore organic, nutrient or other wastes if it is reasonable to assume that they are being assimilated by land and water areas that are already included in the eco-footprint calculation because said lands produce some relevant commodity (e.g., agricultural products or fish). Still other wastes such as widely dissipated airborne and waterborne toxic compounds and ozone-depleting chemicals cannot be translated into land area at all.

The area of a population's theoretical eco-footprint depends on four factors: (1) the population size; (2) the average material standard of living; (3) the average productivity of land/water ecosystems; and (4) the efficiency of resource harvesting, processing, and use. Regardless of the relative importance of these factors and how they interact, *every population has an ecological footprint* and the productive land and water captured by the EFA represents much of the 'natural capital' (productive natural resource base) required to meet that study population's consumptive demands. It is important to recognize that population eco-footprints constitute *mutually exclusive* appropriations of productive capacity. The biocapacity used by one population is not available for use by another. Ultimately, *all human populations are competing for the available productive capacity of Earth.*

Note that ecological footprints can be interpreted in terms of thermodynamic theory (Rees, 2002b). All human activities involve the extraction, consumption, and irreversible dissipation of resources, thus increasing global net entropy. Since contemporary production of renewables is driven by solar energy, a population's sustainable ecological footprint represents the area required continuously to generate photosynthetically a quantity of biomass energy and material (negentropy) equivalent to the amount used and dissipated by the population's consumptive activities.

9.2.1 *Basic Methods*

Population eco-footprints are based on final demand for goods and services. Thus, the first step in calculating the ecological footprint (EF) of a study population is to estimate the total annualized consumption of significant categories of commodities and consumer goods consumed by that population. Data are obtained from national production and trade statistics and other sources such as various United Nations statistical publications. For accuracy, consumption data should be trade-corrected. Thus, the population's consumption of pulses (beans, peas and lentils) can be represented as follows:

$$\text{consumption}_{\text{pulses}} = \text{production}_{\text{pulses}} + \text{imports}_{\text{pulses}} - \text{exports}_{\text{pulses}}$$

The second step is to convert consumption of each item into the land/water area required to produce that item (or to assimilate the wastes generated in its production) by dividing total consumption by world average land productivity or yield for that item. This step gives us the EFs of the individual consumption categories.

In the case of non-organic items such as metals, we use the land sterilized for mines, tailings and smelters as well carbon dioxide output – from the fossil energy used in production – to estimate that component's eco-footprint (see reference to carbon dioxide emissions below). Built-up and damaged lands (usually urban-related) are measured directly. The third step estimates the total EF of the population by summing the footprints for the individual consumption/waste categories. Finally, we can obtain the *per capita* EF of the study population obtained by dividing the total population footprint by population size.

For some wastes such as carbon dioxide emissions, or nutrients such as phosphates and nitrates, it is also possible to calculate the exclusive land/aquatic ecosystem area required for sustainable assimilation and recycling. (Carbon sinks constitute most of the eco-footprint area associated with fossil fuels themselves as well as that for many energy-intensive non-organic products.) In all such cases, the assimilation rate per hectare and year is substituted for yield in the calculations described above.

For basic population ecological footprints, EFA assessors usually use world average yields/assimilation rates in each major land categories (cropland, pasture, forest land, productive marine area, etc.). This simplifies calculations since we do not have to trace all the sources of trade goods and waste sinks nor determine the productivity and assimilative capacities of the corresponding production/assimilation areas. In all cases, we strive to avoid overlap and double-counting. For example, we would not need to estimate the grazing land eco-footprint associated with leather goods because animal hides are usually a by-product of the meat and dairy industry whose contributions are already included.

To facilitate comparisons between countries and to estimate national ecological surpluses or deficits, analysts further adjust the basic footprint calculations to a common scale. For example, if country 'A' uses a large area of relatively unproductive pasture land per capita compared to another country, 'B', that uses more highly productive cropland, then country 'A's eco-footprint will seem disproportionately (unfairly) large compared to country 'B's. To provide a more balanced comparison, we make an 'equivalence adjustment' by converting each land-type component of the basic national EFs into its equivalent area in terms of 'global hectares' where a global hectare is a standardized hectare (ha) of world average productivity (see WWF, 2004). Thus, if country 'A' uses the equivalent of 2 ha of average pasture *per capita*, and average pasture is half as productive as a standard global hectare, then the representation of pasture in country 'A's EF is scaled down to only 1 global hectare *per capita* ($2\,\text{ha} \times 0.5_{\text{equiv}}$).

One of the most interesting and useful applications of nation-level EFA is in the assessment of each country's ecological surplus/deficit. Here we ask the following questions:

What part of a country's net consumption could be accommodated by that country's domestic biocapacity? Could this country be more or less self-sufficient if necessary? We answer these questions by subtracting each country's EF (based on global hectares) from the area of its domestic productive land-base (also converted to equivalent global hectares by factoring in domestic yields). A positive answer implies that the subject country's consumption imposes a load on Earth less than the its total domestic biocapacity, i.e., the country has an ecological surplus. If the answer is negative, then the country's ecological load exceeds its domestic biocapacity. In this case, the relevant population is living, in part, on apparently surplus biocapacity imported from other countries or from the global commons.

9.2.2 *The Eco-Footprints of Nations: Measuring Relative Sustainability*

Figure 9.1 displays the equivalence-adjusted eco-footprints of a selection of the world's nations (based on 2001 data compiled in WWF, 2004). As might be expected, *per capita* EFs are positively correlated with income. The residents of the United States, Australia, Canada, many Western European and other high-income countries each require from 5 to 10 hectares (12–25 acres) of productive land/water to support their consumer lifestyles. By contrast, the citizens of the world's poorest countries have average EFs as low as half a hectare. Even burgeoning China's *per capita* eco-footprint in 2001 was less than 2 hectares. The average human EF is about 2.2 ha.

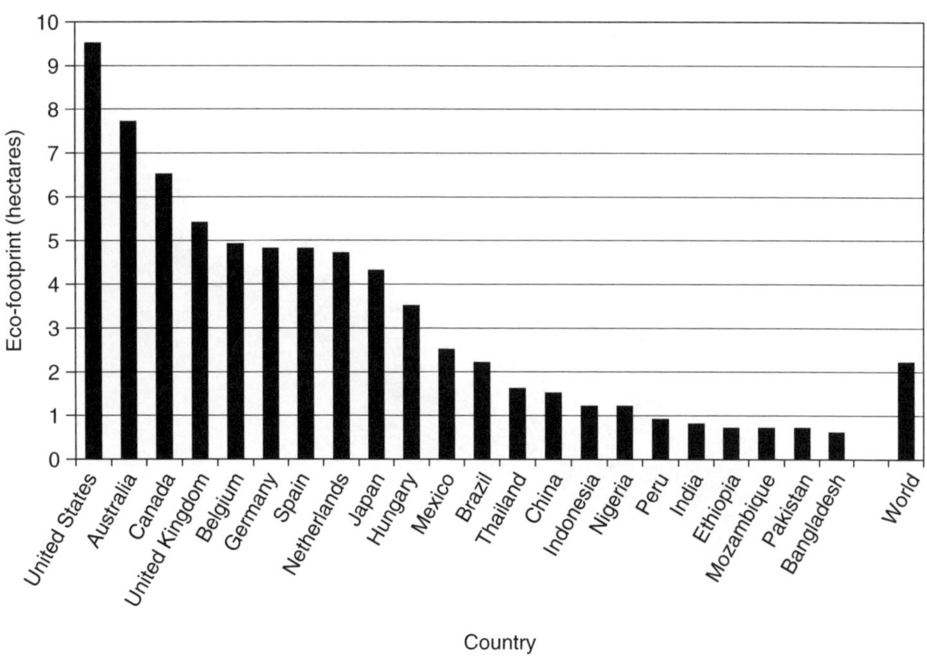

Figure 9.1 *Equivalence-adjusted per capita ecological footprints of selected countries (2001 data)*

Consider these demand data in light of global supply and prospects for sustainability. In 2001, there were only about 10 billion ha of productive cropland, pasture, and forest on Earth and about one billion ha of equivalent fishable shallow ocean, for a total of about 11.1 billion ha for over 6.1 billion people. In short, there were only 1.8 ha of productive ecosystem *per capita* on the entire planet. With an estimated average eco-footprint of 2.2 ha *per capita*, the human population already had a total eco-footprint of almost 13.5 billion ha. This means that by a fairly conservative estimate, humanity had already 'overshot' the long-term human carrying capacity of the Earth by over 20% in 2001 – the whole planet is in deficit. (A population can live in overshoot – i.e., beyond its ecological means – for a considerable period by depleting vital ecosystems and non-renewable resource stocks.)

These aggregate eco-footprint data imply that to bring just the present world population up to, say, North American material standards with prevailing technology would require four additional Earth-like planets! (Figure 9.2). This poses a serious conundrum for those who insist on sustainability through material growth – 'good planets are hard to find.'

The situation is more complex from the sustainability perspective than is suggested by gross overshoot alone. Many high-density high-income countries have eco-footprints several times larger than their domestic territories. These countries are running large 'ecological deficits' with the rest of the world (Rees, 1996). The Netherlands, for

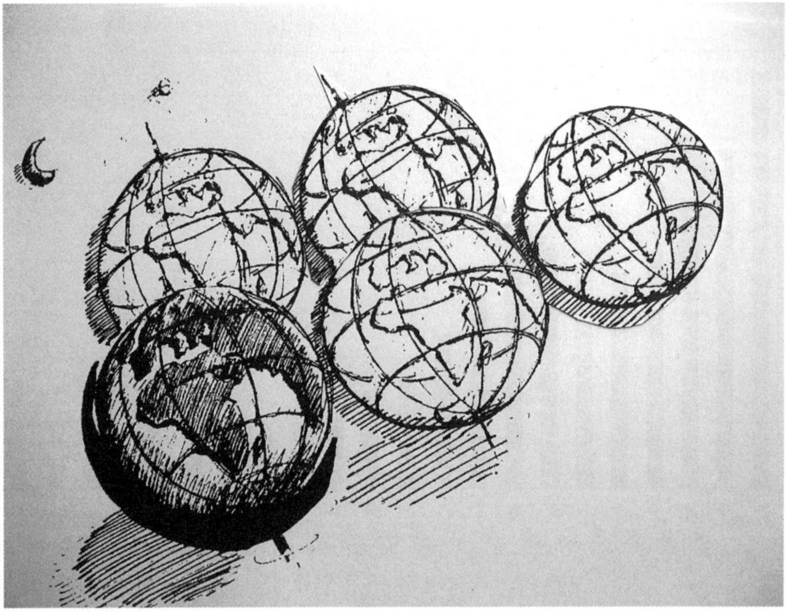

Figure 9.2 *Phantom Orbs – to sustain today's 6.2 billion global villagers at North American levels of energy and material consumption would require four additional Earth-like planets. Reproduced from Gallaugher and Woods (eds), 'Ecological Footprints of Farmed and Harvested Salmon' in The World Summiton Salmon: Proceedings. With permission from the Simon Fraser University. Copyright 2004.*

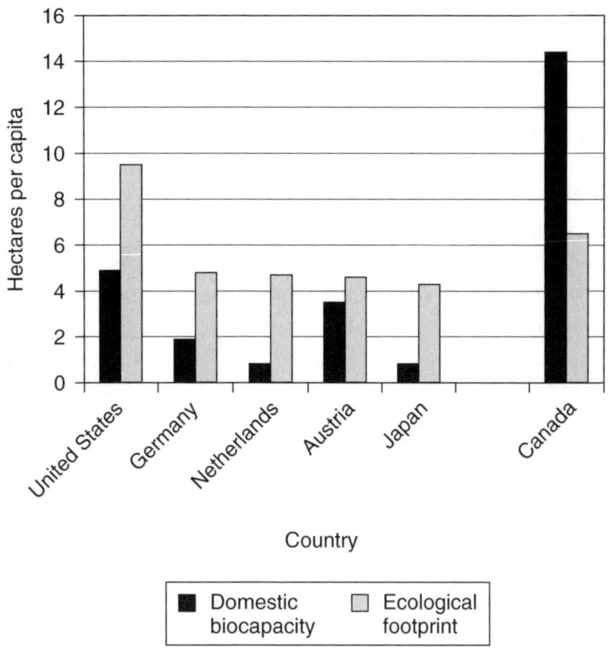

Figure 9.3 *Domestic biocapacity compared to eco-footprints of selected countries (2001 data)*

example, requires almost six times its domestic biocapacity to sustain prevailing levels of net consumption (Figure 9.3, Table 9.1).

As noted above, citizens of such deficit countries live, in part, on life support services imported from other countries and by imposing a disproportionate load on the global commons. Indeed, wealthy market economies like those of the USA, Canada, most Western European countries and Japan appropriate two to five times their equitable share (1.8 ha) of the planet's productive land/water (and 10 to 20 times more *per capita* than the chronically impoverished). By contrast, low-income countries like Bangladesh, Mozambique

Table 9.1 *Overshoot – the ecological deficits of a selection of the world's most advanced economies*

Country	*Total eco-footprint (EF) (millions of global ha)*	*Energy component of EF (millions of global ha and %)*	*Domestic biocapacity (millions of global ha)*	*Overshoot factor (ecological deficit) (%)*
United States	2,736	1,757 (64)	1,411	94
Germany	395	255 (64)	156	153
Austria	37	20 (66)	28	32
Netherlands	75	46 (61)	13	488
Japan	547	356 (62)	102	438
Canada	198	102 (52)	446	–56

Source: based on data from WWF (2004).

and even China, use only a fraction of their equitable population-based allocation. (Note that the prevailing forces of globalization tend to exacerbate rather than level these gross eco-economic inequities.)

Eco-footprinting thus reveals a hidden role of global trade. The enormous purchasing power of the world's richest nations enables them to finance their ecological deficits by extending their eco-footprints deeply into exporting nations and throughout the open ecosphere (Rees, 2002b). Wealthy and powerful nations can now achieve through global commerce what used to require territorial occupation.

The obvious ecological problem is that not all countries can run a biophysical deficit – for every sustainable deficit there must be a permanent surplus somewhere else. Unfortunately, the apparent surpluses of the few large 'under-populated' countries such as Australia and Canada (Figure 9.3, Table 9.1) have already been absorbed into the eco-deficits of other countries. This is why, in net terms, there is no global surplus.

9.3 Inherent Strengths in EFA

9.3.1 The Scientific Merit of EFA

Ecological footprint analysis has gained considerable momentum around the world as both a heuristic device and a practical method of assessing sustainability. This success derives in part from methodological strengths of EFA that are both scientifically well founded and reflect thinking people's intuitive sense of reality. On the technical/scientific side, EFA has several qualities that reinforce its credibility as a sustainability indicator. EFA does the following:

- acknowledges that humans are biophysical beings who make constant metabolic demands on their supportive ecosystems and that all our manufactured capital and related cultural artefacts impose a parallel and much larger industrial metabolism on the ecosphere;
- recognizes the crucial role of natural capital and natural income (biophysical stocks and flows) in economic development and sustainability;
- accepts that the economy is a fully contained, growing, dependent, sub-system of the non-growing ecosphere;
- is closely related conceptually to Odum's embodied energy (emergy) analyses (Hall, 1995) and the 'environmental space' concept of the Sustainable Europe Campaign (Carley and Spapens, 1998);
- accounts for both population size and resource consumption in estimating the appropriated ecosystem area. This aligns EFA closely with Catton's (1980) concept of human 'load' (population times *per capita* consumption).
- corresponds closely to and incorporates all the factors in Ehrlich's and Holdren's (1971) well-known definition of human impact on the environment: I=PAT, where 'I' is impact, 'P' is population, 'A' is affluence (i.e., level of consumption) and 'T' is a technology scalar.

9.3.2 *Popular Acceptance of EFA*

People are intelligent beings capable of responding rationally to new knowledge particularly if it can be shown to be directly relevant to their own circumstances. For this reason, the eco-footprint concept resonates better with the public than do more abstract and impersonal sustainability indicators. In particular, people appreciate the way EFA draws them into reflecting on their personal consumption habits as illustrated by the popularity of EFA-oriented websites that offer simple calculators that visitors can use to estimate their personal eco-footprints. Attributes of EFA that help to communicate biophysical reality to the public include the following:

- The method is conceptually simple and intuitively appealing. Even sceptics recognize that they have a positive ecological footprint.
- EFA personalizes sustainability by focusing on consumption – everyone is a consumer and must ultimately take responsibility for his/her own 'load' on the planet.
- EFA consolidates measurable energy and material flows into a single concrete variable, the corresponding appropriated land/water (ecosystem) area.
- Land itself is a powerful indicator. Everyone understands 'land'. (Popular understanding of the ecological crisis is prerequisite to any politically viable solutions.)
- Eco-footprint estimates can be compared to finite local and global 'supplies' of terrestrial and aquatic ecosystems (i.e., people and populations can compare their demands to available biocapacity).
- The 'ecological deficit' – the difference between domestic bio capacity and a larger eco-footprint – requires little explanation and many people see it as more important than the fiscal deficits with which their governments are often preoccupied.
- EFA appeals to both the ecologically and socially conscious. For example, it reflects gross material inequity but also shows that growth is not a sustainable option to relieve it.
- Perhaps as important as any other factor, 'ecological footprint' is a powerfully evocative metaphor – would people be as quickly captivated by the concept had it been called the 'human impact index' instead?

9.4 Answering the Critics

For all its strengths, ecological footprint analysis is not a perfect tool and the method has attracted its fair share of criticism from both academics and practitioners. Some criticisms stem from real weaknesses of the method or from misunderstandings of its scope and potential. Others, however, seem more to reflect the critic's discomfort with the findings and implications of EFA than they do flaws in the method. The following selection of criticisms are drawn both from the published literature (e.g., several articles in *Ecological Economics* (2000)), from papers that the author has reviewed on behalf of various journals, and from colleagues' direct questions and challenges. The critique can be partitioned into two broad categories: concerns about methods or the concept itself, and concerns about the policy relevance.

9.4.1 Conceptual and Methodological Critiques

1 EFA raises questions of human carrying capacity. We do, in fact, use EFA to compare economic production/consumption to available biocapacity (carrying capacity) as a sustainability test. Economists in particular object to this on grounds that in today's world human carrying capacity is irrelevant and that a nation's domestic bioproductivity should in no way constrain its growth, influence its development policy or affect its citizens' lifestyle choices. This view reflects the view that individual trading regions are open systems that can effectively import 'carrying capacity' from elsewhere. Unfortunately, this argument fails on the global scale – the Earth as a whole is materially closed. Thus, it is simply not possible for all countries to be simultaneous net importers of biocapacity.

2 EFA has an anti-trade bias. It assumes eco-deficits are bad and reduces sustainability to self-sufficiency. EFA shows that many rich countries and urban regions have enormous ecological footprints. Cities typically 'occupy' an ecological space several hundred times larger than their geographic areas and many whole countries draw on the biophysical capacity of an ecosystem area several times their productive domestic land base (Folke *et al.*, 1997; Rees, 2003b). However, these facts do not in themselves express an anti-urban or anti-trade bias. They merely underscore the biophysical dependence of densely populated, high-consuming regions/countries on other countries and the global commons and remind us that such regions appropriate more than their fair share of global biocapacity (Wackernagel and Silverstein, 2000). In fact, the data show that self-sufficiency actually lies beyond the realistic reach of regions/countries with large eco-deficits. *Some* trade is not only a good thing but is absolutely necessary for sustainability in these situations.

All the same, we live an era of accelerating global ecological change and uncertain geopolitics. This suggests that there may be an optimal level of trade-reliance and that perhaps greater regional *self-reliance* might be a good thing. Many people also question a global economic system that enables the already rich to buy sustainability out from under the struggling poor.

Such reasoning is apparently alarming to neo-liberal economists and other expansionists who favour unrestrained growth facilitated by trade. By stimulating discussion of carrying capacity, optimal trade levels and relative dependence/self sufficiency, EFA implicitly challenges the core economic ideology of our time. This does not weaken EFA but may undermine prevailing ideology.

3 EFA ignores technology and the substitutability of manufactured for natural capital. It therefore supports growth pessimists. Some critics argue that EFA 'seems not to be in accordance with mainstream ecological economics that assumes at least some substitution between different types of capital'. Others allege that EFA assumes 'technology will not be able to overcome biophysical limits'.

These assertions are erroneous because eco-footprint analysis *per se* makes *no assumptions whatever* about material substitutions or technological change. Most population EFAs typically represent real-time snapshots of *de facto* energy and material flows at the time of the analysis, whatever the prevailing level of technological sophistication. They represent what *is*, not what should be or what could be. EFA is fully responsive to technological changes or substitutions that might significantly affect a population's eco-footprint. (Bear in mind that some innovations and substitutions actually *increase* the eco-footprint (Rees, 2003a, 2003b.))

It is worth noting that 'EFA supports growth pessimists' is really not a criticism but rather a *conclusion*. As such, it may reveal a hidden fear of the growth optimists that perhaps there is a weakness in their own paradigm.

4 On representing energy consumption. Some critics are uncomfortable with the large contribution that energy makes to typical eco-footprints – about half the global average eco-footprint, and up to two-thirds that in high-income countries (Table 9.1). Others dispute the use of terrestrial carbon sequestration as the measure of society's use of fossil fuel on several grounds, e.g., a carbon sink is not a 'real' land use; carbon sequestration faces real economic constraints ('society will never accept the high costs of terrestrial-based CO_2 strategies'); or that renewable energy would generate a smaller eco-footprint.

Much of this criticism comes from either a misunderstanding of how EFA treats energy or from discomfort generated by having to confront humanity's large energy eco-footprint. The fact that energy consumption looms large in EFA simply represents reality. Energy is absolutely vital to industrial society as the means by which we do everything else. Energy-use creates a large EF primarily because of thermodynamic laws, not because of methodological flaws in EFA.

As for using carbon sink area to represent fossil fuel use, remember first that EFA is an ecosystem-based way of assessing consumption and that ecosystems store carbon in biomass; second, modern society consumes prodigious quantities of fossil energy, releasing huge volumes of carbon dioxide into the atmosphere and thus into the biologically active carbon pool.

Now, carbon dioxide waste is a major factor in atmospheric change and a primary 'forcing mechanism' for climate modification. A consensus has therefore emerged that, for sustainability, large quantities of this greenhouse gas must be sequestered. The terrestrial ecosystem area that would be required to assimilate industrial CO_2 emissions – less the amount routinely taken up by the sea – is therefore a legitimate representation of society's fossil fuel energy footprint. Certainly, too, dedicated carbon-sink forests are very real land use.

Some critics argue that we could reduce the energy eco-footprint if we based it on some form of renewable energy alternative such as biomass fuels. This approach would indeed produce a different eco-footprint and when the use of alternative energy begins to have a serious impact on carbon emissions, any positive result will show up in subsequent EFA studies.

Note, however, that techno-optimists may be disappointed by efforts to replace for fossil fuels. For example, thermodynamic law dictates that the biomass equivalent of fossil fuels would generate a *larger* eco-footprint than the carbon sequestering method and would create other problems. For example, more fossil energy is used to produce a litre of ethanol than is contained in the ethanol so that the relevant energy footprint would have to comprise both the original carbon sink area plus the corn-growing area. The latter is substantial. To grow sufficient maize to provide the ethanol equivalent of just one-third of US automotive fuel would require as much cropland as is needed to feed the entire US population (Pimentel, 2003).

Putting all this together, EFA simply recognizes that carbon emissions are a real contemporary problem, that we have not developed artificial carbon dumps or adequate alternative fuels, that the principal way nature stores carbon in the short term is in biomass – indeed, this is currently the only way to sequester substantial quantities of carbon – and that it is possible for humans to establish at least some dedicated carbon sink forests. In short, the large energy footprint due to excessive carbon dioxide emissions is not a fault of EFA (which is merely the bearer of the bad news) but of over-consumption relative to available biocapacity. In this sense, the large contemporary energy eco-footprint is actually a *robust finding* of the EFA method.

5 EFA assumes that 'land' is being used sustainably. This charge is true and bothersome. Much land and many critical ecosystems are being degraded. Part of the reason we do not account for the unsustainable use of nature is the sheer labour intensity of determining erosion and depletion rates of areas in question. The question does arise, however, of how would one use the data if they were available. Suppose soil degradation were 30 times the rate of renewal (probably close to the world average). If we inflated the arable land component of the eco-footprint to reflect such information, *per capita* eco-footprints would be numbingly large – would anyone then take them seriously? As matters stand, the estimated average EFs, unadjusted for land degradation, show everything needed for a reasoned policy response *in the right direction* without being intimidating or discouraging.

9.4.2 EFA and Sustainability Policy

1 EFA offers no policy guidance. Economists in particular reject EFA on grounds that the method is a simplistic static tool that 'provides little use in directing policy', is 'inadequate for policy design' or 'does not assist in the analysis of sustainability'.

It is true that the EF is a single numerical index so that, in isolation, it can hardly suggest policy directions (the same can be said of GDP/capita). However, the notion that EFA is irrelevant to policy would come as a surprise to the hundreds of government agencies, academics, citizen organizations, NGOs, policy advisers, etc., around the world who are using EFA precisely to identify, highlight or address numerous policy issues pertaining to sustainability. As this chapter reveals, EFA is powerfully indicative of policy choices available to countries that have excessive eco-footprints, including greater regional self-reliance, conserving natural capital, developing dedicated carbon sink forests, investing in alternative energy, working toward greater global equity, etc. Significantly, many of the policy issues and options raised by EF studies (e.g., biocapacity, the risks of excess trade dependence, ecological deficits, self-reliance, etc.) remain invisible to conventional analysis.

Indeed, it is more than a little ironic that economists would reject EFA on grounds that it has little to offer to sustainability policy. Unlike EFA, economists' policy models are *totally abstracted from biophysical reality.* Moreover, according to some economists, they no longer even describe economic reality. The major problem is '*the nearly complete collapse of the prevailing economic theory*... It is a collapse so complete, so pervasive, that the profession can only deny it by refusing to discuss theoretical questions in the first place' (Galbraith, 2000, italics added for emphasis). So much for policy relevance.

2 What about socio-economic factors? Some critics reject the use of EFA in policy analysis because it is 'devoid of any socio-economic factors'. On one level, this is a straw-man partial truth, and to that extent it is irrelevant. The very term, 'ecological footprint analysis' declares that the method is intended to generate a human ecological index, not a social indicator. There is no logical reason why an ecological index should incorporate social factors *per se*.

That said, it remains instructive on socio-economic grounds to compare to the size and composition of per capita eco-footprints at opposite ends of the income distribution (Figure 9.1). The relatively wealthy 25% of the world's population with the largest eco-footprints are responsible for 86% of personal consumption. This bloated share of consumption appropriates virtually all the biophysical capacity of the planet in important categories. Such data show that efforts to relieve poverty and achieve sustainability through material economic growth are futile, an uncomfortable reality for any expansionist to contemplate.

3 Typical EFA studies lack predictive power. This seeming 'criticism' is only partially true and mostly irrelevant. Many useful indices are based on static analyses. For example, average 'life expectancy' is not a predictive indicator but it is a good measure of population health; the 'human development index' is not a predictive tool but it is accepted as a good aggregate indicator of, well, relative 'human development'; 'GDP' is not a predictive tool but it is generally regarded as a fair indicator of aggregate economic activity (but, to be sure, it is a poor indicator of human welfare). One might criticize a dynamic model that fails to make good predictions but it is silly to reject static models and indicators (contemporary 'snapshots') on the same grounds. In any event, analysts can use EFA in simulation studies involving assumed lifestyle changes or advances in technology and thereby predict the effect of such changes on EF size.

4 EFA results are depressing. To quote one critic, '[EFA] is becoming a global aggregated indicator of ecological overshoot and doom.' As shown in a previous section, EFA shows that the world economy is, in fact, in a state of overshoot in that the estimated total global eco-footprint is somewhat larger than the aggregate eco-productive area of the planet. This conclusion is supported by myriad empirical data on everything from the collapse of fish stocks through accelerating landscape/soil degradation to the accumulation of greenhouse gases. If this suggests 'doom' to the critics, it may once again reflect their own subjective fears that the findings may actually be accurate. On the positive side, having a tool that recognizes the danger enables EF analysts to recommend the policies, strategies and behavioural changes required to address the situation precisely to *avoid* ecosystemic collapse and related forms of 'doom'.

9.5 Summary

This chapter argues that ecological footprint analysis provides a robust method of assessing ecological sustainability. EFA is firmly rooted in human ecological reality and clearly reflects the most basic of biophysical laws. The logic of the method flows from the simple fact that the scale of human economic activity must fit within the productive capacity of the ecosphere.

In this light, it is encouraging that the 'footprint' metaphor seems to have captured the public imagination. Because EFA personalizes the sustainability crisis – everyone is a consumer and everyone has an eco-footprint – the human eco-footprint has become one of the most effective and best known indicators of global (un)sustainability.

Most importantly, by focusing on physical resource stocks and flows, and on quantifiable real-world ecosystems capacity, EFA stimulates questions about sustainability that are invisible to conventional economic assessments. In particular, EFA has succeeded in putting the questions of local and global carrying capacity back on the policy table. Hardly any sustainability analyst is not now aware of the 'overshoot' phenomenon as highlighted by EFA. The finding that to raise just the present world population to North American material standards sustainably would require several addition Earth-like planets is frequently cited in the sustainability literature and in public debates around the world.

EFA is certainly an imperfect tool. However, its major weakness may be the inherent conservatism of the method rather than the concerns expressed by economists and techno-optimists. EFA findings, already alarming enough, likely under-estimate rather than over-estimate the total human load. In this light, the real sustainability problem is that the official world remains in the thrall of the perpetual growth myth.

References

Bartlett, A. (1996) The exponential function XI: the new Flat Earth Society, *The Physics Teacher*, **34**, 342–343.

Carley, M. and Spapens, P. (1998) *Sharing the World: Sustainable Living and Global Equity in the 21st Century*, Earthscan Publications, London.

Catton, W. R. (1980) *Overshoot: The Ecological Basis of Revolutionary Change*, University of Illinois Press, Urbana and Chicago.

Ecological Economics (2000) 'Forum on the Ecological Footprint', **32**, 341–394.

Ehrlich, P. and Holdren, J. (1971) Impact of population growth, *Science*, **171**, 1212–1217.

Folke, C., Jansson, A., Larsson, J. and Costanza, R. (1997) Ecosystem appropriation by cities, *Ambio*, **26**, 167–172.

Fowler, C. W. and Hobbs, L. (2003) Is humanity sustainable? *Proceedings of the Royal Society of London, Series B: Biological Sciences*, **270**, 2579–2583.

Galbraith, J. K. (2000) How the economists got it wrong, *The American Prospect*, **11**, 18–20.

Hall, C. (ed.) (1995) *Maximum Power: The Ideas and Applications of H.T. Odum*, University Press of Colorado, Niwot, CO.

Pimentel, D. (2003) Ethanol fuels: energy balance, economics and environmental impacts are negative, *Natural Resources Research*, **12**, 127–134.

Rees, W. E. (1992) Ecological footprints and appropriated carrying capacity: what urban economics leaves out, *Environment and Urbanization*, **4**, 120–130.

Rees, W. E. (1996) Revisiting carrying capacity: area-based indicators of sustainability, *Population and Environment*, **17**, 195–215.

Rees, W. E. (2001) Ecological footprint, concept of, *Encyclopedia of Biodiversity* (S. Levin, Editor-in-Chief), vol. 2, Academic Press, San Diego, pp. 229–244.

Rees, W. E. (2002a) Carrying capacity and sustainability: waking malthus' ghost, in Bell, D.V.J. and Cheung, Y. A. (eds), Theme: 'Introduction to Sustainable Development', *Encyclopedia of Life Support Systems* (EOLSS), Eolss Publishers, Oxford. Available at: (http://www.eolss.net)

Rees, W. E. (2002b) Globalization and sustainability: conflict or convergence, *Bulletin of Science, Technology and Society*, **22**, 249–268.

Rees, W. E. (2003a) Net-pen salmon farming: failing on two fronts—an eco-footprint analysis. in Gallaugher, P. and Wood, L. (eds) *Proceedings of the World Summit on Salmon*, Simon Fraser University, Burnaby, BC, pp. 139–152.

Rees, W. E. (2003b) Understanding urban ecosystems: an ecological economics perspective, in Berkowitz, A., Hollweg, K. S. and Nilon, C. H. (eds) *Understanding Urban Ecosystems*, Springer-Verlag, New York.

Rees, W. E. and Wackernagel, M. (1994) Ecological footprints and appropriated carrying capacity: measuring the natural capital requirements of the human economy, in Jansson, A-M., Hammer, M., Folke, C. and Costanza, R. (eds), *Investing in Natural Capital: The Ecological Economics Approach to Sustainability*, Island Press, Washington, DC.

Simon, J. and Kahn, H. (eds) (1984) *The Resourceful Earth: A Response to 'Global 2000'*, Basil Blackwell, Oxford.

Wackernagel, M. and Rees, W. E. (1996) *Our Ecological Footprint: Reducing Human Impact on the Earth*, New Society Publishers, Gabriola Island, BC, and Philadelphia, PA.

Wackernagel, M. and Silverstein, J. (2000) Big things first: Focusing of the scale imperative with the ecological footprint, *Ecological Economics*, **32**, 391–394.

WWF (2004) *Living Planet Report 2004*, Worldwide Fund for Nature International, Gland, Switzerland.

10

The Sustainable Process Index (SPI)

Michael Narodoslawsky and Anneliese Niederl

10.1 Introduction

Although implementation of sustainable development requires profound changes across the board of human activities, technology will play a key role, especially concerning environmental aspects of sustainability. Measurement of environmental sustainability of technological processes and artefacts therefore becomes a necessary guidance for all those who play an active role in technological progress and the implementation of technologies. As technological progress and technology implementation are a shared societal quest, this includes almost all societal actors and professions. However, as a matter of fact, engineers will continue to play a key role in shaping technologies and therefore must become prime clients for environmental sustainability valuation.

The challenge for the Sustainable Process Index (SPI) is to provide a valuation tool for environmental sustainability applicable to a wide variety of technologies and technology-related environmental impacts, that 'speaks the language of engineers' but at the same time offers secure guidance towards sustainable development. The SPI is not only meant to identify the respective impact on environmental sustainability exerted by different technological alternatives. It also allows the engineer to go into the details of his/her technology and to pinpoint those steps and factors that have the greatest influence on this impact. Thus, the SPI is not only a valuation method but also a tool to optimize technologies, products and their respective application in everyday life. This ability to aid the engineer in his/her creative development work is a precondition for any tool acceptable to this community.

Renewables-Based Technology Edited by J. Dewulf and H. Van Langenhove

The work of engineers requires translating scientific concepts into artefacts fulfilling human aspirations. This task evidently obliges engineers to cut compromises and constantly optimize the outcome of their work. Technological progress is not so much about finding the ultimate solution than about pursuing ever better compromises between contradictory requirements that by tedious work lead us to a net overall gain. Hence engineers are acutely aware that improvements on one front (say, ozone depletion) tend to lead to disadvantages in other areas (say, greenhouse gas emissions) and they want to know about these trade-offs. Thus, a high degree of aggregation, which allows valuating one impact versus another, is a necessity for any engineering tool for evaluation of environmental sustainability. This argument lies at the basis of the SPI concept.

A high degree of aggregation needs a uniform 'target function' for sustainable environmental performance. In order to be attractive to engineers, this target function must also be:

- universally acceptable;
- sound from the vantage point of natural sciences (as they constitute the base for engineering);
- limited (as this allows formulation of clear goals for development).

One very attractive candidate as the basis of measuring environmental sustainable development is the natural income by solar radiation. As a matter of fact our planet is a thermodynamically 'open' system, that is open to the flux of solar radiation to its surface and that emits energy to the universe. However, it is (almost) closed to any material flux from the universe. This means that solar radiation is the sole sustainable natural driving force for all environmental as well as human processes (see Figure 10.1).

Solar radiation is a limited flux (although it will be available practically indefinitely) and its recipient is our planet's surface. This means that all processes, natural as well as induced by human activity, vie for 'their place in the sun', meaning that they require a certain fraction of this limited flux for running and therefore require a certain surface area of our planet. This means that technological processes compete against each other and with natural processes for this limited resource, area.

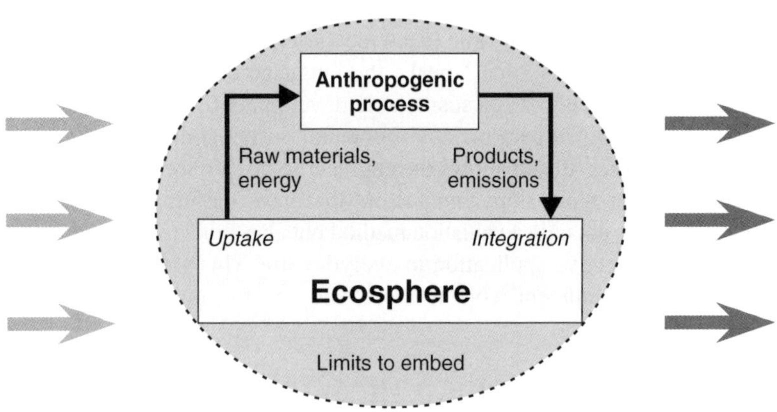

Figure 10.1 *Anthroposphere and ecosphere as co-evolving systems powered by solar energy*

It is noteworthy that this concept actually was developed in parallel from two very different points of departure, from the economic point of view as an 'ecological footprint' (Rees and Wackernagel, 1996, see Chapter 9 in this volume) and the engineering perspective (Narodoslawsky and Krotscheck, 1995), leading to surprisingly similar results.

However, for engineers, the most important aspects of environmental sustainability (and the factors they can influence most effectively) are material fluxes that technological artefacts exchange with their environments. If the SPI can make use of area as a unit of measurement, then we need a concept to link material fluxes to the competition of different processes for area. In other words we need a concept for environmental sustainability taking into account the limitation in the natural income and setting criteria for the exchange of material flows between the technosphere and the environment.

This concept was developed by SUSTAIN (1994). This states that environmental sustainability requires human activities:

- not to alter long-term storage compartments of global material cycles either in quality or in quantity (as human intervention by transferring material from these long-term storage compartments to other, more vulnerable compartments is at the heart of many unsustainable changes to the global environment);
- to keep local flows to the environment within the qualitative and quantitative range of natural variations in environmental compartments (as this optimizes the chance that natural systems can cope with human-induced pressures);
- to preserve (or increase) variety of species, landscapes and habitats (as variety is a key factor for flexible response of natural systems to pressures).

The first two of these criteria allow a direct link between the impact of human-induced material flows and the measure of surface area:

- All natural global cycles are necessarily driven by solar radiation and hence are phenomena that are ultimately linked to our planet's surface. So are the processes of long-term storage (e.g. sedimentation of organic matter to the bottom of oceans as a process to replenish the long-term storage of carbon). Comparing human-induced flows to the environment of these natural flows to long-term storages and measuring the area they require to store the same amount of material as emitted by a human activity reflects how much environment (in the form of surface area of our planet) the activity considered appropriates.
- All flows that emanate from the life cycle of a given technological artefact must necessarily be re-introduced (dissipated) to the environment, causing a local environmental impact. Comparing this flow to natural qualities and rates of replenishment of environmental compartments like (ground)water, (top)soil, and the natural exchange between surface and air allows us to estimate the local impact on the environment. As replenishment of environmental compartments soil and water as well as the exchange between surface and air are inherently linked to surface area, we again have the possibility of calculating how much environment a certain activity appropriates to dissipate all flows back to the environment.

Using these two lines of argument and with some simple algorithms, the SPI allows the translation of material flows extracted from and dissipated to the environment into area. The area that is calculated using the SPI for a given process or life cycle for a technological artefact is the area that is necessary to embed this process or the life cycle sustainably into the ecosphere without compromising its ability to transform the natural income of solar radiation into natural evolution as well as goods and services for human utilization.

One of the major advantages of this approach is that the SPI only uses basic engineering data (the mass and energy balances of processes or the life cycle inventory in particular), on the one hand, and natural qualities (of compartments) and rates (of replenishment of environmental compartments and long-term storage), on the other. Thus, the result of the SPI calculation is not dependent on (ever changing) threshold values or environmental standards. Another interesting feature of the SPI is that it may be 'regionalized' as the values for environmental compartment replenishment rates as well as their qualities may be adapted to specific applications.

For normal engineering work, depicting environmental pressures for every single flow into or out of processes and life cycles in terms of areas as well as an analysis of what exactly causes this pressure (in terms of raw material use or dissipation) is usually sufficient. It focuses the eye of the engineer on those flows and process steps (and those materials) that exert the largest impact and induces the engineer to search for alternatives that appropriate less environment (shown by a smaller area for the process).

The SPI, however, offers also the possibility to compare the impact of human activities to absolute values. Every human has a birthright to access nature's services and the results of natural evolution. As the surface area of our planet is limited, this translates (at least statistically) to a certain fraction of 'planet surface' per person. In an ideally sustainable societal, economical and technological system, this area (approximately 8 hectares per person as a global average) must produce all the products and services consumed by a person. Relating the area necessary for sustainably providing a certain service (say, transportation over 1 km or providing heating for 1 m^2 of housing for a year) to this statistical 'area per inhabitant' gives an indication how much of one's 'personal natural budget' is needed to fulfil a certain aspiration. Although this feature of the SPI is certainly less important for technological development, it might still prove helpful in societal discourse about sustainable development pathways and shaping lifestyles.

10.2 Computation of the SPI

Human activities exert impacts on the environment in different ways. They extract raw materials, use energy, need physical installations, employ people and emit material such as waste or emissions to the environment. Taking these different aspects into account, the total area for sustainable embedding a specific process into ecosphere is therefore given by:

$$A_{tot}=A_R+A_E+A_I+A_S+A_P \quad [m^2] \tag{10.1}$$

where A_R stands for the area necessary to produce raw materials, A_E represents the area requirement to provide process energy, A_I takes into account the area attached to physical installations, A_S is the area required for staff and A_P denotes the area to accommodate products and by-products in the ecosphere. The specific areas will be computed on the basis of mass and energy flows and the infrastructural requirements for the reference period of one year of operation. As the area A_{tot} corresponds to real technical processes, e.g. a biodiesel plant producing 10000 t of biodiesel per year, the area for a single unit of product a_{tot} or a single unit of service, like transport km, may be calculated as:

$$a_{tot} = \frac{A_{tot}}{N} = \frac{1}{y_{tot}} \qquad [\mathrm{m^2a\ unit^{-1}}] \tag{10.2}$$

where N is the capacity of the plant (e.g. kg biodiesel per year) and y_{tot} the 'specific yield' for providing the product/service sustainably. If a certain service requires the input of several products (that are produced by different processes), the specific area per service unit is

$$a_{tot} = \sum_i M_i \cdot a_{tot,i} \qquad [\mathrm{m^2a\ unit^{-1}}] \tag{10.3}$$

where M_i denotes the input flow of product i necessary to provide the service in question and $a_{tot,\,i}$ represents the partial footprint of process i.

This specific area is already a possible comparative measure of sustainability. Let us look, for example, at transport with a car and let us assume an average of 2000 km per person and year. Let us compare using either biodiesel from used vegetable oil or fossil diesel as fuel in a car consuming respectively 5 kg of biodiesel and 4.3 kg fossil diesel per 100 km. The service unit transport of 100 km now corresponds to 3720 MJ of combustion energy. With the specific area of 4.8 $\mathrm{m^2a\ MJ^{-1}}$ and 28.3 $\mathrm{m^2a\ MJ^{-1}}$ for the combustion of biodiesel and fossil diesel, respectively (for detailed calculation of this area, see case study) the area necessary for the service 'transport of 2000 km' to be embedded sustainably into ecosphere is 0.018 $\mathrm{km^2a}$ for biodiesel compared to 0.105 $\mathrm{km^2a}$ for fossil diesel.

We now can apply the 'budget aspect' by relating this area to the area a_{in} statistically available to a person to provide all services and goods in a sustainable way (see Figure 10.2). This area can be estimated to 80 000 $\mathrm{m^2a\ cap^{-1}}$ using the surface of Planet Earth and a population number of 6.4 billion. This finally gives the SPI

$$SPI = \frac{a_{tot}}{a_{in}} \tag{10.4}$$

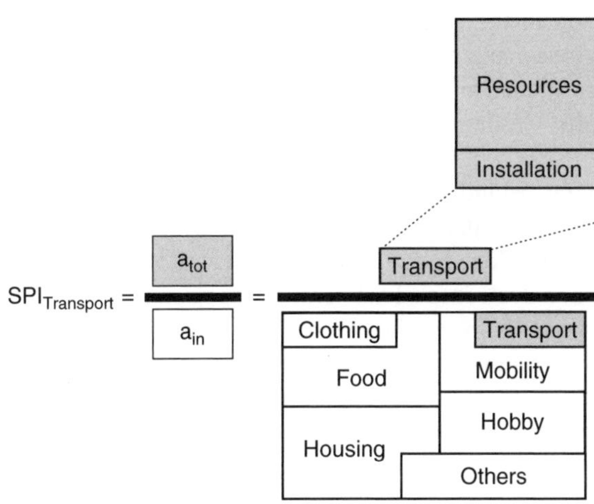

Figure 10.2 *The sustainable process index (SPI) of transport*

SPI, thus, is the fraction of the area per inhabitant related to the provision of a certain service. Its physical meaning is how much of the area theoretically available for a person to guarantee their sustainable subsistence is used up to produce the service in question. The lower the SPI (or a_{tot}), the lower the impact on the ecosphere to provide the service or good. For our case study the $\mathrm{SPI}_{\mathrm{Transport}}$ would be 0.22 when using biodiesel and 1.3 when using fossil diesel as fuel. A SPI of 0.22 means that already 22% of a person's budget is spent on transportation. Whereas a SPI greater than 1 (here 1.3) means that this particular service (e.g. transportation) on its own clearly exceeds the natural person's budget and is not provided in a sustainable way.

Let us now look into the detailed calculation of the individual contributions to A_{tot}.

10.2.1 The Raw Material Area A_R

This area accounts for the sustainable provision of raw materials. The SPI concept calls for a life cycle approach and therefore also includes 'grey' environmental pressures, i.e. pressures that are indirectly caused by the provision of raw materials (like the energy used for planting and harvesting a crop). The computation of the raw material area itself is different for renewable (area A_{RR}) and non-renewable raw materials (area A_{RN}).

Renewable raw materials (basically agricultural, fishery and foresting products) take part in at least on global cycle. For this category of materials, the area required for production is the area needed to convert the building blocks available in the biosphere (e.g. carbon dioxide, water) into biomass which is subsequently fed into the process. This is the area required to produce the necessary amount of the crop. Thus, the specific yield y_R (e.g. the kg of rapeseed that can be harvested in one year in kg $m^{-2}a^{-1}$) and the feed F_R (e.g. rapeseed in kg a^{-1}) that is processed to provide the service in question are the basis for computation of the renewable raw material area A_{RR}:

$$A_{RR} = \frac{F_R}{y_R} \qquad [m^2] \tag{10.5}$$

In this case the area also accounts for the ultimate disposal of the product since it embeds the raw material generation in the global material cycles. From the point of view of sustainable development, renewable resources have thus a distinct advantage over fossil raw materials: They close global material cycles directly during their production and therefore automatically fulfil the first requirement for environmental sustainability given in the introduction.

The same reasoning lying behind renewable raw materials can be applied to fossil organic raw materials. This means that they can be treated as renewable resources albeit with a low rate of regeneration. The reason for this is that at any given time there is a stream from the global carbon cycle into a long-term storage compartment. This is mainly realized by sedimentation onto the ocean bed. When fossil raw materials are used at the same rate at which carbon is removed to long-term storage, there is no alteration of the global cycle itself and no unsustainable accumulation occurs. However, the area to 'store' 1 kg of organic sediment per year is about 500 m^2a (Bolin and Cook, 1983), which means a yield of 0.002 kg $m^{-2}a^{-1}$. This yield is taken as the 'production yield' for fossil raw materials.

For most mineral materials, no global cycles exist and their use is inherently dissipative. For the provision of this kind of raw materials, the pressures exerted by the production of these materials until they reach the doorstep of our industrial process have to be taken into account. The dissipative pressures of these materials will be taken care of in the dissipation area (explained below). Among these raw material pressures, the provision of energy to mine, refine and transport the materials plays a predominant role and is therefore included in the calculation. Thus, if the energy demand E_D to provide 1 kg of a given material to the process in question is known in kWh kg^{-1}, the area requirement can be calculated by:

$$A_{RN} = \frac{F_R \cdot E_D}{y_{EI}} \qquad [m^2] \qquad (10.6)$$

In equation 10.6, F_R is the flow of the mineral raw material in kg a^{-1} and y_{EI} represents the energy yield for industrial energy in kWh $m^{-2}a^{-1}$. As the conversion and upgrading of primary raw materials (like ores) are almost exclusively performed on an industrial scale, this means industrial energy supply y_{EI} takes into account the energy mix (process heat, electricity, mechanical power, etc.) used in industry. It may vary with the geographic context and with the technologies used in the technology mix to supply the energy. However, as a result of numerous case studies in this field, it may be stated that this variation is confined to a relatively small range between 2 (in the case of a high nuclear and fossil energy contribution like in France) and 12 kWh $m^{-2}a^{-1}$ (for an energy system depending primarily on renewable sources like hydropower as in the case of Austria and Norway) (Krotscheck, 1995).

In some cases the precise energy content per mass unit of raw material is not available. In these cases the following procedure is proposed (Krotscheck and Narodoslawsky, 1996), where the price of the raw material is the base of computation

$$E_D = \frac{C_N \cdot 0.95}{C_E} \qquad [kWh\ kg^{-1}] \qquad (10.7)$$

E_D is the energy demand to supply 1 kg of the material in question, C_N is the price of this material (world market price, taxes excluded) and C_E is the price of one kWh of energy (industrial price, taxes excluded). This relation is based on the assumption that energy almost exclusively defines the prices of basic raw materials. Although this seems to be a very crude estimate, it is true for a large number of staple products with only minor deviations from the factor 0.95.

10.2.2 The Energy Supply Area A_E

A_E takes into account the area necessary to provide energy for the process in question. Any energy carrier like coal, oil, or biomass directly used in the process is treated like a raw material. In this case it is only necessary to know how much of an energy carrier is necessary to sustain the process per year and to calculate the raw material area for this energy provision. For non-material-based energies (electric power, solar energy), energy yields in kWh m^{-2} a^{-1} have been established by extensive studies (Krotscheck *et al.*, 2000; Narodoslawsky and Krotscheck, 2003).

The energy supply area is calculated using the energy yield Y_E in kWh m^{-2} a^{-1} and the energy F_E in kWh a^{-1} utilized in the process:

$$A_E = \frac{F_E}{y_E} \qquad [m^2] \tag{10.8}$$

This area varies considerably with the quality of the energy needed (different temperature levels of process heat, electricity or mechanical power, etc.) and the transformation technology. As a rough guideline it can be seen that the higher the quality of the energy service, the higher the area required for supplying it.

Table 10.1 gives a selection of the specific area (corresponding to $1/Y_E$) to produce one kWh of electricity or thermal energy. For electricity provision a mix of different technologies is used. This technology mix may vary with the geographic location and time of the process. Using data from Table 10.1, an appropriate mean 'electricity yield' may be calculated for any given technology mix.

10.2.3 The Area for Installation and Staff A_I, A_S

From the economic point of view, investment goods and installation as well as staff costs are usually important factors in the analysis. Many years of application of the SPI, however, have shown that these factors are much less important in evaluating environmental sustainability, albeit they may not be neglected in specific cases. Due to space limitations in this chapter, these aspects of the SPI will not be described in detail and the reader is kindly referred to the original literature (Krotscheck and Narodoslawsky, 1996; Narodoslawsky and Krotscheck, 1995).

The SPI treats investment goods in the same way as non-renewable resources (of which they are usually made). As these goods serve the process over their full lifetime (and the timeframe of analysis is usually one year), the impact for producing these goods must be depreciated much like in economic analysis. As input data for this pressure, the

Table 10.1 *Specific area for the production of energy*

Type	*Electricity*
Coal-fired plant	316
Natural gas	126,7
Photovoltaics	63,8
Hydropower	11,7
Biomass	43,4
Fuel oil	193
Nuclear power	531,7
Electricity UCPTE[1]	371,6
Electricity Austria	152,3

Note:
All values in m^2 a kWh^{-1}.
[1] Union for the Coordination of Production and Transmission of Electricity.

SPI uses costs (as they are usually known to the engineer). Using averaged industry data, costs are related to construction material input, that is then treated like non-renewable resources. This method is called the 'retropagatory calculation' and renders approximate results within the error margin of the whole evaluation.

Direct land use (e.g. in the form of factory area or infrastructure area like roads) goes directly into the calculation of the total area necessary to embed a technology sustainably. Again the impact of this category tends to be small.

Staff can be factored in by adding the 'statistical area per inhabitant' for every employee to the total area. This partial area, however, is only worthwhile calculating when comparing alternatives with vastly differing workforces to provide the same artefact or service.

10.2.4 The Area for Dissipation of Products A_P

This area is usually a decisive factor in the SPI concept, besides the raw material and energy provision areas. The basic idea behind the calculation of this area is that every flow leaving the process will be eventually dissipated to the environment and dissipation must be related to natural rates of regeneration and natural qualities.

In order to estimate the area allocated to dissipation, the following reasoning is applied: if there is a rate at which the content of a given environmental compartment is renewed, any product stream can be 'diluted' by the newly added mass, until this mass reaches qualities (concentrations) that are equal to the quality of the initial compartment. Therefore, it is necessary to know the rate of renewal of a certain environmental compartment and the actual concentration of different components (e.g. heavy metals, sulphur, chloride, etc.) in this compartment. So the dissipation area can be calculated using the rate of renewal R_c (kg m^{-2}a^{-1}) of the environmental compartment, the actual concentration of the substance $conc_i$ (kg$_i$ kg^{-1}) in the compartment and the product flow $F_{P,i}$ (kg$_i$ a^{-1}) to this compartment (index i describes a certain substance).

The area for dissipating a single component of a certain product flow to a given compartment is:

$$A_{P,c,i} = \frac{F_{P,i}}{(R_c \cdot conc_{ci})} \qquad [\mathrm{m}^2] \qquad (10.9)$$

This calculation must be made for all product flows leaving the process in question. The area $A_{P,S,c}$ assigned to the dissipation of a certain stream 'S' leaving a process is the largest area $A_{P,c,i}$ computed for this stream (for each compartment c water, soil or air) and determines hereby the largest area caused by constituents of the flow:

$$A_{P,S,c} = \max_m (A_{P,c,i}) \qquad [\mathrm{m}^2] \qquad (10.10)$$

The product dissipation area A_P is calculated as the sum of the individual dissipation areas of each compartment:

$$A_p = \sum_c A_{P,S,c} \qquad [\mathrm{m}^2] \qquad (10.11)$$

For dissipation of products into the compartment water, the basis for the assimilation capacity is the seeping rate to the ground water body (usually between 30 and 50% of the precipitation rate per m^2). With an annual precipitation rate of 1200 kg $m^{-2}a^{-1}$ and a seeping ratio of 30%, we get a rate of replenishing the compartment water of 360 kg $m^{-2}a^{-1}$ (Central Europe values) (Krotscheck, 1995). Assuming that there is a concentration of 0.005 mg of cadmium per litre of ground water and 1 kilogram of cadmium is dissipated in a process in Austria per year, an area of 0.56 km^2 (=1/(360*0.005)) according to equation 10.9 is appropriate for 'sustainable' dissipation.

For soil, the process of composting is the base of the calculation. Through composting biomass is converted into a 'soil-like' material that can be used to replenish top soil. Composting needs fresh biomass, so the mass of compost resulting from 1 m^2 of fresh biomass is the basic unit in this case. This 'renewed' compartment is supposed to be 'empty'. The reasoning now is that a flow leaving the process will go to this 'renewed' compartment, which will assimilate it. This compost generation rate can be estimated to be 0.42 kg $m^{-2}a^{-1}$. This is a typical value for Central Europe, where there is a loss in mass during composting of 56% (values for grassland in Austria). An example for the allowable yearly dissipation into the compartment soil for mercury is the following: the natural concentration of mercury in top soil is 1 mg kg^{-1} (Central European value). This makes an allowable yearly dissipation of mercury into the compartment soil of 0.42 mg $m^{-2}a^{-1}$. Further allowable yearly dissipations into the compartment soil, but also into water and air can be found in literature (Krotscheck and Narodoslawsky, 1996).

For the compartment air, flows from the process will be compared to natural emissions of forests per m^2, which are known for most relevant gases. An example for the annual allowable dissipation into air is given with 255 mg $m^{-2}a^{-1}$ for sulphur dioxide SO_2 representing a global mean value (Krotscheck, 1995).

10.3 Case Study: Biodiesel from Used Vegetable Oil

In order to exemplify the computation of the SPI, a life cycle assessment case study 'Biodiesel from Used Vegetable Oil' is enclosed. The case of biodiesel from used vegetable oil is specific in two aspects: first, the source for vegetable oil itself is a renewable agricultural product. Second, the process constitutes a 'second use phase' for this product, utilizing the energy content of the renewable resource vegetable oil at the end of its societal use. The service in question is the combustion energy (in MJ) from biodiesel compared to fossil diesel. Due to its low influence (see above) infrastructure as well as staff is not considered in the setting of system boundaries. The first step in the life cycle of the production of biodiesel is the collection of the raw material used vegetable oil from households and restaurants (see Figure 10.3), which mainly comprises of transport (using fossil diesel). Then this collected raw material is processed in a transesterification step to yield biodiesel. Another transport process (fuel delivery) is included before the biodiesel fuel is combusted in an engine.

For the transesterification process, methanol as transesterification agent and some other chemicals (namely NaOH, KOH and H_2SO_4) are used as raw materials. Furthermore, electricity and fuel oil (light) as thermal energy source are input to the biodiesel process. For details of the inventory, see Table 10.2, and for details on the data used and the detailed life cycle assessment study, see Niederl and Narodoslawsky (2004).

The biodiesel process is a multi-output process and besides biodiesel it also yields glycerol and to a small amount K_2SO_4 as valuable by-products. Glycerol faces an

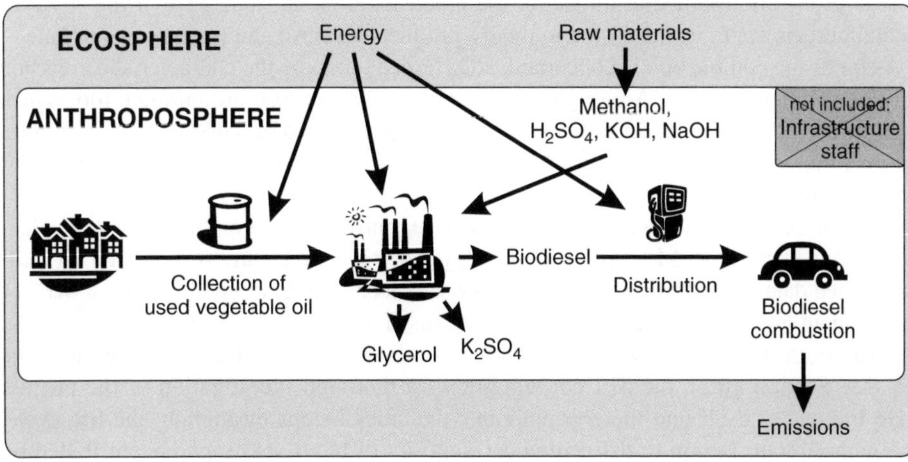

Figure 10.3 System boundaries of the production and usage of biodiesel from used vegetable oil

increasing market as chemical feedstock and potassium sulphate can be used as fertilizer. According to the ISO norm 14040 and due to similar prices for glycerol and biodiesel, mass allocation has been applied at this stage. In Table 10.2 the numbers for the raw material flow F_R and the energy flow F_E are already allocated values corresponding

Table 10.2 Calculation of the partial areas in the case study biodiesel from used vegetable oil

	Raw material	F_R	y_R	A_R
		[g MJ^{-1}]	[g m^{-2} a^{-1}]	[m^2 a MJ^{-1}]
	MeOH	2.65	2.60	1.02
	NaOH	0.047	1.89	0.02
	KOH	0.329	51.36	0.01
	H_2SO_4	0.329	25.44	0.01
Energy		F_E	y_E	A_E
		[g MJ^{-1}]	[g m^{-2} a^{-1}]	[m^2 a MJ^{-1}]
	Fuel oil	0.92	1.47	0.62
		[Wh MJ^{-1}]	[Wh m^{-2} a^{-1}]	[m^2 a MJ^{-1}]
	Electricity [1]	0.799	3.31	0.24
		[tkm MJ^{-1}]	[tkm m^{-2} a^{-1}]	[m^2 a MJ^{-1}]
	Transport [2]	0.009	0.011	0.84
Dissipation		F_P	R_c * $\textbf{conc}_i$	A_P
		[g MJ^{-1}]	[g m^{-2} a^{-1}]	[m^2 a MJ^{-1}]
	NO_x [3]	0.237	0.13	1.81
	Benzene [4]	0.000791	3.6E-03	0.22
A_{tot}				4.79

Notes:
[1] Electricity mix in Great Britain.
[2] Transport includes the collection of the raw material (100 km) and fuel distribution (50 km) in a 16 t HGV using fossil diesel as fuel.
[3] Characteristic emission into the compartment air.
[4] Characteristic emission into the compartment water.

to 87% of the environmental impact of the biodiesel process, since 13% of the environmental burden has to be attributed to the by-products glycerol and potassium sulphate.

As far as the combustion is concerned, NO_x turned out to be the characteristic emission into the compartment air. Although, for example, 0.17 g of SO_x are emitted, too, which would with the yield of 0.255 g m^{-2} a^{-1} need a dissipation area of 0.668 m^2 a, NO_x defines the maximum area (here 1.81 m^2 a MJ^{-1}) needed for dissipation in the compartment air (see equation 10.1).

The total ecological footprint A_{tot} of the service transport based on biodiesel was calculated with 4.79 m^2 a per MJ combustion energy. The same procedure was applied on well-established data on fossil diesel (Suter and Frischknecht, 1996) to yield 28.3 m^2 a MJ^{-1} of energy in an internal combustion engine using fossil diesel.

Apart from this comparison with different kinds of services or the calculation of the SPI (see Section 10.2), the SPI concept allows a thorough investigation of the process (here life cycle) itself and thereby pinpoints the crucial steps throughout the life cycle. The highest contribution to the ecological footprint of biodiesel usage is contributed by the formation of NO_x during combustion. Methanol, mainly based on fossil resources, also has a share of over 20%. Electricity and thermal energy used during the process contribute approximately the same amount as transport for collection of raw material and delivery of biodiesel. An engineer or decision-maker with this information can clearly define which precautions have to be met in order to make the whole process more sustainable. (e.g. use methanol from renewable resources, reduce the pressure caused by transport with better logistics, fuel the trucks for collection and delivery with biodiesel).

10.4 Summary

The SPI is a measure of eco-sustainability of processes, products and service provision. It is based on information about mass and energy flows that are exchanged between ecosphere and anthroposphere. Thus, it allows a comprehensive valuation and aggregation of all relevant emissions and resources.

Since the assessment is based on natural flows and qualities of the environment that remain constant in the foreseeable future, the SPI is independent of legal norms that may vary over time and that in general do not reliably mirror environmental restrictions. Therefore, the results of this evaluation may be used for long-term strategies.

A further advantage is that it uses the 'common currency area/ecological footprint' and therefore can be used for pedagogical and awareness-raising purposes. The ecological footprint developed by Rees and Wackernagel (1996) is a consumption index, whereas the SPI presented in this contribution is based on a life cycle approach. The SPI aggregates resources as well as emissions to the three compartments: air, water and soil.

The high resolution moreover allows discrimination between closely related alternatives. The SPI concept has been applied to valuation and technological optimization of various processes, from transport, bulk material production, to pulp and paper production, to various energy provision technologies, to name just a few. Even regional economies can be measured according to their sustainability by the SPI since it strongly emphasizes the regional context of the process under consideration (Eder and Narodoslawsky, 1996; Steinmüller and Krotscheck, 1997).

With the SPI concept, process engineers have a versatile tool for quick and reliable environmental assessment of processes. On the one hand, this is due to the possibility of using easy accessible engineering data in the valuation process. On the other, the SPI can be used to pinpoint ecological bottlenecks of technologies and therefore may be used for technological optimization in process industry and related technological fields already in an early stage of development.

From the point of view of valuation of technologies on the base of renewable resources, a key advantage of the SPI is that it discerns raw materials according to their origin. Thus, the inherent advantage of renewable resources as being neutral for global material cycles, like the carbon cycle, can be clearly factored into the technological evaluation. The SPI distinguishes itself here clearly from consumption-based valuation concepts. It not only values conventional eco-efficiency in terms of reduced material input to a process, but sends a strong signal concerning the quality of the input to (as well as emissions from) a process.

A major drawback of the SPI concept, especially from the engineer's and process designer's point of view, is that toxicity aspects are not dealt with. This is due to the fact that only the characteristic areas for each compartment (air, water and soil) define the product dissipation area of a process stream. According to the arguments of the SPI concept (relating process flows to natural flows), toxic substances (often in a very low concentration) are not predominantly a sustainability problem. However, the problem-oriented approach (developed by CML – Centrum voor Milieukunde Leiden (Guinée *et al.*, 2002)) representing the most commonly used method in life cycle assessment, considers human and ecotoxicity aspects. In this method six different toxicity potentials (human toxicity, marine and freshwater, marine and freshwater sediment, and terrestrial toxicity) are expressed in kg 1,4 – dichlorobenzene equivalents (kg 1,4-DCB eq.-values for various substances see (Guinée *et al.*, 2002)). This opens the possibility to integrate toxicity aspects in the SPI based on the following. Instead of using different toxicity potentials, the highest is integrated into the SPI. By analogy with the SPI concept, all mass flows are converted to equivalents of the non-natural substance 1,4 – dichlorobenzene. This value now provides information about the toxicity of a process comparable to the information about eco-sustainability provided by the SPI.

For both the ecological and the toxicological assessment a spreadsheet program has been developed (SPIonExcel and TOXonExcel). With most of the relevant yields for an easy calculation of the SPI based on simple mass and energy balances as well as some case studies incorporated, this tool can be obtained from the Institute for Resource Efficient and Sustainable Systems at Graz University of Technology (http://www.rns.tugraz.at).

References

Bolin, B. and Cook, R. B. (1983) *The Major Biochemical Cycles and their Interactions: Scope 21*, John Wiley & Sons, Ltd, New York.

Eder, P. and Narodoslawsky, M. (1996) Input-output-based valuation of the compatibility of regional activities with environmental assimilation capacities, in *Proceedings of the Inaugural Conference of the European Branch of the International Society for Ecological Economics*, International Society for Ecological Economics, Paris.

Guinée, J. B., Gorrée, M., Heijungs, R., Huppes, G., Kleijn, R., de Koning, A., van Oers, L., Sleeswijk, A. W., Suh, S., de Haes, H. A. U., de Bruijn, H., van Duin, R., and Huijbregts, M. A. J. (2002) *Life Cycle Assessment: An Operational Guide to the ISO Standards*, Kluwer, Dordrecht.

Krotscheck, C. (1995) Prozessbewertung in der nachhaltigen Wirtschaft, PhD thesis, Graz University of Technology.

Krotscheck, C., König, F. and Obernberger, I. (2000) Ecological assessment of integrated bioenergy systems using the Sustainable Process Index, *Biomass and Bioenergy*, **18**, 341–368.

Krotscheck, C. and Narodoslawsky, M. (1996) The sustainable process index: a new dimension in ecological evaluation, *Ecological Engineering*, **6**, 241–258.

Narodoslawsky, M. and Krotscheck, C. (1995) The sustainable process index (SPI): evaluating processes according to environmental compatibility, *Journal of Hazardous Materials*, **41**, 383–397.

Narodoslawsky, M. and Krotscheck, C. (2003) What can we learn from ecological valuation of processes with the sustainable process index (SPI)?: The case study of energy production systems, *Journal of Cleaner Production*, **12**, 111–115.

Niederl, A. and Narodoslawsky, M. (2004) Life cycle assessment as an engineer's tool? In *Proceedings of the Conference on Process Integration, Modelling and Optimisation for Energy Saving and Pollution Reduction, PRES 04*, Process Engineering Publisher, Prague.

Rees, W. E. and Wackernagel, M. (1996) *Our Ecological Footprint: Reducing Human Impact on the Earth*, New Society Publishers, Philadelphia, PA.

Steinmüller, H. and Krotscheck, C. (1997) Regional assessment with the sustainable process index (SPI) concept, in *Proceedings of the 7th International Conference on EnvironMetrics*, Innsbruck, Austria.

SUSTAIN (1994) *Forschungs- und Entwicklungsbedarf für den Übergang zu einer nachhaltigen Wirtschaftsweise in Österreich*, TU Graz, Graz.

Suter, P. and Frischknecht, R. (1996) *Ökoinventare von Energiesystemen*, 3rd edn. Bundesamt für Energiewirtschaft, Bern.

Part III
Case Studies

11

Assessment of Sustainable Land Use in Producing Biomass

Helmut Haberl and Karl-Heinz Erb

11.1 Introduction

Biomass, the sum of recent, non-fossil organic material of biological origin, provides the basis of heterotrophic life on Earth. In the process of primary production, plant biomass is formed through photosynthesis, i.e. a set of complex biochemical pathways in which solar energy is fixed by chemical reduction of carbon dioxide (CO_2) to organic carbon. Except for the production of chemolithoautotrophs, biomass can thus be regarded as stored solar energy on which all heterotrophic organisms depend in sustaining their metabolism. Just like all other heterotrophic organisms, humans need biomass as food to sustain their metabolism, i.e. biomass serves as energy supply and provides building blocks for the reproduction and proper functioning of human bodies. Humans also use biomass to sustain their 'exosomatic metabolism', i.e. as feed for domesticated animals, as raw material for various products (wood, wood-products, paper, fibres, etc.), and for the provision of energy in technical installations such as furnaces, steam engines, etc. (technical energy use). The use of biomass as an energy source is not new; wood was the most important energy carrier in the past. In developing countries, biomass still provides around 40% of the technical energy used (Hall *et al.*, 1993a).

Recently, the use of biomass as a source of technical energy as well as raw material for human use is being rediscovered and increasingly promoted (Allgeier *et al.*, 1995; European Commission, 1997; Sampaio-Nunes, 1995) for a variety of reasons: (1) biomass is seen as a renewable resource for which, in contrast to fossil fuels, problems of resource exhaustion seem to be irrelevant; (2) biomass is seen as a CO_2-neutral energy source which, if substituted

Renewables-Based Technology Edited by J. Dewulf and H. Van Langenhove

Table 11.1 *Current and projected future level of global biomass and energy use and global terrestrial net primary production: a compilation of estimates*

	Energy flow [EJ/yr]	*Year*	*Sources*
1 Current global use of energy and biomass			
Biomass used for the provision of technical energy	35–55	mid-1990s	[1,2,3,4]
Global technical energy consumption excluding biomass	350 (376)	1995	[5]
Global human biomass extraction, incl. wood, food feed, etc.	235	1992–94	[6]
2 Scenarios of future global technical biomass energy use (potentials)			
Short-term potential according to *World Energy Assessment*	145	2025	[2]
Mid-term potential according to *World Energy Assessment*	94–280	around 2050	[2]
Long-term potential according to *World Energy Assessment*	132–325	2100	[2]
Range of mid-term potentials/scenarios found in a review	35–450	2050	[1]
WEC/IIASA scenarios mid-term	78–154	2050	[3]
WEC/IIASA scenarios long-term	174–266	2100	[3]
IPCC-SRES scenarios mid-term	52–193	2050	[7]
IPCC-SRES scenarios long-term	67–376	2100	[7]
Potential according to Fischer/Schrattenholzer	370–450	2050	[8]
Potential according to Hoogwijk *et al.*	33–1135	2050	[9]
3 Global terrestrial NPP			
Average from [10] model intercomparison project	2 140	mid-1990s	[10]
NPP estimate by Ajtay *et al.* (1979)	2 460	1970s	[11]
Current 'best guess' according to Saugier *et al.* (2001)	2 440	mid-1990s	[12]

Notes:

[1] Berndes *et al.* (2003) summarizes findings of 17 studies, including some of the below-quoted.
[2] Turkenburg (2000).
[3] Nakicenovic *et al.* (1998).
[4] Hall *et al.* (1993a).
[5] Podobnik (1999), own conversion assuming 1 toe = 41.868 GJ (net caloric value). Value in brackets: estimate of gross caloric value.
[6] Haberl *et al.* (2006). This estimate is based on FAO data for wood harvest, data for agricultural biomass harvest, including grazing, assessed by Wirsenius (2000), an estimate of fibre consumption, and an estimate on the under-representation in FAO statistics of biomass used for energy provision in subsistence economies.
[7] Nakicenovic and Swart (2000).
[8] Fischer and Schrattenholzer (2001).
[9] Hoogwijk *et al.* (2003).
[10] Cramer *et al.* (1999), converted assuming a carbon content of biomass of 47.5% and 18.5 MJ/kg gross calorific value of dry matter biomass.
[11] Ajtay *et al.* (1979), converted assuming 18.5 MJ/kg gross calorific value of dry matter biomass.
[12] Saugier *et al.* (2001), converted as in [10].

for fossil fuels, may help to mitigate global warming; this argument is based on the fact that combustion adds to the atmosphere CO_2 that had previously been absorbed during plant growth; and (3) biomass use is seen as a strategy to reduce a country's dependency on foreign markets for their resource supply, and to promote rural economic development, as many countries have productive areas at their disposal on which they can grow biomass for fibre and energy.

Biomass use plays a significant role in global socio-economic energy supply. Biomass currently contributes some 9–13%, that is 35–55 EJ/yr (1 EJ = 10^{18} Joule) to the global supply of technical energy (see Table 11.1). This figure, however, by far underestimates the importance of biomass for humanity's 'energetic metabolism' (Haberl, 2001a, 2001b). Global human biomass harvest, including crops, by-products, grazing by livestock, fibre consumption and forest products amounted to about 235 EJ/yr around 1993 (Table 11.1). This value includes an estimate of biomass used in subsistence economies for energy provision (Hall *et al.*, 1993b; Scurlock and Hall, 1990) unaccounted for in statistical data such as those of the FAO (FAO, 2002).

Notable future increases in biomass demand are expected. First, the projected growth of world population – world population might be around 8 billion in 2030 and between 7 and 11 billion in 2050 (Lutz *et al.*, 2004) – and likely improvements in human diets are strong driving forces for further increases in the amount of biomass required as food and feed. Second, many energy scenarios predict strong increases in the amount of biomass used for energy provision (Table 11.1). The global potential for biomass energy provision has been estimated to be in the order of magnitude of current global technical energy use, i.e. around 400 EJ/yr (Fischer and Schrattenholzer, 2001), although even a biomass potential of over 1 000 EJ/yr has been reported (Hoogwijk *et al.*, 2003).

11.2 Sustainability Issues Involved in Promoting Biomass Energy

Due to its potential to replace fossil fuels, bioenergy is often regarded as an environmentally favourable option. But bioenergy production can also result in similar environmental problems as intensive agriculture: undesired effects of fertilizers and pesticides on humans and ecosystems, increased erosion, replacement of valuable ecosystems with monocultures, low energy return on investment, etc. Although some of these problems may be overcome or at least mitigated through careful planning (Rosillo-Calle and Hall, 1992), there exist some almost unavoidable impacts of biomass use that have to be addressed. Through the process of biomass harvest, humans alter patterns and processes in ecosystems, thus significantly affecting important biogeochemical processes and biodiversity. The magnitude of such impacts essentially depends on the amount of biomass harvested, compared to the amount of biomass produced by ecosystems each year.

The magnitude of current and potential future human uses of biomass becomes apparent when remembering that global terrestrial net primary production (NPP) is currently estimated to be between 2 140 and 2 460 EJ/yr (Table 11.1). NPP is the net amount of primary biomass production after the costs of plant respiration – the energy needed for the plant's metabolism – are included. The figures presented in Table 11.1 suggest that humans currently harvest about 10% of global NPP. Taking into account harvest losses and roots of harvested plants, 'human consumption' of NPP has been estimated to be about 415 EJ/yr (uncertainty range 288–533 EJ/yr or 14–26% of terrestrial NPP; Imhoff *et al.*, 2004). Moreover, present and historical land use may already have reduced the Earth's terrestrial productivity by over 10% (Vitousek *et al.*, 1986).

Annual NPP equals the amount of biomass produced per year and is the upper limit of the amount of biomass that could, hypothetically, be used without exceeding renewal rates. In practice, much less biomass can be used, among others because aggregate NPP includes below-ground NPP, most of which is hardly useable, if at all. Moreover, photosynthetically fixed energy is not only vital for humans, on the contrary: 'NPP provides the basis for maintenance, growth, and reproduction of all heterotrophs (consumers and decomposers); it is the total food resource on Earth' (ibid.). A large fraction of annual NPP is consumed in heterotrophic food chains, i.e. provides the energy required to keep alive the Earth's wild-living animals, fungi and micro-organisms, and is thus necessary to guarantee ecosystem functioning and resilience. Biomass not consumed by heterotrophs may accumulate in the ecosystem, thus absorbing carbon from the atmosphere – in this case vegetation acts as a carbon sink. Human harvest of biomass often leads to a reduction in vegetation's carbon uptake (Haberl *et al.*, 2003; Leemans *et al.*, 1996; Schlamadinger *et al.*, 1997b). Moreover, in ecosystems dominated by perennial plants, above all forests, standing crop (biomass stock) can be considerably (up to 30 times) larger than NPP. In such ecosystems it is possible to harvest considerably more biomass than is being renewed each year – in this case biomass use may not even be renewable.

Therefore, the following sustainability issues should be considered in promoting biomass as a renewable energy and material resource:

1. *The effect of biomass use on land demand.* How much land is required, i.e. how large is the ecological 'footprint' of a proposed biomass utilization scheme? Does this land demand compete with others, such as land required for agriculture?
2. *Quality of land use.* How intensively is the land used? This can be addressed by assessing the 'human appropriation of NPP' (HANPP, defined below; see Haberl, 1997; Vitousek *et al.*, 1986) caused by a biomass utilization scheme.
3. *Contribution to mitigating the accumulation of CO_2 in the atmosphere.* Does a biomass utilization scheme entail land-cover changes that result in carbon flows into the atmosphere? Does it reduce the vegetation's carbon sink strength?

In addressing these questions it will often be necessary to compare different options, e.g. by assessing how much land is needed or HANPP caused per unit of energy gained. In this case, it is important to relate these impacts to the net amount of energy produced, i.e. to take the Energy Return on Investment (EROI; see Hall *et al.*, 1986) of biomass supply options to be compared into account. This is important because some kinds of biomass use, for example, liquid biofuels, may have unfavourable energy input to output ratios below 1:3 (Giampietro *et al.*, 1997), so the amount of fossil energy required to produce them may be considerable.

11.2.1 The 'Footprint' of Biomass Use

Biomass production requires land, therefore the promotion of bioenergy might significantly increase human land demand (Nonhebel, 2004). Above-ground NPP represents an upper limit to the amount of biomass which can be harvested each year without exceeding renewal rates. NPP varies between almost nil in deserts or arctic/high alpine systems, around 15–25 $MJ/m^2/yr$ in humid, temperate regions and more than 30 $MJ/m^2/yr$ in moist tropical forests (Saugier *et al.*, 2001). The amount of area required per unit of biomass energy gained strongly depends on productivity (i.e. yields per area unit) of the managed ecosystem, which in turn depends on a host of factors such as crop, climate, soil, technology, etc. The determination of the amount of land required for a specific biomass utilization scheme or plan to be evaluated is straightforward when the origin of the biomass is known and data on yields and energy input:output ratios are available:

$$\text{Area requirement [ha]} = \frac{\text{Net energy supplied [GJ]}}{\text{Yield [GJ/ha]}} \tag{11.1}$$

This formula requires that yields are expressed as net energy gained per unit area. For example, let us assume the following case: annual yields of a short-rotation plantation could be 5 tons dry matter per hectare, energy input : output ratio 1:20, and a gross caloric value of the biomass of 19.5 GJ/t. Gross yield is then $5\times19.5=97.5$ GJ/ha, net yield 92.6 GJ/ha. Supplying 1 PJ/yr of this biomass would require 10 799 ha, i.e. about 108 km^2 of land.

Yields of different biomass production systems vary greatly, and a great deal of the variability in the exemplary figures compiled in Table 11.2 stems from differences in

Table 11.2 *Biomass energy yields: a range of exemplary estimates*

	Energy inputs (EROI) considered [Y/N]	*Yield [GJ/ha/yr]*	*Area required for 1 EJ/yr [Mha]*	*Sources*
Perennial crops (willow, eucalyptus, switchgrass)	Y	155–244	4.1–6.5	[1]
Poplar short rotation plantations (Netherlands, Portugal)	Y	86–689	1.5–11.6	[2]
Wood extensive	Y	73	13.7	[3]
Wood intensive	Y	188	5.3	[3]
Straw as by-product	Y	43*	(23.3)	[3]
Miscanthus	Y	254	3.9	[3]
Rape oil	Y	58	17.2	[3]
Methyl ester of rapeseed	Y	47	21.3	[3]
Ethanol from sugar cane, sugar beet, wood, wheat, maize	Y	16–93	10.8–62.5	[3]
Methanol from wood	Y	117	8.5	[3]
Biogas	Y	50*	(20)	[3]
Hybrid poplar, Midwest USA (1990)	N	208	4.8	[4]
Switchgrass, Midwest USA (1990)	N	158	6.3	[4]
Sorghum, Midwest USA (1990)	N	320	3.1	[4]
Sugar cane for energy provision, South-east USA (1990)	N	324	3.1	[4]

Notes:
[1] Turkenburg (2000).
[2] Nonhebel (2002).
[3] Stöglehner (2003).
[4] Hall *et al.* (1993a) (values are given net of harvest and storage losses).

All estimates refer to gross calorific values, except when explicitly stated. An asterisk marks biomass energy carriers which also may be assumed not to require additional area because they stem from biomass residues (area requirement in brackets).

yields caused by a host of natural (soil, climate, etc.) and socio-economic (management, technology, etc.) factors. Note that the above quoted formula assumes that energy required for biomass production is of the same quality as the biomass energy produced which is not necessarily the case (Hall *et al.*, 1986). For biomass systems with high EROI this may be negligible, but for biomass systems with unfavourable EROI (e.g., liquid biomass fuels) these aspects should be considered in order to obtain accurate and meaningful results.

One option to reduce land demand is to increase productivity. But high productivity can seldom be achieved in a sustainable manner; in most cases highly productive systems require large fossil fuel inputs, either directly, or embodied in fertilizers, pesticides, etc., so taking this trade-off into account is important. Moreover, intensive systems often result in the above-mentioned detrimental effects of intensive agriculture.

Assessment of the area requirements of by-products, e.g. residues from agriculture, forestry or organic waste, may require the allocation of area requirements to more than one product. In energy studies of cogeneration systems, several approaches have been proposed (Fritsche *et al.*, 1992). For biomass residues, the following variants seem applicable:

1 One can allocate all inputs to the main product, in this case the by-product is assumed to come free. This is justified if the use of the byproduct does not alter the efficiency (e.g., yield) with which the main product is gained. For example, if wheat straw is to be used which would otherwise be discarded, it is justifiable to use this approach. In this case the calculation of an area equivalent of the energy needed to gain the by-product may be warranted (e.g., based on the ecological footprint approach, see Wackernagel and Rees, 1996; see Chapter 9).
2 If the utilization of the by-product reduces the efficiency with which the main product is gained, it is justified to allocate to the production of the by-product the additional amount of inputs (in this case: area) needed to keep the production level of the main product constant. For example, using straw-rich cereal variants may greatly increase straw output at the expense of grain yields, thus requiring additional area if grain output is to be kept constant (Haberl *et al.*, 2003). In this case it would be sensible to allocate to the by-product gained the additional amount of area required to keep cereals production constant. Again, energy inputs required to collect the straw and to extend the cultivated area should be considered.

If no information on the origin of the biomass or no yield estimates are available, it may be to some extent helpful to use average global yields (as is done in ecological footprint studies, see Wackernagel and Rees, 1996), but the accuracy of such assessments is far lower than that of assessments based on actual yield figures.

Strategies aiming at the large-scale reinforcement of biomass utilization have to be analyzed objectively and critically from the land demand perspective. Tenfold increases in biomass provision, as suggested by some authors, will interact with other land uses. This is often neglected in bioenergy studies (Berndes *et al.*, 2003). The expansion of bioenergy could imply pressures on forests and unmanaged ecosystems, and even competition for land availability between food and energy provision (Leemans *et al.*, 1996). Large-scale bioenergy scenarios revealed that energy crop production could have far-reaching consequences such as deforestation and land degradation. Therefore, biomass promotion strategies should be underpinned by assessments of the economics of scarce land resources and the competition for land between food and energy crop production.

11.2.2 Intensity of Land Use: Human Appropriation of NPP (HANPP)

A useful indicator for the intensity with which the above calculated areas are used is the 'human appropriation of net primary production' (HANPP) (Vitousek *et al.*, 1986; Wright, 1990). HANPP has been defined as the difference between the NPP of potential vegetation (Tüxen, 1956), i.e. the amount of biomass energy that would be available in an ecosystem without human intervention, and the proportion of the NPP of the actually prevailing vegetation remaining in the ecosystem after human harvest has been subtracted (Haberl, 1997; see Figure 11.1).

This definition of HANPP considers changes in the availability of NPP for ecological processes induced by (1) alterations of the productivity of vegetation that result from land use; and (2) extraction of NPP from ecosystems through biomass harvest. HANPP is thus the difference between NPP_0, the NPP of potential vegetation, and NPP_t, the part of the NPP of actual vegetation (NPP_{act}) remaining in ecosystems after harvest. HANPP can be expressed as material (kg dry matter), substance (kg carbon) or energy flow (Joule) or as a percentage of NPP_0.

HANPP indicates how intensively a defined area of land is being used in terms of ecosystem energetics (Haberl *et al.*, 2004c). With reference to a given territory, it reveals how much energy is diverted by humans as compared with the trophic energy potentially available in the ecosystem. HANPP is thus a measure of how strongly human use of a defined land area affects its primary productivity, and how much of the NPP is diverted to human uses and is thus not available for processes within the ecosystem. In other words, HANPP is an indicator of land-use intensity, and it may have significant effects on biodiversity discussed below.

Assessing the HANPP caused by a defined biomass utilization scheme or plan requires information regarding the origin of the biomass. Only then is it possible to calculate or estimate the required variables, i.e. NPP_0 (NPP of potential vegetation), NPP_{act} (NPP of actually prevailing vegetation) and NPP_h (harvest). HANPP is then defined as follows:

$$HANPP = NPP_0 - NPP_t \text{ with } NPP_t = NPP_{act} - NPP_h \quad (11.2)$$

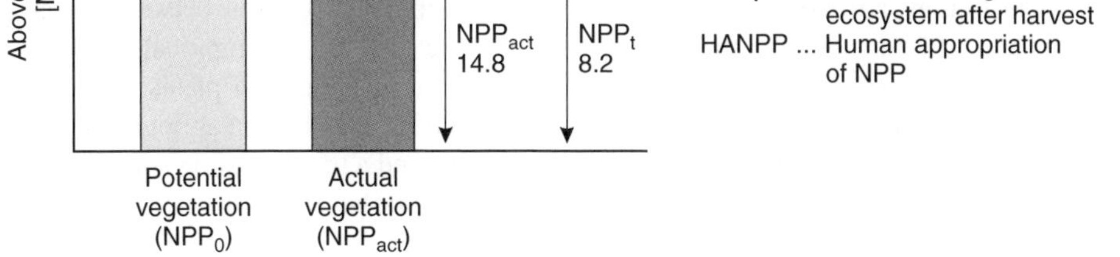

Figure 11.1 *Definition of the human appropriation of net primary production (HANPP)*
Note: Figures refer to Austria 1990, according to Haberl *et al.* (2001).

In principle, HANPP may be negative if $NPP_{act} > NPP_o$ and NPP_h is low or zero, but these conditions are rare. A full-blown description of methods to assess HANPP is beyond the scope of this chapter. Methodological details can be found in the literature (Haberl *et al.*, 2001; Schandl *et al.*, 2002), so we here present only a first overview. Estimates of NPP_0 can be obtained either by applying literature values on the productivity of the type of vegetation under similar climatic and soil conditions (e.g., from Cannell, 1982; a useful compilation of NPP data can be found at http://daac.ornl.gov/NPP/npp_home.html). Another possibility would be to use suitable models, such as DGVMs (Dynamic Global Vegetation Models). A first rough estimate can be obtained by using Lieth's 'Miami model' (Lieth, 1975) which requires only annual mean temperature and precipitation as inputs; more precise calculations of course require more detailed modelling efforts (e.g., McGuire *et al.*, 2001; Sitch *et al.*, 2003). For agricultural areas NPP_{act} can best be approximated by using harvest indices (see Krausmann, 2001; Haberl *et al.*, 2001, for more references) which estimate NPP based on commercial harvest data. NPP_h is usually well known.

HANPP should be considered in sustainability assessments of biomass utilization schemes for at least two reasons: (1) HANPP is an indicator of the intensity with which land is used in producing biomass. Limiting the assessment to the amount of area required without taking into the account how intensively this area is used obviously leaves out a highly important aspect. (2) There are theoretical reasons as well as empirical findings suggesting that HANPP may be an important driver of biodiversity loss. The theoretical background behind this notion is the so-called species–energy hypothesis (Brown, 1981; Hutchinson, 1959; Wright, 1983) which holds that species numbers in ecosystems depend on the availability of trophic energy. If humans remove energy from ecosystems, species numbers would therefore, in turn, be bound to decline (Wright, 1990).

A problem with the species–energy hypothesis, however, is that the mathematical relation between species diversity and energy flows is contested. Some argue that the relation between energy flow (E) and species richness (S) follows a monotonous function (Currie and Paquin, 1987; Lennon *et al.*, 2000; Turner *et al.*, 1988; Weiher, 1999; Wright, 1983; Wright *et al.*, 1993), others favour unimodal ('hump-shaped') species-energy curves (Rapson *et al.*, 1997; Rosenzweig, 1992; Rosenzweig and Abramsky, 1993; Tilman, 1980). A literature review of studies that relate productivity (NPP per unit area) to species diversity found that linear and unimodal patterns seem to be about equally frequent (Waide *et al.*, 1999). Moreover, the effect of human-induced changes in ecosystem energetics on biodiversity may be different from the effect of natural variations in productivity.

At present, only indirect tests of the species–energy hypothesis are possible. According to this hypothesis, HANPP should be correlated with species loss, for which practically no data exist because the potential species richness of current landscapes is unknown. Indirect tests may be carried out, however, by analyzing the relation between species diversity and NPP remaining in the system (NPP_t). In one such test we empirically analyzed relations between species diversity of seven taxonomic groups (vascular plants, bryphytes, orthopterans, gastropods, spiders, ants, and ground beetles) and NPP_t in agricultural landscapes in East Austria (Haberl *et al.*, 2004b). The study used a transect of 38 squares sized 600×600 m in East Austria. Both a linear and a quadratic polynomial function were fitted to the data. The decision which model to select was based on the Akaike Information Criterion (AIC). The study found a highly significant correlation between NPP_t and species richness ($0.13 < r^2 < 0.76$, depending on taxon). The AIC confirmed that the relation between NPP_t and species richness was linear. Results from a second study on bird species richness in Austria support these findings (Haberl *et al.*, 2005). In order to consider possible effects of spatial scale, this study considered 4 different plot sizes: 0.25×0.25 km, 1×1 km, 4×4 km, and 16×16 km. The study showed that NPP variables explained bird species richness better than all landscape heterogeneity indicators considered. Consistent with the species–energy hypothesis highly significant, monotonous (nonlinear) positive correlations between NPP_t and bird species numbers were found.

Although these empirical results do not suffice to establish beyond doubt that HANPP results in species loss, they clearly warrant further investigation. Moreover, as long as strong evidence supports the notion that HANPP may reduce species diversity (Haberl *et al.*, 2004a; 2005; Wright, 1987), bioenergy strategies that entail considerable increases in HANPP should be envisaged with caution, according to the precautionary principle.

11.2.3 Impacts of Biomass Use on Carbon Flows

The substitution of bioenergy for fossil fuels is promoted as an effective measure to reduce net CO_2 emissions. However, stocks and flows of biomass in ecosystems play an important role in the global carbon cycle: ecosystems absorb carbon from the atmosphere, store it (temporarily) in living and non-living compartments (e.g., standing crop, soil organic matter), before releasing it back to the atmosphere. Human harvest of biomass may release carbon earlier than it would have been released in the absence of human activities. Moreover, human-induced land-cover changes induced by establishing biomass production systems (e.g., agriculture, forestry) affect, and often reduce, the amount of carbon stored in ecosystems. Whether or not the substitution of bioenergy for fossil fuels is actually beneficial with respect to atmospheric carbon critically depends on two factors: (1) how much CO_2 from fossil fuels is saved?; and (2) how much additional carbon is emitted due to bioenergy provision and utilization?

In order to assess the net effect of a bioenergy utilization scheme, it is thus important to take its whole life cycle into account. Energy is required to cultivate, harvest, and process biomass. All these energy flows, and the CO_2 emissions resulting from them, must be considered. Industrial agriculture relies heavily on energy inputs from outside the production system, e.g. in the form of fertilizers, pesticides or mechanical power. Assessments of energy inputs for corn production in the USA (Pimentel *et al.*, 1991), and of global agricultural energy inputs in the 1980s (Smil, 1991) demonstrate that energy inputs per unit of biomass produced may be considerable (around 3 J output per 1 J input), so CO_2 emissions of the biomass production cycle must not be neglected.

Furthermore, biomass harvest considerably alters the carbon balance of ecosystems, affecting not only fluxes of carbon such as NPP or net ecosystem production (NEP), but also stocks of carbon such as standing crop or soil organic carbon. Land use generally results in significant reductions of carbon stocks in vegetation (Houghton, 1995; Schimel, 1995; Watson *et al.*, 2000). This effect can be considerable. For example, in its above-ground compartment alone, Austria's currently prevailing vegetation stores about 64% less carbon than potential vegetation, i.e. the vegetation expected to prevail in the absence of human-induced land-cover change (Erb, 2004). Past and present land use has reduced the amount of carbon stored by vegetation on Austria's territory by about 633 MtC, around 40 times Austria's current annual carbon emissions from fossil fuel combustion. The main reason for this reduction can be found in the (historic) conversion of pristine ecosystems (mainly forests) to managed ecosystems, such as agricultural areas and grasslands. In addition, forest management substantially reduces above-ground carbon stocks (23–38%, depending on forest type), due to (1) changes in species composition and (2) reduction of the average stand age. Although these figures refer to Austria, the situation is probably similar not only in central Europe, but in most temperate regions with similar land-use patterns. Furthermore, land conversion and biomass harvest can considerably contribute to a loss of soil organic matter, leading to significant net CO_2 releases into the atmosphere (Lal, 2002; 2004; Pulleman *et al.*, 2000). Thus, bioenergy strategies must be evaluated carefully, taking into account the full process chains and all related carbon flows and stock changes in a comprehensive manner (Apps and Price, 1996).

Comprehensive system approaches to carry out such evaluations exist (Schlamadinger *et al.*, 1997a). Their model takes into account a wide array of influencing parameters, such as changes in ecosystem carbon stocks (e.g., standing crop, soil organic carbon), land productivity, taking into account land use history, the energy needed for land management activities (e.g., mechanical power or fertilizers; in forestry, the energy input to output ratio is of minor importance (ca. 1%); however, for agricultural bioenergy production, the ratio might be significant, see above) and for biomass conversion processes (e.g., refinement), and storage of C in biomass products and artefacts (e.g., construction wood in buildings). Moreover, it evaluates the substitution effects of carbon from fossil fuel combustion, and additional substitution effects due to reductions in 'embodied' energy by the use of less energy-intensive biomass materials (e.g., construction wood requires less energy for its production than steel, fulfilling the same purpose). It is assumed that, with current technology, 1 ton of C in wood fuel displaces only about 0.6 tons of C in fossil fuels, although higher ratios are feasible if innovative technologies such as wood gasification are used (Schlamadinger and Marland, 1996b). On the other hand, energy input for fabrication of goods from wood requires only 3–14 GJ/t, whereas plastics or metal manufacturing requires considerably more energy (e.g. plastics 60–80, steel 20–25, aluminum 190 GJ/t). Such a broad approach is required to assess overall carbon effects of biomass strategies and to evaluate potential trade-offs in implementing carbon strategies.

Measures to increase bioenergy use are in an inherent conflict with strategies that aim at increasing carbon storage in terrestrial vegetation. Carbon storage can best be 'maximized' when ecosystems are allowed to remain undisturbed or (re)grow to maturity (e.g., climax vegetation), whereas harvest of biomass will result in a reduction of on-site carbon storage. Trade-offs between these two options have been discussed widely (Hall, 1991; Kirschbaum, 2003; Marland and Schlamadinger, 2002; Rosillo-Calle and Hall, 1992; Winjum *et al.*, 1998). One major argument in favour of bioenergy use is the time frame: fossil fuel substitution has the potential to be carried out indefinitely, thus replacing – at least in the calculation – fossil-fuel borne carbon emissions over long periods of time, while carbon sequestration strategies are time-limited and effective only until the ecosystems reach maturity. Then, no further net-carbon gain will take place, while fossil fuels may still be in use. Furthermore, carbon sinks suffer from 'non-permanence', i.e. they could be reversed due to future land use changes, forest fires, etc.

For the evaluation of the carbon-performance of bioenergy systems the initial conditions of the ecosystems used for biomass production are particularly important (Marland and

Schlamadinger, 1995; Schlamadinger and Marland, 1996b). One of the major underlying factors of the net carbon effect of bioenergy strategies is ecosystem productivity. Other important initial system conditions are land-cover type (e.g., forest or agricultural land), initial carbon stock on site, and harvest regimes (Thornley and Cannell, 2000). In cases with low ecosystem productivity measures to promote carbon storage in forests (e.g., through afforestation) performed better than biomass utilization strategies (Marland and Marland, 1992). Forest productivity, the efficiency with which wood is used to produce energy, prior use of land converted to forest management, and the time scale under consideration are decisive variables for the carbon performance of any bioenergy strategy.

One case study found that harvest of mature old-growth forests results in net emissions of carbon due to reductions of carbon stocks in vegetation even if the carbon stored in artefacts is taken into account. Even over a time period of over 100 years this biomass scheme did not result in negative net C emissions. Only 51% of the initial carbon stock were stored in the recovered forest stand. Storage offsite (e.g. in buildings and other structures) did not offset these substantial losses (Harmon *et al.*, 1990). A similar site-specific model which included the substitution effect of forestry products with respect to fossil fuel use, found a net reduction in C flows into the atmosphere after 100 years (Schlamadinger and Marland, 1996b; Schlamadinger *et al.*, 1997b). The same model, however, did not return to its initial value for over 100 years when calculated not for 1 hectare, but for a mosaic of 100 hectares with a 60-year harvest cycle. These analyses highlight the importance of the time perspective of biomass systems: increases in biomass supply from existing forests – e.g., through increases in exploitation rates – will in most cases result in net carbon emissions in the beginning which are then compensated in subsequent periods after a more or less elongated 'payback period'. In many scenarios, carbon emissions were found to prevail in the first stages of exploitation, due to carbon stock depletion, whereas C substitution effects gained importance later.

Note, however, that the above-mentioned findings are based on bottom-up models which take neither land-use dynamics on the landscape level, nor feedbacks within the overall socio-economic energy system into account. A top-down approach which considers the socio-economic energy system and land use dynamics from a systemic perspective may be of additional significance for the evaluation of renewable biomass strategies (Haberl, 2001a). Such an approach considers recent and historic land use dynamics and associated carbon flows. It is well known that currently the vegetation of the global temperate and boreal zones acts as a net sink for atmospheric carbon, mainly due to an increase in carbon stocks in forests (Janssens *et al.*, 2003; Kauppi *et al.*, 1992; Myneni *et al.*, 2001; Sedjo, 1992). This recent role of vegetation as a sink for atmospheric carbon probably represents a recovery from past carbon losses and might be understood as reversals of past land conversions (Schimel *et al.*, 2001; Caspersen *et al.*, 2000; Sabine *et al.*, 2004).

This observed reforestation can mostly be explained by changes in the socio-economic energy system, i.e. as an indirect consequence of the introduction of fossil energy some 100–200 years ago (Haberl, 2001b). As human use of fossil fuels does not depend on land productivity, the exploitation of this energy source allows society to energetically 'subsidize' agriculture. During the industrialization of agriculture significant increases in such energetic subsidies allowed for tremendous increases in agricultural yields (Krausmann, 2001). Consequently, more output could be achieved on smaller areas, ultimately resulting in an economic optimization process in agriculture which led to a concentration of cropland farming in favourable regions and abandonment of marginal soils. These abandoned areas were subsequently reforested or allowed to regrow. The availability of fossil fuels additionally decreased the demand for firewood, thus helping to reduce wood harvests (Erb, 2004).

It is therefore questionable to what extent a global, large-scale replacement of fossil fuels with biomass, i.e. a reversal of historic trends, could work. In particular, the interaction between a possible utilization of bioenergy potentials such as those compiled in Table 11.1, and the food system deserves further study. High yields are not only decisive with regard to the period of time in which net carbon gains can be achieved by replacing fossil fuels with bioenergy (Schlamadinger and Marland, 1996a), high yields in food production are also required in order to set aside a sufficient area of productive land on which to grow bioenergy. At present, land suitable for agriculture is already scarce in most of the developing world, and where it is abundant in industrial countries (Döös, 2002), this abundance entirely hinges on high yields which are currently obtained through large inputs of fossil energy. Such feedbacks and possible competing uses of land certainly merit more attention than they have received so far (but see Leemans *et al.*, 1996).

Therefore, the overall contribution of biomass strategies to a reduction of atmospheric carbon has to be critically assessed, in combination with many other environmental land use and socio-economic factors. The promotion of biomass as a renewable energy and material source requires taking into account scenarios of future changes in land use, the rates of ecosystem productivity and how they are sustained, and the manner and efficiency with which biomass products are produced, refined and used.

11.3 Recommendations

Considerable potentials exist to increase global bioenergy use. Some even claim that these potentials might of the same order of magnitude as current global technical energy use. Fully exploiting these potentials, however, can have ecologically detrimental effects. Besides impacts that can be avoided or at least mitigated through careful planning – e.g., undesired effects of fertilizer or pesticide use, erosion, monocultures, etc. – biomass harvest in any case requires the use of area for human purposes, results in HANPP, and directly or indirectly entails a release of carbon into the atmosphere that may even exceed the amount of carbon saved through substitution of biomass for fossil fuels. The assessment of sustainability issues involved with using biomass for energy provision should therefore ask the following questions:

- How much area is required per unit of net energy gained?
- How intensively is this area used in terms of HANPP, how much NPP is appropriated per unit of net energy gained?
- How much carbon is released into the atmosphere per unit of net energy gained?

Policies aimed at promoting the use of biomass for energy provision should therefore aim at the highest possible efficiency in biomass use, i.e. biomass should be used as sparingly as possible. Moreover, the utilization of biomass from residues (i.e., agricultural crop residues, forest residues, manure, organic wastes) should be given priority over biomass utilization schemes that require the additional harvest of biomass. There exist considerable

Table 11.3 Estimates of global energy potentials from biomass residues

Residues	*Global energy potential [EJ/yr]*	*Sources*
Agricultural residues	10–32	[1]
Forest residues	10–16	[1]
Animal residues/manure	9–25	[1]
Organic waste	1–3	[1]
Total according to [1]	30–76	[1]
Residues from agriculture	53	[2]
Residues from forestry	35	[2]
Total according to [2]	88	[2]
Residues from maize, wheat, rice, sugar cane	35	[3]
Dung (gross amount produced)	41	[3]
Forest residues	36	[3]
Total according to [3]	112	[3]

Notes:
[1] Hoogwijk *et al.* (2003).
[2] Fujino *et al.* (1999).
[3] Hall *et al.* (1993a).

potentials for such a strategy of 'cascade utilization of biomass' (Fraanje, 1997; Haberl and Geissler, 2000; Haberl *et al.*, 2003; Lal, 2004). On a global level, biomass residues could yield some 30–112 EJ/yr (Table 11.3). However, also those schemes should be carefully evaluated with regard to their effects on ecosystem functioning, as a portion of agricultural residues, for instance, is required to abate soil erosion and sustain soil fertility. However, as the exploitation of these biomass residues does not result in HANPP increases, it should be given priority over schemes that require the harvest of additional amounts of biomass. Biomass strategies, developed under consideration of integrative land use concerns, could then indeed contribute substantially to gains in sustainability.

11.4 Summary

Projected future population growth, likely improvements in human diet, and possibly also increased use of biomass for energy provision could result in considerable increases in human use of biomass over the next decades. At present, biomass contributes some 35–55 EJ/yr to global socio-economic energy supply; in total, humanity uses around 235 EJ/yr for the provision of food, feed, biomass-based goods, and energy which is around 10% of global terrestrial NPP. Through past and present land use, humanity has probably already lowered global terrestrial productivity by another 10%, thus leaving ever smaller amounts of biomass in ecosystems. There is evidence that this 'human appropriation of net primary production' (HANPP) may result in biodiversity loss.

According to one study, about 36% of the Earth's bioproductive area is currently entirely dominated by man, 37% is partially disturbed and only about 27% undisturbed

(Hannah *et al.*, 1994); another study comes to the similar conclusion that 83% of the terrestrial surface is under direct human influence (Sanderson *et al.*, 2002). From an ecological point of view, the amount of area required for any biomass utilization scheme should therefore be an important criterion.

The substitution of fossil fuels by biomass can result in a net reduction in CO_2 emissions, thus eventually helping to mitigate global warming. However, there are many circumstances under which the simple assumption that biomass is CO_2-neutral must be scrutinized. This chapter discusses methods taking these issues into account, and outlines some of the drawbacks that have to be considered in analysing the net C consequences of biomass use.

References

Ajtay, G. L., Ketner, P. and Duvigneaud, P. (1979) Terrrestrial primary production and phytomass, in Bolin, B., Degens, E. T., Kempe, S. and Ketner, P. (eds) *The Global Carbon Cycle*, John Wiley & Sons, Ltd, Chichester, pp. 129–182.

Allgeier, H. J., Caratti, G. and Sandberg, O. (1995) Towards a European bio-energy strategy, in Chartier, P., Beenackers, A. A. C. M. and Grassi, G. (eds), *Biomass for Energy, Environment, Agriculture and Industry*, Vol. 1, Pergamon Press and Elsevier, Oxford, pp. 11–19.

Apps, M. J. and Price, D. T. (1996) *Forest Ecosystems, Forest Management and the Global Carbon Cycle*, Springer, Heidelberg.

Berndes, G., Hoogwijk, M. and van den Broek, R. (2003) The contribution of biomass in the future global energy supply: a review of 17 studies, *Biomass and Bioenergy*, **25**, 1–28.

Brown, J. H. (1981) Two decades of homage to Santa Rosalia: toward a general theory of diversity, *American Zoologist*, **21**, 877–888.

Cannell, M. G. R. (1982) *World Forest Biomass and Primary Production Data*, Academic Press, London.

Caspersen, J. P., Pacala, S. W., Jenkins, J. C., Hurtt, G. C., Moorcroft, P. R. and Birdsey, R. A. (2000) Contributions of land-use history to carbon accumulation in U.S. forests, *Science*, **290**, 1148–1151.

Cramer, W., Kicklighter, D. W., Bondeau, A., Moore III, B., Churkina, G., Nemry, B., Ruimy, A., Schloss, A., Kaduk, J. and The Participants of the Potsdam NPP Model Intercomparison (1999) Comparing global models of terrestrial net primary productivity (NPP): overview and key results, *Global Change Biology*, **5**, S1–15.

Currie, D. J. and Paquin, V. (1987) Large-scale biogeographical patterns of species richness of trees, *Nature*, **329**, 326–327.

Döös, B. R. (2002) The problem of predicting global food production. *Ambio*, **31**, 417–424.

Dvorak, M., Ranner, A. and Ber, H.-M. (1993) *Atlas der Brutvögel Österreichs: Ergebnisse der Brutvogelkartierung 1981–1985 der Österreichischen Gesellschaft für Vogelkunde*, Federal Environment Agency, Vienna.

Erb, K.-H. (2004) Land-use related changes in aboveground carbon stocks of Austria's terrestrial ecosystems, *Ecosystems*, **7**, 563–572.

European Commission (1997) *Energy for the Future: Renewable Sources of Energy*, White Paper for a Community Strategy and Action Plan, European Commission, Luxembourg.

FAO (2002) FAO-STAT 2001 Statistical Database, FAO. Available at: www.fao.org

Fischer, G. and Schrattenholzer, L. (2001) Global bioenergy potentials through 2050, *Biomass and Bioenergy*, **20**, 151–159.

Fraanje, P. J. (1997) Cascading of pine wood, *Resources, Conservation and Recycling*, **19**, 21–28.

Fritsche, U. R., Leuchtner, J., Matthes, F. C., Rausch, L. and Simon, K. H. (1992) *Gesamt-Emissions-Modell Integrierter Systeme (GEMIS)*, Version 2.0, Hessisches Ministerium für Umwelt, Energie und Bundesangelegenheiten, Wiesbaden.

Fujino, J., Kenji, Y. and Yamamoto, H. (1999) Biomass-balance table for evaluating bioenergy resources, *Applied Energy*, **63**, 75–89.

Giampietro, M., Ulgiati, S. and Pimentel, D. (1997) Feasibility of large-scale biofuel production, *BioScience*, **47**, 587–600.
Haberl, H. (1997) Human appropriation of net primary production as an environmental indicator: implications for sustainable development, *Ambio*, **26**, 143–146.
Haberl, H. (2001a) The energetic metabolism of societies, Part I: accounting concepts, *Journal of Industrial Ecology*, **5**, 11–33.
Haberl, H. (2001b) The energetic metabolism of societies, Part II: empirical examples, *Journal of Industrial Ecology*, **5**, 71–88.
Haberl, H., Amann, C., Eisenmenger, N., Erb, K.-H., Fischer-Kowalski, M., Krausmann, F. and Weisz, H. (2006) The energetic metabolism of societies: energy inputs of the EU-15 and the USA, *Journal of Industrial Ecology*, **10**(4) in print.
Haberl, H., Erb, K.-H., Krausmann, F., Adensam, H. and Schulz, N. B. (2003) Land-use change and socioeconomic metabolism in Austria, Part II: land-use scenarios for 2020, *Land Use Policy*, **20**, 21–39.
Haberl, H., Erb, K.-H., Krausmann, F., Loibl, W., Schulz, N. B. and Weisz, H. (2001) Changes in ecosystem processes induced by land use: human appropriation of net primary production and its influence on standing crop in Austria, *Global Biogeochemical Cycles*, **15**, 929–942.
Haberl, H., Erb, K.-H., Plutzar, C., Fischer-Kowalski, M. and Krausmann, F. (2004a) Human appropriation of net primary production (HANPP) as indicator for pressures on biodiversity, in Moldan, B., Hak, T. and Bourdeau, P. (eds) *Assessment of Sustainability Indicators*, SCOPE, Island Press, Washington, DC.
Haberl, H. and Geissler, S. (2000) Cascade utilisation of biomass: how to cope with ecological limits to biomass use, *Ecological Engineering*, **16**, S111–121.
Haberl, H., Plutzar, C., Erb, K.-H., Gaube, V., Pollheimer, M., and Schulz, N. B. (2005) Human appropriation of net primary production and avifauna diversity in Austria, *Agriculture, Ecosystems and Environment*, **110**, 119–131.
Haberl, H., Schulz, N. B., Plutzar, C., Erb, K.-H., Krausmann, F., Loibl, W., Moser, D., Sauberer, N., Weisz, H., Zechmeister, H. G. and Zulka, P. (2004b) Human appropriation of net primary production and species diversity in agricultural landscapes, *Agriculture, Ecosystems and Environment*, **102**, 213–218.
Haberl, H., Wackernagel, M., Krausmann, F., Erb, K.-H. and Monfreda, C. (2004c) Ecological footprints and human appropriation of net primary production: a comparison, *Land Use Policy*, **21**, 279–288.
Hall, C. A. S., Cleveland, C. J. and Kaufmann, R. K. (1986) *Energy and Resource Quality: The Ecology of the Economic Process*, Wiley Interscience, New York.
Hall, D. O. (1991) Biomass energy, *Energy Policy*, **19**, 711–737.
Hall, D. O., Rosillo-Calle, F., Williams, R. H. and Woods, J. (1993a) Biomass for energy: supply prospects, in Johansson, T. B., Kelly, H., Reddy, A. K. N., and Williams, R. H. (eds), *Renewable Energy: Sources for Fuels and Electricity*, Earthscan, London. pp. 653–698.
Hall, D. O., Scurlock, J. M. O., Bolhàr-Nordenkampf, H. R., Leegood, R. C. and Long, S. P. (eds) (1993b) *Photosynthesis and Production in a Changing Environment: A Field and Laboratory Manual*, Chapman & Hall, London.
Hannah, L., Lohse, D., Hutchinson, C., Carr, J. L. and Lankerani, A. (1994) A preliminary inventory of human disturbance of world ecosystems, *Ambio*, **23**, 246–250.
Harmon, M. E., Ferrell, W. K. and Franklin, J. F. (1990) Effects on carbon storage of conversion of old-growth forests to young forests, *Science*, **247**, 699–702.
Hoogwijk, M., Faaij, A., Broek, R. v. d., Berndes, G., Gielen, D. and Turkenburg, W. (2003) Exploration of the ranges of the global potential of biomass for energy, *Biomass and Bioenergy*, **25**, 119–133.
Houghton, R. A. (1995) Land-use change and the carbon cycle, *Global Change Biology*, **1**, 275–287.
Hutchinson, G. E. (1959) Homage to Santa Rosalia, or why are there so many kinds of animals? *The American Naturalist*, **93**, 145–159.
Imhoff, M. L., Bounoua, L., Ricketts, T., Loucks, C., Harriss, R. and Lawrence, W. T. (2004) Global patterns in human consumption of net primary production, *Nature*, **429**, 870–873.
Janssens, I. A., Freibauer, A., Ciais, P., Smith, P., Nabuurs, G.-J., Folberth, G., Schlamadinger, B., Hutjes, R. W. A., Ceulemans, R., Schulze, E. D., Valentini, R. and Dolman, A. J. (2003) Europe's terrestrial biosphere absorbs 7 to 12% of European anthropogenic CO_2 emissions, *Science*, **300**, 1538–1542.

Kauppi, P. E., Mielikäinen, K. and Kuusela, K. (1992) Biomass and carbon budget of European forests, 1971 to 1990, *Science*, **256**, 70–74.

Kirschbaum, M. U. F. (2003) To sink or to burn? A discussion of the potential contributions of forests to greenhouse gas balances through storing carbon or providing biofuels, *Biomass and Bioenergy*, **24**, 297–310.

Krausmann, F. (2001) Land use and industrial modernization: an empirical analysis of human influence on the functioning of ecosystems in Austria 1830–1995, *Land Use Policy*, **18**, 17–26.

Lal, R. (2002) Soil carbon dynamics in cropland and rangeland, *Environmental Pollution*, **116**, 353–362

Lal, R. (2004) Soil carbon sequestration to mitigate climate change, *Geoderma*, **123**, 1–22.

Leemans, R., Amstel, A. v., Battjes, C., Kreileman, E. and Toet, S. (1996) The land cover and carbon cycle consequences of large-scale utilizations of biomass as an energy source, *Global Environmental Change*, **6**, 335–357.

Lennon, J. J., Greenwood, J. J. D. and Turner, J. R. G. (2000) Bird diversity and environmental gradients in Britain: a test of the species-energy hypothesis, *Journal of Animal Ecology*, **69**, 581–598.

Lieth, H. (1975) Modeling the primary productivity of the world, in Lieth, H. and Whittaker, R. H. (eds) *Primary Productivity of the Biosphere*, Springer, Berlin, pp. 237–264.

Lindemann, R. L. (1942) The trophic-dynamic aspect of ecology, *Ecology*, **23**, 399–418.

Lutz, W., Sanderson, W. C. and Scherbov, S. (2004) The end of world population growth, in Lutz, W., Sanderson, W. C., and Scherbov, S. (eds) *The End of World Population Growth in the 21st Century: New Challenges for Human Capital Formation and Sustainable Development*, Earthscan, London, pp. 17–83.

Marland, G. and Marland, S. (1992) Should we store carbon in trees? *Water, Air and Soil Pollution*, **64**, 181–195.

Marland, G. and Schlamadinger, B. (1995) Biomass fuels and forest-management strategies: how do we calculate the greenhouse-gas emissions benefits? *Energy*, **20**, 1131–1140.

Marland, G. and Schlamadinger, B. (2002) Forests for carbon sequestration or fossil fuel substitution? A sensitivity analysis, *Biomass and Bioenergy*, **13**, 389–397.

McGuire, A. D., Sitch, S., Clein, J. S., Dargaville, R., Esser, G., Foley, J. A., Heimann, M., Joos, F., Kaplan, J., Kicklighter, D. W., Meier, R. A., Melillo, J. M., Moore III, B., Prentice, I. C., Ramankutty, N., Reichenau, T., Schloss, A., Tian, H., Williams, L. J. and Wittenberg, U. (2001) Carbon balance of the terrestrial biosphere in the twentieth century: analyses of CO_2, climate and land-use effects with four process-based ecosystem models, *Global Biogeochemical Cycles*, **15**, 183–206.

Myneni, R. B., Dong, J., Tucker, C. J., Kaufmann, R. K., Kauppi, P. E., Liski, J., Zhou, L., Alexandreyev, V. and Huges, M. K. (2001) A large carbon sink in the woody biomass of Northern forests, *Proceedings of the National Academy of Sciences*, **98**, 14784–14789.

Nakicenovic, N., Grübler, A. and McDonald, A. (1998) *Global Energy Perspectives*, Cambridge University Press, Cambridge.

Nakicenovic, N. and Swart, R. (2000) *Special Report on Emission Scenarios*, Intergovernmental Panel on Climate Change (IPCC), Cambridge University Press, Cambridge.

Nonhebel, S. (2002) Energy yields in intensive and extensive biomass production systems, *Biomass and Bioenergy*, **22**, 159–167.

Nonhebel, S. (2004) Renewable energy and food supply: will there be enough land? *Renewable and Sustainable Energy Reviews*, **9**(2), 191–201.

Pimentel, D., Dazhong, W. and Giampietro, M. (1991) Technological changes in energy use in US agricultural production, in Gliessman, S. R. (ed.) *Agroecology: Researching the* Technological Basis for Sustainable Agriculture. Springer, New York, pp. 305–321.

Podobnik, B. (1999) Toward a sustainable energy regime: a long-wave interpretation of global energy shifts, *Technological Forecasting and Social Change*, **62**, 155–172.

Pulleman, M. M., Bourma, J., van Essen, E. A. and Meijles, E. W. (2000) Soil organic matter content as a function of different land use history, *Soil Science Society of America Journal*, **64**, 689–693.

Rapson, G. L., Thompson, K. and Hodgson, J. G. (1997) The humped relationship between species richness and biomass – testing its sensitivity to sample quadrat size, *Journal of Ecology*, **85**, 99–100.

Rosenzweig, M. L. (1992) Species diversity gradients: we know more and less than we thought, *Journal of Mammalogy*, **73**, 715–730.

Rosenzweig, M. L. and Abramsky, Z. (1993) How are diversity and productivity related? In Ricklefs, R. E. and Schluter, D. (eds) *Species Diversity on Ecological Communities*, The University of Chicago Press, Chicago, pp. 52–65.
Rosillo-Calle, F. and Hall, D. O. (1992) Biomass energy, forests and global warming, *Energy Policy*, **20**, 124–136.
Sabine, C. L., Heiman, M., Artaxo, P., Bakker, D. C. E., Chen, C. T. A., Field, C. B., Gruber, N., Quéré, C. L., Prinn, R. G., Richey, J. E., Romero-Lankao, P., Sathaye, J. A. and Valentini, R. (2004) Current status and past trends of the global carbon cycle, in Field, C. B. and Raupach, M. R. (eds) *The Global Carbon Cycle. Integrating Humans, Climate, and the Natural World*, Island Press, Washington, DC.
Sampaio-Nunes, D. (1995) The role of biomass in the European energy policy, in Chartier, P., Beenackers, A. A. C. M. and Grassi, G. (eds) *Biomass for Energy, Environment, Agriculture and Industry*, Vol. 1, Pergamon Press and Elsevier, Oxford, pp. 20–30.
Sanderson, E., Jaiteh, M., Levy, M., Redford, K., Wannebo, A. and Woolmer, G. (2002) The human footprint and the last of the wild, *BioScience*, **52**, 891–904.
Saugier, B., Roy, J. and Mooney, H. A. (2001) Estimations of global terrestrial productivity: converging toward a single number? In Roy, J., Saugier, B., and Mooney, H. A. (eds), *Terrestrial Global Productivity*, Academic Press, San Diego, pp. 543–557.
Schandl, H., Grünbühel, C. M., Haberl, H., and Weisz, H. (2002) *Handbook of Physical Accounting: Measuring Bio-Physical Dimensions of Socio-Economic Activities*, Social Ecology Working Paper 73, Vienna.
Schimel, D. (1995) Terrestrial ecosystems and the carbon cycle, *Global Change Biology*, **1**, 77–91.
Schimel, D. S., House, J. I., Hibbard, K. A., Bousquet, P., Ciais, P., Peylin, P., Braswell, B. H., Apps, M. J., Baker, D., Bondeau, A., Canadell, J., Churkina, G., Cramer, W., Denning, A. S., Field, C. B., Friedlingstein, P., Goodale, C. L., Heimann, M., Houghton, R. A., Melillo, J. M., Moore III, B., Murdiyarso, D., Noble, I. R., Pacala, S. W., Prentice, I. C., Raupach, M. R., Payner, P. J., Scholes, R. J., Steffen, W. and Wirth, C. (2001) Recent patterns and mechanisms of carbon exchange by terrestrial ecosystems, *Nature*, **414**, 169–172.
Schlamadinger, B., Apps, M. J., Bohlin, F., Gustavsson, L., Jungmeier, G., Marland, K., Pingoud, K. and Savolainen, I. (1997a) Towards a standard methodology for greenhouse gas balances of bioenergy systems and comparison with fossil energy systems, *Biomass and Bioenergy*, **13**, 359–375.
Schlamadinger, B., Canella, L., Marland, G. and Spitzer, J. (1997b) Bioenergy strategies and the global carbon cycle, *Science et Géologie Bullétin*, **50**, 157–182.
Schlamadinger, B. and Marland, G. (1996a) Full fuel cycle carbon balances of bioenergy and forestry options, *Energy Conversion and Management*, **37**, 813–818.
Schlamadinger, B. and Marland, G. (1996b) The role of forest and bioenergy strategies in the global carbon cycle, *Biomass and Bioenergy*, **10**, 275–300.
Scurlock, J. M. O. and Hall, D. O. (1990) The contribution of biomass to global energy use (1987), *Biomass*, **21**, 75–81.
Sedjo, R. A. (1992) Temperate forest ecosystems in the global carbon cycle, *Ambio*, **21**, 274–277.
Sitch, S., Smith, B., Prentice, I. C., Arneth, A., Bondeau, A., Cramer, W., Kamplan, J. O., Levis, S., Lucht, W., Sykes, M. T., Thonicke, K. and Venevsky, S. (2003) Evaluation of ecosystem dynamics, plant geography and terrestrial carbon cycling in the LPJ dynamic global vegetation model, *Global Change Biology*, **9**, 161–185.
Smil, V. (1991) *General Energetics, Energy in the Biosphere and Civilization*, John Wiley & Sons, New York.
Stöglehner, G. (2003) Ecological footprint: a tool for assessing sustainable energy supplies, *Journal of Cleaner Production*, **11**, 267–277.
Thornley, J. H. M. and Cannell, M. G. R. (2000) Managing forests for wood yield and carbon storage: a theoretical study, *Tree Physiology*, **20**, 477–484.
Tilman, D. (1980) Resources: a graphical-mechanistic approach to competition and predation, *American Naturalist*, **166**, 362–393.
Turkenburg, W. C. (2000) Renewable energy technology, In Goldemberg, J. (ed.), *World Energy Assessment: Energy and the Challenge of Sustainability*, United Nations Development Programme (UNDP), United Nations Department of Economic and Social Affairs, World Energy Council (WEC), New York, pp. 219–272.
Turner, J. R. G., Lennon, J. J. and Lawrenson, L. A. (1988) British bird species distribution and the energy theory, *Nature*, **335**, 539–541.

Tüxen, R. (1956) Die heutige potentielle natürliche Vegetation als Gegenstand der Vegetationskartierung, *Angewandte Pflanzensoziologie*, **13**, 5–42.
Vitousek, P. M., Ehrlich, P. R., Ehrlich, A. H. and Matson, P. A. (1986) Human appropriation of the products of photosynthesis, *BioScience*, **36**, 363–373.
Wackernagel, M. and Rees, W. E. (1996) *Our Ecological Footprint: Reducing Human Impact on the Earth*. New Society Publishers, Gabriola Island, BC.
Waide, R. B., Willig, M. R., Steiner, C. F., Mittelbach, G., Gough, L., Dodson, S. I., Juday, G. P. and Parmenter, R. (1999) The relationship between productivity and species richness, *Annual Review of Ecology and Systematics*, **30**, 257–300.
Watson, R. T., Noble, I. R., Bolin, B., Ravindranath, N. H., Verardo, D. J. and Dokken, D. J. (2000) *Land Use, Land-Use Change, and Forestry: A Special Report of the IPCC*, Cambridge University Press, Cambridge.
Weiher, E. (1999) The combined effects of scale and productivity on species richness, *Journal of Ecology*, **87**, 1005–1011.
Winjum, J. K., Brown, S. and Schlamadinger, B. (1998) Forest harvests and wood products: sources and sinks of atmospheric carbon dioxide, *Forest Science*, **44**, 272–284.
Wirsenius, S. (2000) *Human Use of Land and Organic Materials: Modeling the Turnover of Biomass in the Global Food System*, Chalmers University, Göteborg, Sweden.
Wright, D. H. (1983) Species-energy theorie: an extension of the species-area theory, *Oikos*, **41**, 495–506.
Wright, D. H. (1987) Estimating human effects to global extinction, *International Journal of Biometeorology*, **31**, 293–299.
Wright, D. H. (1990) Human impacts on the energy flow through natural ecosystems, and implications for species endangerment, *Ambio*, **19**, 189–194.
Wright, D. H., Currie, D. J. and Maurer, B. A. (1993) Energy supply and patterns of species richness on local and regional scales, in Ricklefs, R. E. and Schluter, D. (eds) *Species Diversity in Ecological Communities*, The University of Chicago Press, Chicago, pp. 66–76.

12

Assessment of the Forest Products Industries

Klaus Richter, Frank Werner and Hans-Jörg Althaus

12.1 Introduction

Wood is the world's most important renewable material and regenerative fuel. Like stone, wood has been used as a material since prehistoric times, and wood-based fuels have formed the main source of energy for millennia. Technological innovation such as wood-based panels (particle board, Oriented Strand Board (OSB) Medium Density Fibreboard (MDF), plywood, etc.) or engineered wood products (glued laminated timber, Laminated Veneer Lumber (LVL), I-joists (I-shaped composite beam), etc.) have contributed to the fact that even today wood is still one of the leading construction materials in the world. Wood is also the most important raw material for the pulp and paper industry and plays a significant role in the transport and packaging sector.

The forest products sector is estimated to contribute about 1% of world GDP and it comprises 3% of international merchandise trade (FAO, 2003). While there is increasing interest in the range of non-timber forest products and environmental services from forests, timber remains the primary economic product from forests. In 2000, the world's total roundwood harvest exceeded $3.3 \times 10E9$ m^3. Only 50% of this quantity is used as industrial roundwood (including logs, wood residues, chips and particles), while the other half is fuelwood. World production of processed wood products has been increasing since the 1960s for each of the four main product categories: sawn wood, pulp, paper and wood-based panels, with paper and panels showing the highest rates of growth. Major consuming regions are the USA and Europe, with over 55% of the world consumption,

Renewables-Based Technology Edited by J. Dewulf and H. Van Langenhove

followed by China and Japan. Global trade in forest products has increased in the past few decades in both value and volume terms, but the vast majority of all wood-based production is still destined for consumption in the domestic markets of producing countries. This implies that domestic trade is more important than international (mostly intra-regional) trade. The value of world trade in the main categories of wood products is estimated at approximately US$ 140×10^9 in 1997, with paper accounting for nearly half of this.

Apparently, the significance of fuelwood has not decreased despite development efforts over the past few decades, aimed at reducing dependence on fuelwood. Globally, 7–11% of energy consumption is met by wood incineration but developing countries account for 90% of global fuelwood use (FAO, 2003).

Forests and forest products have become prominent topics on the international agenda dealing with sustainable development, as is expressed in actual discussions on themes like, e.g., loss of forest cover and forest degradation in the tropics and boreal forests, overaged forests in Europe due to decreasing roundwood prices, conservation of biodiversity, the contributions of forestry and wood industry to GDP and employment in marginalized regions, the role of forests and wood products for the mitigation of climate change, and the environmental consequences of conventional materials substituting for wood products (Robledo and Werner, 2003). Beside general concepts, different forestry and wood-specific criteria and indicator systems have been developed to monitor sustainability aspects and to facilitate decision-making for consumers, companies and national governmental bodies.

This chapter focuses on sustainability criteria and metrics for the forestry and wood processing industry in Europe. Tropical forestry and the sustainability assessment of tropical timber imports – about 17% of the total European wood consumption – are beyond the scope of this chapter.

The main advantages of wood lie in its favourable strength-to-weight ratio combined with good vapour permeability and adequate thermal insulation, its appealing, individual appearance, and the ease in which it is worked with proven techniques. Wood as a commodity is a relatively low priced material and stands out for its regenerative nature. But as a biogenic material, wood is also subject to biological and photochemical degradation. Outdoor applications may require constructive and/or chemical wood protection. Its hygroscopic nature can lead to changes in dimension ('swelling and shrinking') dependent on the moisture content of the environment. This can cause deformation and cracks if wood is not applied in a carefully considered way.

Wood can be used as solid wood, particles, and fibres or even on a molecular basis. The intrinsic characteristics and the threefold use options of wood as a material, fuel and chemical resource allow for an almost 100% utilization of this resource. Even after the first use phase of wood-based products, the material characteristics of wood (theoretically) still allow for a variety of options for reuse, recycling (paper recycling!) or for use as energy carrier. Reuse and recycling of wood as a material must be considered as down-cycling (except reuse in a very strict sense): diameters of wood pieces and fibre length decrease while 'unwanted' contaminants (e.g., from wood treatment or coating) increase with each processing step. The more often wood is reprocessed, the more limited are its potential applications.

12.2 Metrics and Criteria to Assess the Sustainability of Forestry

12.2.1 History of the Term 'Sustainability'

The term 'sustainability' is historically closely linked to forestry and is by no means an invention of recent times. Conceptually, it was developed in the early eighteenth century, when population growth and industrialization had increased pressure on the resource wood to an extent threatening further supply. For the first time in Europe, forest management was ruled in a way that permitted a continuous, stable and 'sustainable' use (Carlowitz, 1713).

Today, sustainable management of wood resources is defined in a broader sense. One of the most widely accepted definitions for European forests was set up in 1993 by the Ministerial Conference for the Conservation of Forests in Europe in Helsinki:

> The stewardship and use of forests and forest lands in a way, and at a rate, that maintains their biodiversity, productivity, regeneration capacity, vitality and their potential to fulfil, now and in the future, relevant ecological, economic and social functions, at local, national and global levels, and that does not cause damage to other ecosystems.
>
> (MCPFE, 1993)

Several criteria and indicator systems have been developed to operationalize this definition for specific purposes.

12.2.2 Existing Criteria and Indicator Systems for Forestry in Europe

In general terms, criteria and indicator systems for sustainable forest management have to be divided into national systems and systems for operational forestry entities, some of the systems being developed as standards of forest certification and labelling schemes.

At the national level, several generic criteria and indicator systems have been designed in the follow-up to the UNCED 'Earth Summit' in Rio 1992 (e.g. ITTO, 1999; MCPFE, 1998a; MPCI, 1995). The relevant system for Europe was set up by the Ministerial, Conference for the Conservation of Forests in Europe in Lisbon, containing the following six criteria (MCPFE, 1998a) with a set of indicators to each as guidance for the planning and implementation of forestry interventions:

- maintenance and appropriate enhancement of forest resources and their contribution to global carbon cycles;
- maintenance of forest ecosystem health and vitality;
- maintenance and encouragement of productive functions of forests (wood and non-wood);
- maintenance, conservation and appropriate enhancement of biological diversity in forest ecosystems;
- maintenance and appropriate enhancement of protective functions in forest management (notably soil and water);
- maintenance of other socio-economic functions and conditions.

These criteria and respective indicators have been adopted by the European countries as a reference for their national forest inventories, as a basis on which national forest resource management can be monitored and planned. The case study in Section 12.2.4 presents structure and results of the Swiss national forest inventory.

Much more political turbulence was caused by criteria and indicator systems for operational forestry entities, particularly in relation with product labelling and certification. The concept of forest certification was conceived by environmental non-governmental organizations in the late 1980s, mainly as an alternative to outright total boycotts of tropical timber. Besides the Earth Summit in Rio, the Uruguay Round of the General Agreement on Trade and Tariffs (GATT) stipulated a no discrimination of products by sources, which was further the key driver for the elaboration of internationally accepted forest certification schemes. The concept of certification quickly grew to encompass temperate and boreal forests and was intended as a market-based tool for promoting more sustainable forest management practices.

Several certification schemes were developed in the mid-1990s. The pioneering one was the Forest Stewardship Council (FSC) certification and labelling scheme, which was elaborated in a large stakeholder consultation (FSC, 2000). This scheme now serves as a template which is or has been specified in working groups for national forest conditions and practices.

Some stakeholders – particularly small woodland owners – remained sceptical about FSC certification; they believed that certification is not economically viable for most family forest situations. Private property rights were also an issue: few private owners want outsiders influencing their decision-making. European private forest landowners, together with the industry and trade, have thus created the Pan-European forest certification (PEFC) framework as an alternative to the FSC. This PEFC framework relies on criteria and indicators that have been elaborated by the Ministerial Conference on the Protection of Forests in Europe held in Lisbon in 1998 (MCPFE, 1998b).

Both the FSC standard and the PEFC framework do not differ that much with regard to the criteria and indicators of sustainable forest management. Both schemes also have described procedures for group certification to address the concerns of small forest land owners. Still, the two schemes differ with regard to stakeholder involvement, representation, and requested monitoring and certification procedures. Negotiations are currently ongoing on the mutual acceptance of FSC and PEFC certificates and on the integration of further national standardization initiatives under one of the 'two umbrella' labels.

Some other stakeholders, especially forest landowner associations and industry and trade organizations prefer alternative certification mechanisms, such as the International Organizations for Standardization's ISO 14001 environmental management system (adapted to forestry companies based on ISO 14061), or intergovernmental arrangements like the proposed European Eco-Management and Audit Scheme (EMAS). However, these certification schemes do not permit the labelling of a product and have not been of much relevance for the forest sector in Europe so far.

12.2.3 Current Status of Certification

In 2004 176 million hectares of forests were certified worldwide, representing less than 4% of the world's forest and the vast majority being situated in the Commonwealth of Independent States, Europe and North America. PEFC has 27 independent national forest certification schemes as members, of which to date 16 have been approved to be in conformity with the PEFC criteria and indicators. These 16 schemes account for over

52 million hectares of certified forests producing millions of tons of certified timber to the market place, making PEFC the world's largest certification scheme. The other national member schemes are at various stages of development and are working towards mutual recognition under the PEFC processes (PEFC, 2004).

According to FSC standards, 27 million hectares have been certified in Europe in the past 10 years while several thousand products are produced using FSC certified wood. FSC operates through its network of national initiatives in more than 60 countries (FSC, 2004). The FSC certification scheme is the only one with a worldwide coverage.

12.2.4 Case Study: The Swiss National Forest Inventory

In 1993–95, the second Swiss National Forest Inventory was established, adopting the Helsinki definition of sustainable forest management (Brändli, 1999). The primary motivation of this project was to set up an integral information system with forestry, geophysical and socio-economic aspects as a basis for the national forest policy. Furthermore, changes were to be detected compared to the first inventory from 1983–1985. To this end, a combination of methods was used in the second inventory (1993–1995). Sampling followed a double sampling design based on aerial photos interpretation and terrestrial sample plots. Further information was obtained from interviewing the local forest services, from external data sources and models describing the site conditions, and from studies of roundwood transportation systems and the effects of game browsing on tree growth. Static models were used for the evaluation of the following complex forest characteristics: the volume of standing and harvested timber, tree growth, the labour and cost involved in timber felling and extraction, the sustainability of forest regeneration, the protection provided by the forest against avalanches and rockfall, its recreational value, and the biotope values of the stands and forest edges. Furthermore, a dynamic model was developed which yields prognoses of the future development of each single tree depending on management scenarios.

Table 12.1 illustrates the different criteria and indicators that are used to monitor and assess the sustainable development of the forests in Switzerland. Based on this sustainability assessment, priorities in Swiss forestry policy have been defined and a forestry action programme 2006–2015 has been established (BUWAL, 2004).

Table 12.1 *Result of the second national forest inventory as evaluation of the state of Swiss forests with respect to the Helsinki definition of sustainable forestry*

	Criteria and indicators	*Status 1993–95*	*Trend within 10 years*
1	Forest resources	++	++
	Forest area, forest structure	+	+
	Growing stock	++	++
	Carbon storage	++	++
2	Health and vitality	□	–
	Damage to trees and stands	□	–
	Stand stability	□	o
	Unregulated felling	□	– –
	Damage by grazing	□	o
	Damage by game	–	¤

Table 12.1 *(Continued)*

Criteria and indicators	*Status 1993–95*	*Trend within 10 years*
3 Productive functions	– –	–
Growth/removals	– –	¤
Costs of harvest/yield	– –	–
Accessibility	□	+
Age composition	–	–
Proportion of regeneration area	– –	–
Planning/exploitation	– –	¤
4 Biological diversity	¤	+
Proportion of conifers in broadleaf areas	–	+
Proportion of exotic tree species	¤	o
Number of tree species	¤	+
Proportion of natural regeneration	+	++
Stand density	–	–
Proportion of dead trees	+	¤
Proportion of old-growth forests	+	+
Disturbance of special sites	–	¤
Construction of forest roads	□	–
5 Protective functions	□	o
Actual effectiveness	+	+
Damages, stability	–	–
Tree species, regeneration	–	o
Accessibility	+	+
Tending (forest maintenance)	□	–
6 Other functions	+	+*
Accessible forest area per capita	+	o
Suitability for recreation	+	+*

Key: + positive – negative □ indifferent o unchanged ¤ no data of first inventory * slight change only
Notes: The status of the forests after the second inventory (1993–95) is evaluated based on several criteria and indicators, and the development (trend) compared to the first inventory (1983–85) is shown. 'Indifferent' indicates that the status allows no clear interpretation; 'unchanged' indicates that the trend is not significant.
Source: Brändl (1999).

12.3 Metrics and Criteria for Assessing the Sustainability of the Wood Industry

For the discussion of sustainability metrics and criteria for forests products and industries, company-oriented criteria, product-related criteria and regional/national approaches have to be distinguished.

12.3.1 Company-Oriented Criteria

In general, every entrepreneurial activity has its environmental impacts. Critical aspects of wood processing companies can be, for example, their electricity and thermal energy consumption including respective on-site emissions, impacts (from heat and chemical emissions) on surface water (particularly in the case of pulp and paper production), transport of raw materials, impacts related to the storage and use of ancillary products such as solvents, adhesives, coatings or agents for chemical wood protection, and operational health. Such issues challenge companies not only to achieve maximal economic efficiency and technological effectiveness but also to satisfy internal and external stakeholder interests in terms of legal compliance and social legitimacy (Freeman, 1984; Pfeffer and Salancik, 1978; Thompson, 1967).

Environmental management systems (EMS) based on the internationally renowned ISO 14001 or on the Eco-Management and Audit Scheme (EMAS) developed for the European community allow a systematic monitoring of the environmental relations of a company. Both the ISO standard and EMAS do not provide industry-sector specific metrics and criteria nor 'sustainable thresholds' or minimal requirements; they define general requirements and procedures to be specified with regard to the specific environment of a company.

Environmental management systems based on ISO 14001 or in accordance with EMAS have also been implemented in wood processing industries, often complementing already existing quality management systems based on the ISO standard 9000 (Gayk, 1996; Ruddell and Stevens, 1998).

It is difficult to evaluate the penetration of EMS in the wood processing and paper industry because respective statistical data are not available. However, the experiences with EMS in other industrial sectors might also hold true for the wood processing industry. The implementation of an EMS is the more likely the bigger the company's size, the larger its international market penetration and the more relevant the perceived environmental impacts (Ruddell and Stevens, 1998; Babakri *et al.*, 2003).

The reader interested in EMS in the wood processing industry is referred to environmental reports and declarations, which usually can be found on the Internet for large European pulp and paper mills or for wood-based board manufacturers.

A further noteworthy development in this context is the extension of originally forestry-related labels to the whole chain of custody, i.e. to wood processing industries and even wood recycling. For example, FSC or the Swiss Q-label (based on PEFC-criteria) specify criteria throughout the wood processing chain. The focus of these extended labels lies on criteria such as the wood origin (of certified forests), legal compliance, and compliance with sector-specific social standards (FSC, 2004; Mosimann, 1999).

12.3.2 Product-Oriented Criteria

In Europe and North America, environmental concerns have increasingly entered public discussion and thus also the marketing strategies of many producers. Since the late 1980s, industrial sectors that have been exposed to public environmental concerns, such as the cement, aluminium or plastics industries, have systematically assessed and communicated the environmental relevance (and advantages) of their products. The forest and wood-based industries were more reluctant to accept quantitative analyses of their energy and mass flows. The common judgement prevailed that wood inherently is an environmentally sound material. This opinion – although correct if the sustainable managed resource is considered – ignores the fact that most modern wood products are combinations of different materials. Adhesives, paints, impregnation agents, connectors, fasteners, plastic overlays and other types of ancillaries are used to produce the high-quality and reliable wood-based products demanded by the market. Also transportation and energy generation, process-specific emissions such as formaldehyde or VOCs (volatile organic compound), the compensation of heat losses during the use phase, or emissions from waste wood incineration can contribute to the environmental relevance of wood products.

To get a clearer view, many initiatives have been undertaken – mainly in Europe and North America – to assess the environmental relevance of wood products compared to conventional products. International efforts have constantly improved the life cycle inventory database for wood products (Lippke *et al.*, 2004; Werner *et al.*, 2003; Hischier, 2005). Such data inventories and product comparisons are based on a methodology called life cycle assessment (LCA) and are based on the standards ISO 14040ff. In LCA 'from cradle to grave', all material and energy flows (from and to nature) related to raw materials extraction, energy generation, production, transportation, use, maintenance, and disposal/recycling of a product are inventoried. An LCA 'from cradle to gate', on the other hand, includes only the flows related to the processes necessary for the production of a good or service, thus excluding use, maintenance and disposal. The resulting emissions and resource consumption are usually assessed with an effect-oriented approach, determining their potential impact to different 'environmental problems' such as 'resource depletion', 'greenhouse effect', 'photosmog', 'acidification', etc. (Guinée *et al.*, 2002). The resulting emissions and resources can also be aggregated further to a single-score 'Eco-indicator points' by assessing and weighting their impact on the safeguards 'human health', 'ecosystem quality', and 'resource quality' (Goedkoop and Spriensma, 2000).

Up to now, wood-related comparative studies have included windows, railway sleepers, floorings, electricity poles, doorframes, paper, constructive elements in landscape architecture, roof and wall constructions, insulation materials, etc. (e.g. Jönsson *et al.*, 1994; Künniger and Richter, 1997, 1998, 2000; Frühwald *et al.*, 2000; Murphey and Hillier, 1999; Potting and Blok, 1996; Richter *et al.*, 1996; Richter, 1998; Wegener *et al.*, 1994; Werner, 1998; Werner and Richter, 1997, 2001; Werner *et al.*, 1997).

Figure 12.1 illustrates the outcome of an LCA of doorframes made of solid wood, particleboard and steel in an exemplary way. In this study, the production, maintenance and disposal of one doorframe were inventoried and assessed according to an effect-oriented classification.

The results of these studies can be summarized as follows:

- Properly applied wood products tend to have environmentally favourable profiles, compared to functionally equivalent products of competing materials.
- Fossil fuel consumption, the potential contributions to climate change and quantities of solid waste tend to be lower for wood products than for alternative products.
- Impregnated wood products tend to be seen more critically in relation to toxicity issues than their substituting products.
- Incineration of wood products can lead to higher acidification and/or eutrophication effects than conventional products, although thermal energy can be recovered.
- While composite products like fibreboards or particleboards make use of a greater portion of logs than solid wood products, the embodied energy associated with the production of wood-based composites and the respective environmental impacts is generally quite high.

Life cycle assessment can be used not only for the environmental comparison of products but also for the in-depth analysis of the environmental profile of a specific product. We

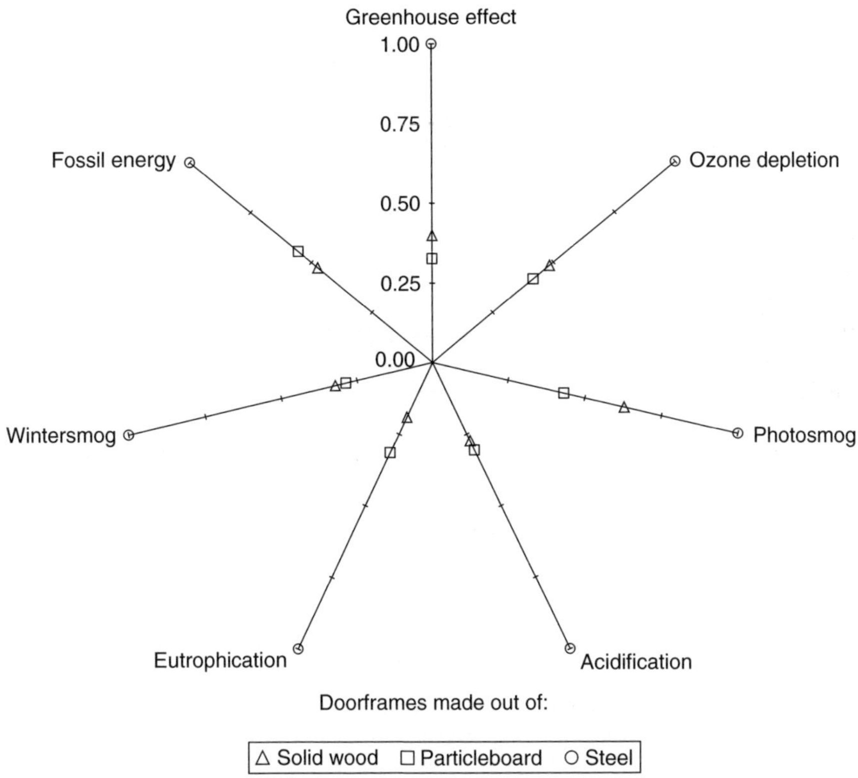

Figure 12.1 *Impact assessment of a life cycle analysis of doorframes. Results relative to the impact of the doorframe made of steel show that both wood-based products cause less environmental impact*
Source: Werner *et al.*, (1997).

choose different qualities of paper from the ecoinvent database (Ecoinvent Centre, 2004), for which the importance of the precursory chains can be analysed. The ecoinvent processes refer to a system from cradle to gate. A cut-off approach is applied to waste material which is recycled. This implies that recycled material used in the production of a new product carries no environmental burden of its primary production. The European average datasets for newsprint basically differ in the content of recycled pulp (de-inked pulp, DIP). The European average datasets for recycled graphical paper differ in the de-inking, which is done in one case and not in the other. Figure 12.2 illustrates the contributions of the precursory chains to the total profiles of the various papers for the three safeguard subjects of the ecoindicator 99 (HA) assessment method.

The energy demand (heat and electricity) dominates the impacts on human health and natural resources in all the examples. The precursory wood chain contributes significantly to the impacts on ecosystem quality for the two newsprints. This is mainly due to land transformation and use. On the other hand, the wood chain has a positive effect on human health, due to the reduction of the global warming potential from carbon stored in the paper. This stored carbon has to be rated as CO_2 emission if the whole life cycle (including incineration or disposal) of the paper is considered. The comparison of the two

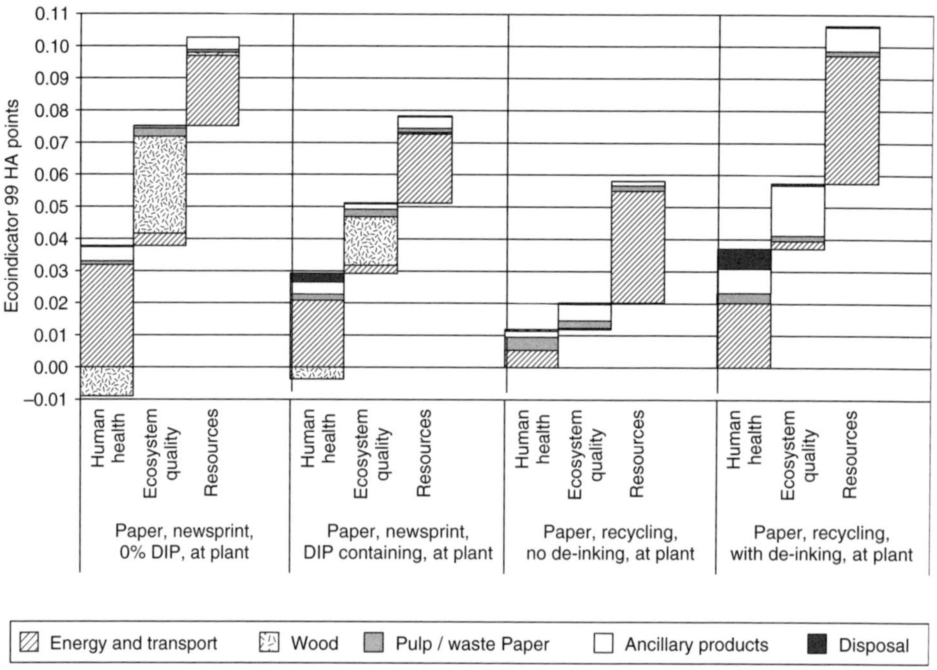

Figure 12.2 *Impact assessment of different papers (Data from Ecoinvent Centre, 2004). The Ecoindicator 99 is a scientifically well-defined impact assessment method. End-point modelling provides effects on safeguard subjects (human health, ecosystem quality, resource quality) and single score results reflecting mental models of decision makers. HA stands for the Hierarchist, average perspective which is proposed as default. Negative impacts denote CO_2 uptake of growing trees from air which is considered as negative emission*
Source: Ecoinvent Centre (2004); Goedkoop and Spriensma (2000).

different newsprints shows that the effects of the wood chain decrease considerably while the effects of production waste treatment increase with an increasing amount of recycled pulp. However, the overall environmental effect of using recycled pulp for newsprint is positive under the given system boundaries (no substitution of e.g. fossil fuels through waste incineration/no burdens associated with recycled input materials). There is no effect of the wood chain in the example of recycled graphical paper, because no fresh wood is used in the production. The comparison of the two examples shows that de-inking leads to higher impacts from the production of ancillary products, mainly from potato starch, and from disposal, especially from the landfilling of the ashes from de-inking sludge incineration. The main difference between the two recycled graphical papers, however, comes from the energy consumption – electricity as well as heat – which is about twice as high with de-inking.

12.3.3 Wood Sector-Oriented Criteria

Apart from plant-specific and product-specific considerations, also wood-sector oriented research has been done to redirect wood (and paper) flows into more sustainable regional or national flow patterns in accordance with various criteria.

Resource efficiency referring to the principles of resource cascading and adequate fit (Sirkin and ten Houten, 1994) has been used as the criterion in material flow analysis (Baccini and Brunner, 1991; Ayres and Simonis, 1994) to define more sustainable wood flows in the Netherlands (Fraanje, 1997; Hekkert *et al.*, 2000; Lafleur and Fraanje, 1997), Switzerland (Binder *et al.*, 2004; Müller, 1998) or in Finland (Korhonen *et al.*, 2001). Other studies have combined material flow analysis with life cycle assessment to estimate the environmental impacts of the national forestry and wood industry (Seppälä *et al.*, 1998).

Another issue is the sector-model based discussion on the sustainability of paper (and wood) recycling, incineration or landfilling (Ekvall, 1999; Finnveden and Ekvall, 1998; Hekkert *et al.*, 2000; Karjalainen *et al.*, 2001; Leach *et al.*, 1997; Turner *et al.*, 1977; Dijkgraaf and Vollebergh, 2004; Anonymous, 1996). Again, resource consumption and aggregated environmental impacts – and sometimes (social) costs – are used as sustainability indicators in these studies. It is beyond the scope of this chapter to elaborate on the findings of these studies. The interested reader is referred to the references. Instead, a further system-based approach to forestry and wood industry is highlighted in the following: the contribution of forestry and wood products to the mitigation of climate change.

A wood-sector oriented approach is also necessary for the assessment of forestry and wood products with regard to their possibilities to mitigate climate change. The common metric applied in such studies is the greenhouse gas (GHG) potential in CO_2-equivalent. To determine the GHG potential, climate impacting gas emissions are weighted according to their relative impact compared to CO_2 (IPCC, 2001). Wood products – particularly those with a long service life – can play multiple roles in the mitigation of climate change:

- *Wood products as carbon pools*: carbon will remain stored in wood products for their life span.
- This transfer of carbon implies the *substitution of non-wood products*: wood products tend to require considerably less energy than comparable non-wood products; therefore, less carbon is emitted for a certain application.
- *Substitution of industrial and post-consumer waste wood for fossil fuels*: waste wood can substitute for fossil fuels.

Ample research on the role of wood and wood products in mitigating climate change is available (Nabuurs and Sikkema, 2001; Marland and Marland, 1992; Marland and Schlamadinger, 1998; Matthews *et al.*, 1996; Pingoud and Lehtilä, 2002; Pingoud *et al.*, 2001; Sedjo, 2002; Skog and Nicholson, 1998; Thompson and Matthews, 1989; Kohlmaier *et al.*, 1998; IPCC, 2000; Hashimoto *et al.*, 2002; Buchanan and Levin, 1999; Börjesson and Gustavsson, 2000; and Sikkema and Nabuurs, 1995). However, the interplay of different GHG effects of wood products taking into account national boundaries has only been investigated in a recent study (Werner *et al.*, 2006). This study estimates the GHG effect of an increased use of wood in the construction sector in Switzerland, excluding effects in the forest: the scenario of a future consumption of wood assumes a yearly growth rate of the building economy of 1%, and an increase of the market share of wood products of 2% every 10 years up to 2030. This leads to an additional use of +0.81 Mio. m^3 wood per year, which is assumed to constantly run up to the year 2130. For the quantification of the mitigation effects, LCA data on GHG emissions of 15 wood products and their substitutes are taken as proxies for the important groups of building products used in construction and interior works. These data are linked to the forecasted wood flows for each group of building products in a cohort-model. The analysis reveals that the GHG emission mechanisms are complex, as different effects with different temporal dynamics overlap (Figure 12.3).

The pool effect is of minor importance as the carbon pool in wood products from a fixed area of sustainably managed forests can only be increased up to an equilibrium state which is reached when the decay of wood products just counteracts new additions to this carbon pool. The substitution effects associated with the thermal use of industrial and post-consumer waste wood and the replacement of 'conventional' energy-intense materials are much more relevant, especially in the long term (–0.36 Mio t CO_2-e). For construction materials, the Swiss share of the GHG effect related to the material substitution is relatively high (–0.261 Mio t CO_2-e) as mainly nationally produced materials are substituted. For products used in interior works, the Swiss share of the GHG effect related to material substitution is rather small because mainly imports are substituted, such as ceramic tiles or steel.

National and international GHG substitution effects of a political measure have to be analyzed carefully for the definition of an effective national strategy to meet the Kyoto Protocol reduction commitment. Compared to the annual Swiss average GHG emissions (53 Mio t CO_2-e), the effects of an increased use of wood would lead to a reduction of 1.8% to 2.3%.

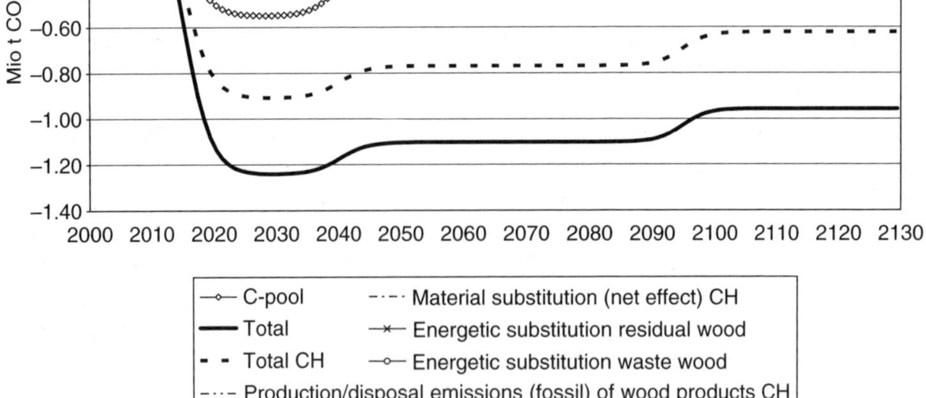

Figure 12.3 Additional GHG emissions of an increased use of wood in Switzerland (2000–2130)

12.4 Scope for Action

Relatively high wood harvesting costs, the resulting economic pressure and increasing environmental (and social) consumer awareness have led to a number of mainly internationally co-ordinated sustainability monitoring and assessment tools, addressing the different interests and scopes for action of the different actors throughout the forestry-wood chain. This is not only true for national forest inventories or inventories of forest companies but also for the different labelling schemes. Current efforts aim at optimizing the use of remotely sensed data for physical forest inventories and/or satellite image-based biomass estimations, the extension of wood labels to cover also processed wood products, and the definition, quantification and marketing of other forest-related environmental services.

Industries based on non-renewable raw materials have become increasingly active in recent decades with respect to challenging traditional wood markets, not only with environmental arguments. This tendency is likely to continue, and self-righteousness of the wood industry related to the environmental relevance of their products would be misconceiving actual market challenges. LCA-based environmental product assessments have shown favourable environmental profiles for wood products. However, data availability and representativeness of existing forestry and wood-product data inventories could be improved if forest owners and wood processors made their production data publicly available.

Economic aspects of wood products over their life cycle could be covered with life cycle costing (LCC), although this tool has barely been applied in this context. The assessment of social aspects over the life cycle of a product has not been conducted so far, as no suitable methodology is at hand. The same is true for a consistent weighting of all three aspects to one integrated sustainability assessment of (wood) products. A further methodological challenge is related to the assessment of land use of forestry and wood products, mainly due to the disputable reference use, the dependency of the assumed timeframe and future land use scenarios.

12.5 Summary

Wood is an important raw material all over the world and will be of vital importance for the transformation of our world towards sustainability. Production, use and disposal processes along the forest–wood–product chain have to be designed and performed in accordance with the management rules of a sustainable use of wood to take full advantage of the beneficial basic values of wood (renewability, low embodied fossil energy, low mass intensity, CO_2-neutrality, multi-purpose use). If improperly used, wood-based products forfeit their credits. Continuous education and information are necessary means to enable a sustainable use of forest products; the metrics, criteria and methods presented in this chapter are useful tools to support the respective decision-making and communication. Simultaneously, the favourable role of forest products with regard to sustainable development needs to be recognized in the political discussion and framework setting to give the required incentives for an increased and responsible use of bio-based resources.

References

Anonymous (1996) Supply of recovered wood and paper is an impetus for recycling, *Industrial Uses*, **6**, 30–31.

Ayres, R. U. and Simonis, U. E. (1994) *Industrial Metabolism: Restructuring for Sustainable Development*, United Nations University Press, New York.

Babakri, K. A., Bennett, R. A. and Franchetti, M. (2003) Critical factors for implementing ISO 14001 standard in United States industrial companies, *Journal of Cleaner Production*, **11**, 749–752.

Baccini, P. and Brunner, P. H. (1991) *Metabolism of the Anthroposphere*, Springer, Berlin.

Binder, C. R., Hofer, C., Wiek, A. and Scholz, R. W. (2004) Transition towards improved regional wood flows by integrating material flux analysis and agent analysis: the case of Appenzell Ausserrhoden, Switzerland, *Ecological Economics*, **49**, 1–17.

Börjesson, P. and Gustavsson, L. (2000) Greenhouse gas balances in building construction: wood versus concrete from life-cycle and forest land-use perspectives, *Energy Policy*, **28**, 575–588.

Brändli, U. B. (1999) *Swiss National Forest Inventory: Results of the Second Sampling 1993–1995*, (eds, Brassel, P. and Brändli, U. B.), Eidgenössische Forschungsanstalt für Wald, Schnee und Landschaft (WSL), Bern, pp. 357–373 (in German).

Buchanan, A. H. and Levin, S. B. (1999) Wood-based building materials and atmospheric carbon emissions, *Environmental Science & Policy*, **2**, 427–437.

BUWAL (2004) *Forest Program Switzerland: Action Program 2006–2015*. Bundesamt für Umwelt, Wald und Landschaft, Bern (in German).

Carlowitz, H. C. ([1713] 2000) *Sylvicultura oeconomica: die naturmässige Anweisung zur wilden Baum-Zucht*, TU Bergakademie/Acad. Buchh, Freiberg.

Dijkgraaf, E. and Vollebergh, H. R. J. (2004) Burn or bury? A social cost comparison of final waste disposal methods, *Ecological Economics*, **50**, 233–247.

Ecoinvent Centre (2004) ecoinvent data v1.1. Final reports ecoinvent 2000 No. 1–15. Swiss Centre for Life Cycle Inventories, Dübendorf. Available at: www.ecoinvent.ch

Ekvall, T. (1999) Key methodological issues for life cycle inventory analysis of paper recycling, *Journal of Cleaner Production*, **7**, 281–294.

FAO (2003) *Trade and Sustainable Forest Management – Impacts and Interactions; Analytic Study of the Global Project GCP/INT/775/JPN: Impact Assessment of Forest Products Trade in the Promotion of Sustainable Forest Management*, Food and Agriculture Organization (FAO), Rome.

Finnveden, G. and Ekvall, T. (1998) Life-cycle assessment as a decision-support tool – the case of recycling versus incineration of paper, *Resources, Conservation and Recycling*, **24**, 235–256.

Fraanje, P. J. (1997) Cascading of pine wood, *Resources, Conservation and Recycling*, **19**, 21–28.

Freeman, R. E. (1984) *Strategic Management: A Stakeholder Approach*, Pitman, Boston.

Frühwald, A., Scharai-Rad, M. and Hasch, J. (2000) *Environmental assessment of wood-Based Materials: Final Report, Complemented for Particleboard and OSB* Universität Hamburg, Hamburg (in German).

FSC (2000) *FSC Principles and Criteria for Forest Stewardship*, Forest Stewardship Council, Bonn.

FSC (2004) Available at: http://www.fsc.org/keepout/content_areas/77/52/files/ABU_REP_70_2004_10_total_FM_and_FM_COC_continent.pdf

Gayk, B. (1996) The ecobalance as an instrument for the economic/environmental management of a company – experiences of the wood construction sector, *Holz Roh Werk.*, **54**, 233–234, (in German).

Goedkoop, M. and Spriensma, R. (2000) *The Eco-indicator 99: A Damage-Oriented Method for Life Cycle Impact Assessment, Methodology Report*, Pré Consultants B.V., Amersfoort.

Guinée, J. B., Gorrée, M., Heijungs, R., Huppes, G., Kleijn, R., de Koning, A., van Oers, L., Wegener Sleesewijk, A., Suh, S., Udo de Haes, H. A., de Brujin, H., van Duin, R., Huijbregts, M. A. J., Lindeijer, E. W., Roorda, A. A. H., van der Ven, B. L. and Weidema, B. P. (2002) *Life Cycle Assessment: An Operational Guide to the ISO Standards*, Kluwer Academic Publishers, Dordrecht.

Hashimoto, S., Nose, M., Obara, T. and Moriguchi, Y. (2002) Wood products: potential carbon sequestration and impact on net carbon emissions of industrialized countries, *Environmental Science and Policy*, **5**, 183–193.

Hekkert, M. P., Joosten, L. A. J. and Worrell, E. (2000) Analysis of the paper and wood flow in the Netherlands, *Resources, Conservation and Recycling*, **30**, 29–48.
Hischier, R., Althaus, H.-J. and Werner, F. (2005) Developments in wood and packaging materials life cycle inventories in Ecoinvent, *International Journal of LCA*, **10**, 50–58.
IPCC (ed.) (2000) *Land Use, Land-Use Change, and Forestry: A Special Report of the IPCC*, Cambridge University Press, Cambridge.
IPCC (ed.) (2001) *Climate Change 2001: The Scientific Basis*, Intergovernmental Panel on Climate Change (IPCC) Cambridge University Press, Cambridge.
ITTO (1999) *Manual for the Application of Criteria and Indicators for Sustainable Management of Natural Tropical Forests: Part A/National Indicators*, International Tropical Timber Organization (ITTO), Yokohama.
Jönsson, A., Tillman, A. M. and Svensson, T. (1994) *Livscykelanalys av golvmaterial*, Chalmers Tekniska Högskola, Göteborg.
Karjalainen, T., Zimmer, B., Berg, S., Welling, J., Schwaiger, H., Finér, L. and Cortijo, P (2001) *Energy, Carbon and Other Material Flows in the Life Cycle Assessment of Forestry and Forest Products: Achievements of the Working Group 1 of the COST Action E9*, European Forest Institute (EFI), Joensuu.
Kohlmaier, G. H., Weber, M. and Houghton, R. A. (1998) *Carbon Dioxide Mitigation in Forestry and Wood Industry*, Springer, New York.
Korhonen, J., Wihersaari, M. and Savolainen, I. (2001) Industrial ecosystems in the Finish forest industry: using the material and energy flow model of a forest ecosystem in a forest industry system, *Ecological Economics*, **39**, 145–161.
Künniger, T. and Richter, K. (1997) *Environmental Comparison of Electricity Poles Made from Impregnated Wood, Reinforced Concrete and Corrosion-Protected Steel*, EMPA, Dübendorf (in German).
Künniger, T. and Richter, K. (1998) *Environmental Assessment of Railway Sleepers in Switzerland: Sleepers Made of Prestressed Concrete, Profiled Steel and Creosote-Treated Beech Wood* Research and Work Report 115/38, EMPA, Dübendorf (in German).
Künniger, T. and Richter, K. (2000) *Environmental Assessment of Constructions of Garden and Landscape Architecture*, EMPA, Dübendorf (in German).
Lafleur, M. C. C. and Fraanje, P. J. (1997) Towards sustainable use of the renewable resource wood in the Netherlands – a systematic approach, *Resources, Conservation and Recycling*, **20**, 19–29.
Leach, M. A., Bauen, A. and Lucas, N. J. D. (1997) A systems approach to materials flows in sustainable cities; a case study of paper, *Journal of Environmental Planning and Management*, **40**, 705–723.
Lippke, B., Wilson, J., Pérez-Garcia, J., Bowyer, J. and Meil, J. (2004) CORRIM: life cycle environmental performance of renewable building materials, *Forest Products J.*, **54**, 8–19.
Marland, G. and Marland, S. (1992) Should we store carbon in trees? *Water, Air, and Soil Pollution*, Special issue natural sinks of CO_2, **64**, 181–195.
Marland, G. and Schlamadinger, B. (1998) Forests for carbon sequestration or fossil fuel substitution? A sensitivity analysis, *Biomass and Bioenergy*, **13**, 389–397.
Matthews, R. W., Nabuurs, G.-J., Alexeyev, V., Birsey, R. A., Fischlin, A., Maclaren, J. P., Marland, G. and Price, D. T. (1996) In Price, D. T. (ed.) *Forest Ecosystems, Forest Management and the Global Carbon Cycle*, Springer Berlin, pp. 293–301.
MCPFE (1993) *Ministerial Conference on the Protection of Forests in Europe, 16–17 June 1993 in Helsinki*, documents (see also www.mcpfe.org).
MCPFE (1998a) *Pan-European Criteria and Indicators Adopted as Annex 2 to the Resolution L2 of the Ministerial Conference on the Protection of Forests in Europe (MCPFE) Held in Lisbon in 1998*, MCPFE.
MCPFE (1998b) *Pan-European Operational Level Guidelines (PEOLG) Adopted as Annex 1 to the Resolution L2 of the Ministerial Conference on the Protection of Forests in Europe (MCPFE) Held in Lisbon in 1998*, MCPFE.
Mosimann, E. (1999) Q-Label jetzt an Holzverarbeiter, *Wald und Holz*, **3**, 20–21.
MPCI (1995) *The Montreal Process*, Montreal Process Working Group, Montreal.
Müller, D. (1998) Modellierung, Simulation und Bewertung des regionalen Holzhaushaltes, PhD dissertation Eidgenössische Technische Hochschule (ETH), Zürich.
Murphey, R. J. and Hillier, W. (1999) An introduction to life cycle assessment (LCA) of painted timber components, *JOCCA – Surface Coatings International*, **82**, 482–487.

Nabuurs, G. J. and Sikkema, R. (2001) International trade in wood products: its role in the land use change and forestry carbon cycle, *Climatic Change*, **49**, 377–395.

PEFC (2004) http://www.pefc.org/internet/html/about_pefc.htm

Pfeffer, J. and Salancik, G. R. (1978) *The External Control of Organizations: A Resource Dependence Perspective*, Harper and Row, New York.

Pingoud, K. and Lehtilä, A. (2002) Fossil carbon emissions associated with carbon flows of wood products, *Mitigation and Adaptation Strategies for Global Change*, **7**, 63–83.

Pingoud, K., Perälä, A.-L. and Pussinen, A. (2001) Carbon dynamics in wood products, *Mitigation and Adaptation Strategies for Global Change*, **6**, 91–111.

Potting, J. and Blok, K. (1996) Life-cycle assessment of four types of floor covering, *Journal of Cleaner Production*, **3**, 201–214.

Richter, K. (1998) In Kohlmaier, G. H., Weber, M. and Houghton, R. A. (eds) *Carbon Dioxide Mitigation in Forestry and Wood Industry*, Springer, Berlin, pp. 219–248.

Richter, K., Künniger, T. and Brunner, K. (1996) *Environmental Assessment of Window Constructions of Different Frame Materials (Without Glazing)*, EMPA-SZFF-Forschungsbericht, Schweizerische Zentralstelle für Fenster- und Fassadenbau (SZFF), Dietikon (in German).

Robledo, C. and Werner, F. (2003) *Wood Research: Knowledge and Concepts for Future Demands*, Research and Work Report 115/50, EMPA, Duebendorf, pp. 199–210.

Ruddell, S. and Stevens, J. A. (1998) The adoption of ISO 9000, ISO 14001, and the demand for certified wood products in the business and institutional furniture industry, *Forest Products Journal*, **48**, 19–26.

Sedjo, R. A. (2002) Wood materials used as a means to reduce greenhouse gases (GHGS): an examination of wooden utility poles, *Mitigation and Adaptation Strategies for Global Change*, **7**, 191–200.

Seppälä, J., Melanen, M., Jouttijärvi, T., Kauppi, L. and Leikola, N. (1998) Forest industry and the environment; a life cycle assessment study from Finland, *Resources, Conservation and Recycling*, **23**, 87–105.

Sikkema, R. and Nabuurs, G.-J. (1995) In Solberg, B. (ed.), *Life-cycle Analysis: A Challenge for Forestry and Forest Industry* EFI Proceedings No 8, European Forest Institute, Hamburg, pp. 169–175.

Sirkin, T. and ten Houten, M. (1994) The cascade chain: a theory and tool for achieving resource sustainability with applications for product design, *Resources, Conservation and Recycling*, **10**, 213–277.

Skog, K. E. and Nicholson, G. A. (1998) Carbon cycling through wood products: the role of wood and paper products in carbon sequestration, *Forest Products Journal*, **48**, 75–83.

Thompson, D. A. and Matthews, R. W. (1989) *The Storage of Carbon in Trees and Timber*, Forestry Commission Research Information Note 160, Edinburgh.

Thompson, J. (1967) *Organizations in Action*, McGraw-Hill, New York.

Turner, R. K., Grace, R. and Pearce, D. W. (1977) In Walter, I. (ed.), *Resource Conservation: Social and Economic Dimensions of Recycling*, Longman, London, pp. 296–343.

Wegener, G., Beudert, M., Frühwald, A., Dreiner, K. and Scharai-Rad, M. (1994) *Evaluation of Wood Compared to Other Material with Regard to the CO_2-balance*, Holzabsatzfonds, Bonn (in German).

Werner, F. (1998) LCI s of insulation material – a comparison, in *10th European Conference and Technology Exhibition 'Biomass for Energy and Industry'*, C. A. R. M. E. N, Würzburg, pp. 180–183.

Werner, F., Künniger, T., Althaus, H.-J., Richter, K. and Jungbluth, N. (2003) *Life Cycle Inventories of Wood as Fuel and Construction Material, Dübendorf, November 2002*, Centre for Life Cycle Inventories, Dübendorf.

Werner, F. and Richter, K. (1997) *Environmental Assessment of Parquet Floorings: Consideration of Mosaic Parquet, 2-layered Prefabricated and 3-layered Prefabricated Parquets*, EMPA/ISP-Forschungsbericht, Dübendorf, Heimberg (in German).

Werner, F. and Richter, K. (2001) *Innovative Fassaden; Tagung Baden-Baden, 7. und 8. November 2001, VDI-Bericht 1642*, VDI-Gesellschaft Bautechnik, Düsseldorf, pp. 243–253.

Werner, F., Richter, K., Bosshart, S. and Frischknecht, R. (1997) *Comparative LCA of Doorframes Made of Solid Wood, Particleboard and Steel*, EMPA/ETH-Forschungsbericht, Dübendorf (in German).

Werner, F., Taverna, R., Hofer, P. and Richter, K. (2006) Greenhouse gas dynamics of an increased use of wood in buildings in Switzerland, *Climatic Change*, accepted.

13

Assessment of the Energy Production Industry

Modern Options for Producing Secondary Energy Carriers from Biomass

André Faaij

13.1 Introduction

Current energy supplies in the world are dominated by fossil fuels (some 80% of the total use of over 400EJ per year). Nevertheless, about 10–15% (or 45 ± 10 EJ) of this demand is covered by biomass resources, making biomass by far the most important renewable energy source used to date. On average, in the industrialized countries biomass contributes some 9–13% of the total energy supplies, but in developing countries this is as high as a fifth to one third. In quite a number of countries biomass even covers between 50 to 90% of the total energy demand. A large part of this biomass use is, however, non-commercial and is used for cooking and space heating, generally by the poorer part of the population. This also explains why the contribution of biomass to the energy supply is not exactly known; non-commercial use is poorly mapped. In addition, some traditional use is not sustainable because it may deprive local soils of needed nutrients, can cause indoor and outdoor pollution and can result in poor health. It may also contribute to GHG emissions and affect ecosystems. Part of this use is commercial though, i.e. the household fuel wood in industrialized countries and charcoal and firewood in urban and industrial areas in developing countries, but there are almost no data on the size of those markets. An estimated 9 ± 6 EJ is covered by this category.

Modern, commercial, energy production from biomass (such as in industry, power generation or transport fuels) makes a lower, but still very significant contribution (some

Renewables-Based Technology Edited by J. Dewulf and H. Van Langenhove

7 EJ/yr in 2000), and this share is growing. It is estimated that end of the 1990s, some 40 GWe biomass-based electricity production capacity was installed worldwide (good for 160 TWh/year) and 200 GW heat production capacity (>700 TWh/year). Total production of biofuels (mainly ethanol produced from sugar cane and surpluses of corn and cereals and to a far lesser extent biodiesel from oil-seed crops) amounted to some 18 billion litres per year. This equals about 0.5 EJ as transport fuel (around 2000), but worldwide production of biofuels (especially bio-ethanol) is growing rapidly (Turkenburg, 2000).

The (technical) potential contribution of bio-energy to the future world's energy supply could be very large. In theory, energy farming on current agricultural land could contribute over 800 EJ, without jeopardizing the world's food supply. Organic wastes and residues could possibly supply another 40–170 EJ, with uncertain contributions from forest residues and potentially a very significant role for organic waste, especially when bio-materials are used on a larger scale. In total, the upper limit of bio-energy potential could be over 1000 EJ (per year). This is considerably more than the current global energy use of 400 EJ (Hoogwijk *et al.*, 2003).

Latin America and Eastern Europe clearly are promising regions, also Oceania and East and NE Asia jump out as potential biomass production areas in the longer term. This can be explained in particular by the projected demographic developments (declining population in China after 2020) and fast technological progress in agriculture, leading to substantial productivity increases. This analysis also shows that a large part of the technical potential for biomass production may be developed at low production costs in the range of 2 U$/GJ.

Major transitions are, however, required to exploit this bio-energy potential. In particular, improving agricultural efficiency in developing countries (i.e. increasing crop yields per hectare) is a key factor. It is uncertain to what extent and how fast such transitions can be realized. Under less favourable conditions, the bio-energy potential could be low as well.

Also, it should be noted that technological developments (in conversion as well as large-scale transport chains) can dramatically improve the competitiveness and efficiency of bio-energy, potentially a driver as such to start exploiting the potential of bio-energy. This chapter will further focus on key technologies for bio-energy production.

13.2 Technology Overview

Conversion routes for producing energy carriers from biomass are plentiful. Figure 13.1 illustrates the main conversion routes that are used for the production of heat, power and transport fuels or are under development. Below, we will discuss the most important technologies, which are or have been deployed so far with respect to their role, status and generic performance levels. We will also discuss some of the key options which are under development and that could play an important role in the coming decades. Some emphasis is on the European context, since a wide array for developing modern bio-energy technologies has been pursued in this region.

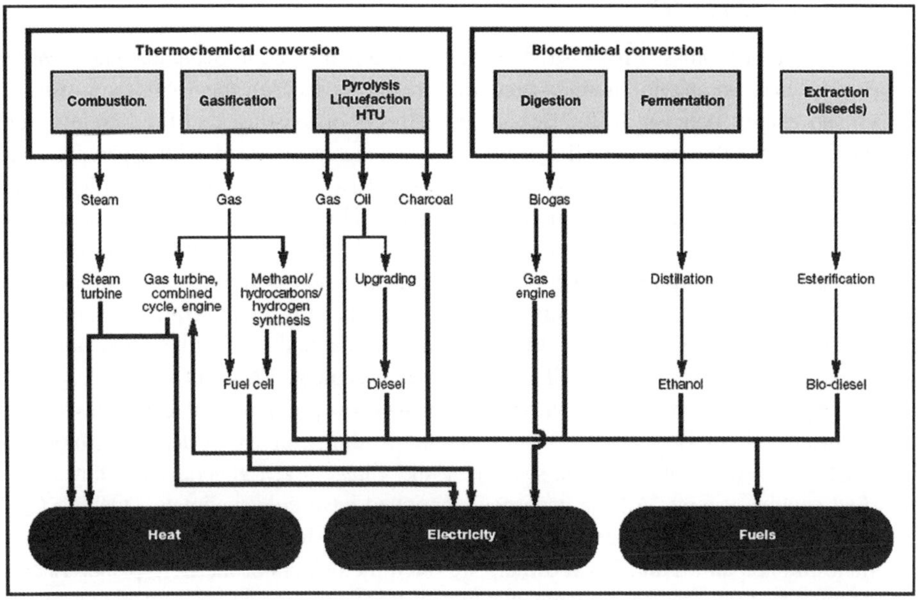

Figure 13.1 *Main conversion options for biomass to secondary energy carriers*
Note: Some categories represent a wide range of technological concepts as well as capacity ranges at which they are deployed, which are dealt with further in the main text.
Source: Turkenburg (2000).

13.2.1 Combustion

Domestic Heating

The classic application of biomass combustion is heat production for domestic applications. This is still a major market for biomass for domestic heating in countries such as Austria, France, Germany and Sweden. Use of wood in open fireplaces and small furnaces in houses is generally poorly registered, but estimated contributions to meet heat demand are considerable in the countries mentioned. Traditional use of wood generally has a low efficiency (sometimes as low as 10%) and generally is accompanied by considerable emissions of dust and soot. Technology development has led to the application of greatly improved heating systems, for example, which are automated, have catalytic gas cleaning and make use of standardized fuel (such as pellets). The efficiency benefit compared to open fireplaces is considerable: open fireplaces may even have a negative efficiency over the year (due to heat losses through the chimney), while advanced domestic heaters can obtain efficiencies of 70–90% with greatly reduced emissions. The application of such systems is especially observed in Scandinavia, Austria and Germany. In Sweden in particular, a significant market has developed for biomass pellets, which are fired in automated firing systems (van Loo and Koppejan, 2002).

District Heating and CHP

The application of biomass fired district heating is widely applied in Scandinavian countries and Austria. In Scandinavia, biomass-fired CHP really took off in the 1980s as a result of climate and energy policies. In the first stages, retrofits of existing coal-fired boilers were popular. Over time, the scale of CHP systems shows an increasing trend, with apparent advantages as higher electrical efficiencies and lower costs. This was also combined with a developing biomass market, allowing for more competitive and longer-distance supplies of biomass resources (especially forest residues) (Hillring, 2002). In the 1990s Denmark developed a major programme for utilizing straw. Various technical concepts were developed and used, such as the so-called cigar burners, combined with efficient straw baling equipment, transport and storage chains. Other innovations were needed to deal with the difficult combustion characteristics of straw such the high alkali and chlorine content. This led to complex boiler concepts, e.g. involving two-stage combustion, but also new pre-treatment techniques such as straw washing (Nikolaisen, 1998). Austria, another leading country in deploying biomass-fired CHP, focused on smaller-scale systems at the village level, generally combined with local fuel supply systems. All the countries mentioned have colder climates, making CHP economically attractive. Furthermore, the involvement of local communities has proved important. Municipalities and forest owners are often the owners of the CHP plants. Energy costs of those systems are usually somewhat higher. Local societal support is generally strong though, especially due to the employment and expenditures that benefit the local community. High labour costs also led to high degrees of automation though, with unmanned operations typical of many of the newer facilities (Serup, 1999).

Combustion of Biomass for the Production of Electricity

Larger-scale combustion of biomass for the production of electricity (plus heat and process steam) is applied commercially worldwide. Many plant configurations have been developed and deployed over time. Basic combustion concepts include pile burning, various types of grate firing (stationary, moving, vibrating), suspension firing and fluidized bed concepts. The key sector for the application for biomass combustion for power generation is the paper and pulp industry for the combustion of black liquor and waste incineration. Conventional boilers for the combined production of power and process steams and recovery of pulping chemicals are common technology for the pulp and paper sector. Waste incinerators were widely deployed starting in the 1980s in countries like Germany and the Netherlands, combined with very stringent emission standards. Typical technology deployed is large-scale (i.e. around 1 Mtonne capacity per plant per year) moving grate boilers (which allow mass burning of very diverse waste properties), low steam pressures and temperatures (to avoid corrosion) and extensive flue gas cleaning. Typical electrical efficiencies are between 15 to over 20% and more efficient designs (reaching some 30% electrical efficiency) have now been commissioned. Mass burning became the key waste-to-energy technology deployed in Europe, but is also relatively expensive with treatment costs in the range of 50–150 Euro/tonne (offset by tipping fees) (Faaij *et al.*, 1998). Typical capacities for stand-alone biomass combustion plants (typically using wood such as forest residues, as fuel) range between 20–50 MWe, with related electrical efficiencies in the 25–30% range. Such plants are only economically viable when fuels are available at low cost or when a carbon tax or feed-in tariff for renewable electricity is in place. In recent years, advanced combustion concepts have penetrated the market (Faaij, 2004). The application of fluid bed technology and advanced gas cleaning allows the efficient and production of electricity (and heat) from biomass. On a scale of about 50–80 MWe, electrical efficiencies of 30–40% are possible (van Loo and Koppejan, 2002; Department of Energy, 1998; van den Broek *et al.*, 1996). Finland is at the cutting edge of the field with development and deployment of BFB and CFB boilers with high fuel flexibility, low costs, high efficiency and deployed on a large scale. One of the latest plants realized in Finland has a capacity of 500 MWth and is co-fired with a portfolio of biomass fuels, partly supplied by water transport. Some of the main characteristics and generic emission data are summarized in Tables 13.1 and 13.2 respectively for various combustion concepts discussed.

Co-combustion

Co-combustion of biomass, in particular in coal-fired power plants is the single largest growing conversion route for biomass in many EU countries (e.g. in Spain, Germany, and the Netherlands to name but a few). The advantages of co-firing are apparent: the overall electrical efficiency is high due to the economies of scale of the existing plant

Table 13.1 *Some main characteristics of key combustion concepts currently deployed*

Combustion concept	*Key advantages*	*Key disadvantages*	*Key applications*
Underfeed stokers	Low cost at smaller capacities, easy control.	Needs high quality fuel with respect to composition and size.	<6 MWth Heating; buildings, district heating.
Grate furnaces	Relatively cheap and less sensitive to slagging.	Different fuels cannot be mixed for certain grate-type. Non-homogeneous combustion conditions (resulting in higher emissions).	Wide range; widely applied, especially for waste fuels and residues. Typical capacity some 20–30 MWe; waste incinerators up to 1–2 Mtonne capacity/yr.
Dust combustion	High efficiency, good load control and low NO_x.	Small particle size required, wearing of lining.	Less common; especially suited for i.e. sawdust
BFB furnaces	Highly flexible for moisture contents, low NO_x, high efficiency and decreased flue gas flow.	High capital and operational costs, sensitive to particle size (requires pre-treatment), high dust load in flue gas, some sensitivity for ash slagging. Erosion of equipment.	>20 MWth; deployed for flexible fueled plants and more expensive fuels (e.g. clean wood).
CFB furnaces	High fuel flexibility with respect to moist and other properties, low NO_x, easy adding additives (e.g. for emission control), high efficiency and good heat transfer.	High capital and operational costs, sensitivity to particle size (requires pre-treatment), high dust loads in flue gas, inflexible for part load operation. Very sensitive to ash slagging, erosion of equipment.	>30 MWth; deployed for flexible fueled plants and more expensive fuels. Of CFB and BFB combined, some 300 installations build worldwide.

Table 13.2 Generic emission data for key biomass combustion options

Combustion option	NO_x (mg/MJ)	Particulates (mg/MJ)	Tar (mg/MJ)	CO (mg/MJ)	UHC as CH4 (mg/MJ)	PAH (μg/MJ)
FB boiler	170	2	n.m.	0	1	4
Grate-fired boiler	110	120	n.m.	1850	70	4040
Stoker burner	100	60	n.m.	460	4	9
Modern wood stove	60	100	70	1730	200	30
Traditional wood stove	30	1920	1800	6960	1750	3450
Fireplace	n.m.	6050	4200	6720	n.m.	110

Notes: Values concern wood firing only and West European conditions (and emissions standards). N.m. implies 'not measured'.

(usually around 40%) and investments costs are low to negligible when high quality fuels as pellets are used. Also, directly avoided emissions are high due to the direct replacement of coal. Combined with the fact that many coal-fired power plants in operation are fully depreciated, this makes co-firing usually a very attractive GHG mitigation option. In addition, biomass firing leads to a lowering of sulphur and other emissions (see e.g. Meuleman and Faaij, 1999). For some countries though, biomass firing is also a route to avoid the need to invest in additional flue gas cleaning equipment needed for 100% coal firing. Generally, relatively low co-firing shares are deployed with very limited consequences for boiler performance and maintenance. Because many plants are now equipped with some co-firing capacity, interest in higher co-firing shares (e.g. up to 40%) is rising. Technical consequences for e.g. feeding lines and boiler performance are more severe though and current development efforts focus on those issues (Van Loo and Koppejan, 2002). Power plants capable of firing natural gas or coal with various biomass streams have been built in Denmark (e.g. the Avedore plant) with the benefit of gaining economies of scale as well as reduced fuel supply risks. In Denmark straw is a common fuel. The chlorine-rich and alkaline-rich straw caused problems in conventional combustion systems through increased corrosion and slagging. In multi-fuel systems, however, straw can be used for raising low temperature steam, after which the steam is superheated by fossil fuels. This approach almost eliminates the problems mentioned (Nikolaisen *et al.*, 1998).

13.2.2 *Gasification*

Gasification as the general method used to convert a range of solid fuels into combustible gas or syngas received considerable attention in the 1980s worldwide. Gasification converts biomass into fuel gas, which can be cleaned prior to combustion (e.g. in a gas turbine; when integrated with a combined cycle this leads to a BIG/CC Biomass Integrated Gasification/Combined Cycle plant) or further conversion. The fuel gas can also be cleaned to syngas quality, from which fuels like methonal, Fischer-Tropsch liquids and hydrogen can be produced.

Smaller-Scale Gasification

At the end of the 1980s and the beginning of the 1990s, small-scale gasification received major support. Downdraft or updraft, fixed bed gasifiers with capacities of less than a 100 kWth up to a few MWth were developed and tested for small-scale power and heat generation using diesel or gas engines. Heat production using (small) gasifier proved to be commercially established. Finland in particular was successful in the 1980s in deploying smaller-scale gasifiers for heat production (Bioneer). Nevertheless, gasification for production of heat finds a strong competitor in combustion. A key concept pursued for a long time was the use of agricultural residues close to their source, thus minimizing transport distances. A wide array of concepts for gasifiers, gas cleaning and system integration for such concepts was proposed and tested in a wide variety of conditions. Technology was also exported to many developing countries with support from international bodies like the World Bank. The key drivers here were rural development and electrification. So far, despite major efforts, investment and a large number of demonstration units, the concept of small-scale gasification linked to gas or diesel engines has not taken off. Small (fixed bed) gasifiers coupled to diesel/gas engines (typically for 100–200 kWe systems with an approximate, modest, electrical efficiency of 15–25%) are commercially available on the market. Especially in India, successful implementation has been realized. However, the critical demands of small-scale gasifiers to fuel quality (preferably standardized and hence using more expensive fuel such as pellets), the required careful operation and high costs, especially for effective gas cleaning given the severe emission standards in the EU, have so far hindered their wide deployment (Kaltschmitt *et al.*, 1998; Stassen, 1995). Possibly, in the longer term, standardized gasification systems ('pre-packaged') using fuel cells and micro-turbines could mean a breakthrough for small scale electricity production from biomass, but such systems need further development and will depend on cheap and reliable fuel cells and again, major advances in small scale gas cleaning.

Larger-Scale (CFB) Biomass Gasification

Larger gasifiers (i.e. over several tens of MWth capacity) are generally associated with (Circulating) Fluidized Bed concepts. At atmospheric pressure (ACFB) gasifiers are used for the production of (raw) fuel gas and process heat (e.g. in Italy, Austria, Sweden and Germany) but not in very large numbers. Biomass Integrated Gasification/Combined Cycle (BIG/CC) systems combine flexibility with respect to fuel characteristics with a high electrical efficiency. Electrical efficiencies around 40% (LHV basis) are possible on a scale of about 30 MWe in the shorter term (Consonni and Larson, 1994a, 1994b; Faaij *et al.*, 1997). BIG/CC became the centre of attention in EU and various national programs in the first half of the 1990s. The promise of this technology, allowing for high electrical efficiency at modest scales combined with modest capital costs, resulted in a variety of research and demonstration initiatives. Furthermore, the CFB technology principally allows for high fuel flexibility and, inherent to the BIG/CC concepts, low emissions to air can be obtained. Demonstration projects were launched in various countries and for various gasification concepts: in Brazil a GEF/World Bank-supported project was set up to demonstrate a 30 MWe ACFB BIG/CC unit fired with cultivated Eucalyptus (Elliott and Booth, 1993). In the same period in Sweden, the first (PCFB) BIG/CC unit (the BIOFLOW pilot-project), based on a pressurized gasification process gained several thousands of hours of operational experience. An atmospheric BIG/CC system was commissioned in 2000 in Yorkshire, UK, but operations stopped just a few years later. An important project in the USA is the demonstration of the indirect FERCO gasification concept at the existing Burlington power station (Morris *et al.*, 2005). In addition, a variety of national initiatives were launched, aimed at pre-commercial or demonstration units of BIG/CC technology (in particular in the EU). However, in practice the realization of the demonstration projects proved to be difficult. Costs of first generation units proved to be very high. The first generation of BIG/CC systems shows high unit capital costs. Depending on the scale, price levels of 5000–3500 Euro/kWe are quoted (Faaij *et al.*, 1998b), which is still far from the desired 1500–2000 Euro/kWe, which could bring BIG/CC into the competitive area. Various technological issues (e.g. concerning pre-treatment and tar removal) still need to be resolved. Later in the 1990s, many utilities involved faced the consequences of the rapid liberalization process in the energy sector and expensive demonstration activities proved to be hard to pursue. Various demonstration units (such as ARBRE and BIOFLOW) were taken out of operation recently. In total, co-firing and proven combustion technology (which also develops over time) are generally favoured by the risk-weary energy sector. This has led to the remarkable situation of a stalling development of a technology that, in the somewhat longer term, is capable of producing power from biomass at competitive price levels. At the somewhat larger scale (over 100 MWe) and considering the ongoing improvement of gas turbine technology, the cost reduction potential of BIG/CC systems is considerable, as has been evaluated by numerous studies (Williams and Larson, 1996; Faaij *et al.* 1998b; Solantausta *et al.*, 1996). The combination of high electrical efficiencies with relatively low unit capital costs can make the use of cultivated biomass as feedstock economically feasible for many areas in the world. So far, however, development has been slow.

Gasification for Co-Firing

Gasification is also a route towards large co-firing shares of existing (coal-fired) power plants, avoiding the need for additional solid fuel feeding lines and allowing for better control of the combustion process. Successful deployment of (A)CFB gasifiers has recently been shown in co-firing schemes (e.g. Lahti in Finland and Amer in the Netherlands) (van Loo and Koppejan, 2002). An interesting alternative to fuel gas produced through biomass gasification is its use in existing (or new) natural gas-fired combined cycles. In this way, economies of scale are utilized, resulting in low costs and (very) high overall efficiencies (currently up to 60% for NG fired combined cycles), combined with a safe fuel supply since one can vary the share of fuel gas and natural gas fired (Rodrigues *et al.*, 2003). So far, this option has not been demonstrated anywhere in the world, but research efforts are increasing and this could prove to be of major importance on short term given that co-firing opportunities at existing coal-fired power plants are increasingly utilized already.

Production of Methanol, Hydrogen and Synthetic Hydrocarbons Via Gasification

Partly as a result of the oil crises, interest in biomass-derived syngas for the production of transport fuels (such as methanol) had been undertaken in the 1980s. Pressurized gasification for methanol production from biomass was tested and developed in France and Sweden. Kemira (a Finnish company active in production of fertilizers) installed a large-scale CFB gasifier in Oulu (Finland), to produce syngas for an ammonia factory (which was shut down). Also noteworthy is the installed gasification capacity (entrained flow) at Schwarze Pumpe (former East Germany) to produce methanol from waste streams, which is a major industrial experience with this technology. Low energy prices caused the interest in advanced gasification technologies for large-scale applications to fall (Kaltschmitt *et al.*, 1998). Renewed attention in using gasification technology to produce transport fuels, in particular, Fischer-Tropsch diesel and hydrogen later arose. Although this seems a viable development given the techno-economic potential of such concepts (see Table 13.3), the technological challenges remain and are likely to be more complex than for BIG/CC concepts because gas cleaning needs to be even more effective in order to protect downstream catalytic gas processing equipment. Once clean syngas is available, the known process technology to produce methanol, Fischer-Tropsch-liquids (via Fischer-Tropsch processes), Dimethyl ether (DME) and hydrogen can be applied. So far, no commercial biomass-fed facilities are operational either in Europe or the rest of the world. The main challenges in this area are gas cleaning, scale-up of processes and process integration. Overall, energetic efficiencies of relatively 'conventional' production facilities, could be close to 60% (on a scale of about 400 MWth input). Deployment on a large scale (e.g over 1000 MWth) is required to gain the necessary economies of scale. In total, however, this (set of) option(s) has a strong position from both an efficiency and economic perspective (Hamelinck and Faaij, 2002; Hamelinck *et al.*, 2004; Tijmensen *et al.*, 2002; Williams *et al.*, 1995). More recent technological concepts, such as liquid phase methanol production and once-through Fischer-Tropsch synthesis (combined with electricity generation) and new gas separation technology offer potentially lower production costs and a higher overall efficiency. More research, demonstration and development activities over a prolonged period of time are, however, needed to reach such a situation. In countries like Germany, the Netherlands and Sweden, interest in developing advanced gasification for syngas production is on the rise again and is playing an important role in long-term RD&D strategies.

13.2.3 Production of Bio-Oils: Pyrolysis and Liquefaction Processes

Pyrolysis converts biomass at temperatures around 500 °C in the absence of oxygen to liquid (bio-oil), gaseous and solid (charcoal) fractions. With flash pyrolysis techniques (fast pyrolysis) the liquid fraction can be maximized (up to 70% of the thermal biomass input). Bio-oil contains about 40 weight percentage of oxygen and is corrosive and acidic. Crude bio-oil can in principle (after some modifications and only for better quality oils) be used for firing engines and turbines. The oil can also be upgraded (e.g. via hydrogenation) in order to reduce the oxygen content. But upgrading comes with both economic and energy penalties. Pyrolysis and upgrading technology are largely in the demonstration phase (Bridgewater, 1998). Liquefaction (conversion under high pressure) and HTU, or hydro thermal upgrading (a process originally developed by Shell and in pre-pilot phase), converts biomass at a high pressure in water and moderate temperatures to bio-crude (Naber *et al.*, 1997).

Up to now, pyrolysis (and even more liquefaction options) have been less well developed than gasification. Major attention since the end of the 1980s/beginning of the 1990s was

Table 13.3 *Global overview of current and projected performance data for the main conversion routes of biomass to power and heat and summary of technology status and deployment in the European context*

Conversion option	*Process*	*Typical capacity range*	*Net efficiency (LHV basis)*	*Investment cost ranges (Euro/kW)*	*Status and deployment in Europe*
Biogas production	Anaerobic Digestion	Up to several MWe	10–15% (electrical)		Well-established technology. Widely applied for homogeneous wet organice waste streams and waste water. To a lesser extent used for heterogeneous wet wastes such as organic domestic wastes.
	Landfill gas	Generally several 100s kWe	Gas engine efficiency		Very attractive GHG mitigation option. Widely applied in EU and in general part of waste treatment policies of most countries.
Combustion	Heat	Domestic 1–5 MWth	From very low (classic fireplaces) up to 70–90% for modern furnaces.	~100/kWth 300–700/kWth for larger furnaces	Classic firewood use still widely deployed in Europe, but decreasing. Replacement by modern heating systems (i.e. automated, flue gas cleaning, pellet firing) in e.g. Austria, Sweden, Germany ongoing for years.
	CHP	0.1–1 MWe 1–10 MWe	60–90% (overall) 80–100% (overall)		Widely deployed in Scandinavia countries, Austria, Germany and to a lesser extent France. In general increasing scale and increasing electrical efficiency over time.
	Stand alone	20–100s MWe	20–40% (electrical)	2.500–1600	Well-established technology, especially deployed in Scandinavia; various advanced concepts using fluid bed technology giving high efficiency, low costs and high flexibility commercially deployed. Mass burning or waste incineration goes with much higher capital costs and lower efficiency; widely applied in countries like the Netherlands, Germany a.o.

	Co-combustion	Typically 5–20 MWe at existing coal fired stations. Higher for new multifuel power plants	30–40% (electrical)	~250 + costs of existing power station	Widely deployed in many EU countries. Interest larger biomass co-firing shares and utilization of more advanced options (e.g. by feeding fuel gas from gasifiers) is growing in more recent years.
Gasification	Heat	Usually smaller capacity range around 100s kWth	80–90% (overall)	Several 100s/kWth, depending on capacity	Commerically available and deployed; but total contribution to energy production in the EU is very limited.
	CHP gas engine	0.1–1 MWe	15–30%	3.000–1.000 (depends on configuration)	Various systems on the market. Deployment limited due to relatively high costs, critical operational demands and fuel quality.
	BIG/CC	30–100 MWe	40–50% (or higher; electrical efficiency)	5.000–3.500 (demos) 2.000–1.000 (longer term, larger scale)	Demonstration phase at 5–10MWe range obtained.. Rapid development in the 1990s has stalled in recent years. First generation concepts prove capital intensive
Pyrolysis	Bio-oil	Generally smaller capacities are proposed of several 100s kWth	60–70% heat content of bio-oil/feedstock.		Not commercially available; mostly considered a pre-treatment option for longer distance transport.

Source: Based on a variety of literature sources (i.e. Van Loo and Koppejan, 2002; Van den Broek *et al.*, 1996; Kaltschmitt *et al.*, 1998; Faaij *et al.*, 1998; Department of Energy, 1998).

Notes: Due to the variability of data in the various references and conditions assumed, all cost figures should be considered as indicative. Some key assumptions for the estimated production cost ranges are given in footnotes; generally they reflect European conditions.

especially caused by the potential deployment of this technology on small scale in rural areas and as feedstock for the chemical industry. Reducing transport costs because of the higher energy density of bio-oil compared to untreated biomass was used as another key argument.

Although considerable experience was gained over time, still, few successful demonstration units were realized, (prime examples were shown by Dynamotive in Canada and Fortum, a Finnish Oil company). Actual market implementation is so far negligible. Pyrolysis now receives increasing attention as a pre-treatment step for long-distance transport of bio-oil that can be used in further conversion (e.g. efficient power generation or (entrained flow) gasification for syngas production).

13.2.4 Fermentation: Production of Ethanol

Ethanol from Sugar and Starch

Production of ethanol via fermentation of sugars is a classic conversion route, which is applied to sugar cane, maize and cereals on a large scale, especially in Brazil, the United States and France. Sweden and Spain have more modest production levels of ethanol. Ethanol is generally mixed with gasoline, which, at low percentages, can be done without adaptations to the current vehicle fleet. Ethanol has the advantage that it lowers NO_x and dust emissions to some extent compared to gasoline use only. The US and European programmes are particularly used for converting surpluses of food crops to a useful (by-) product. Ethanol production from food crops like maize and cereals is, however, far from competitive when compared to gasoline and diesel prices and it is not likely this will change in the longer term.

Ethanol from Ligno-Cellulosic (Woody) Biomass

Hydrolysis of ligno-cellulosic biomass can open the way towards low cost and efficient production of ethanol from ligno-cellulosic biomass. The development of various hydrolysis techniques has gained major attention in the past eight years or so, particularly in Sweden and the United States. However, cheap and efficient hydrolysis processes are still under development and some fundamental issues need to be resolved. The conversion is more difficult than for sugar and starch because from ligno-cellulosic materials, first sugars need to be produced via hydrolysis. This can be done through acid treatment or via enzymatic pathways, respectively being relatively expensive and inefficient and technologically unproven. In addition, the pre-treatment of woody biomass materials for further processing is a technical challenge. Assuming, however, that those issues are resolved and ethanol production is combined with efficient electricity production from unconverted wood fractions (lignine in particular), ethanol costs could come close to current gasoline prices (Lynd, 1996): as low as 12 Euroct/litre assuming biomass costs of about 2 Euro/GJ. Overall system efficiencies could go up to about 70% (LHV). For the agricultural sector and agro-food industry this technology is attracting interest already to boost the competitiveness of existing production facilities (e.g. by converting available crop and process residues), which provides drivers for both industry and agriculture to support this technology.

13.2.5 Digestion

Biogas

Anaerobic digestion of biomass has been demonstrated and applied commercially for a variety of feedstocks such as organic domestic waste, organic industrial wastes, manure, sludges, etc. Digestion has a low overall electrical efficiency (roughly 10–15%, strongly depending on the feedstock) and is particularly suited for wet biomass materials. Digestion has been deployed for a long time in the food and beverage industry to process wastewater with high loads of organic matter. Currently, advanced, large-scale, systems for wet industrial wastestreams are applied in many countries. Countries like Denmark and the Germany are in a strong position with advanced digestion systems used for processing various wet wastestreams (Braber, 1995).

Landfill gas utilization

A specific source of biogas is landfills. The production of methane-rich landfill gas from landfill sites makes a significant contribution to atmospheric methane emissions. In many situations the collection of landfill gas and production of electricity by converting this gas in gas engines are profitable and the application of such systems has become widespread. The benefits are obvious: useful energy carriers are produced from gas that would otherwise contribute to a build-up of GHG in the atmosphere (Faaij *et al.*, 1998). This makes landfill gas utilization, in general, a very attractive GHG mitigation option and it has been widely adopted throughout the EU and North America.

13.2.6 Extraction and Production of Esters from Oilseeds

Oilseeds, like rapeseed, can be extracted and converted to esters and are well suited to replace diesel. Rapeseed production and subsequent esterification and distribution are an established technology in Europe. Significant quantities of RME are produced in the EU (concentrated in Germany, France and to a lesser extent in Austria and Italy). RME, however, requires substantial subsidies to compete with diesel, also in the longer term. Subsidies in Europe generally consist of a combination of farm subsidies (e.g. for producing non-food crops) and tax exemption of the fuel itself. The latter implies about a factor of 3–4 subsidy compared to conventional diesel or gasoline production costs (see Table 13.4). Key drivers for the implementation of RME schemes are rural employment and the flexible nature of the crop because it can easily replace conventional food crops when desired. Energy balances of RME fuel chains are less favourable when compared to perennial crops and net energy production per hectare is lower (Kaltschmitt *et al.*, 1996, Biewinga and Van der Bijl, 1996). Energy balances and economic performance can be improved to some extent, particularly by using straw for efficient heat and power production (International Energy Agency, 1994).

Table 13.4 *Global overview of current and projected performance data for the main conversion routes of biomass to fuels*

Concept	*Energy efficiency (HHV) + energy inputs*		*Investment costs (Euro/kWth input capacity)*		*O&M (% of inv.)*	*Estimated production costs (Euro/GJ fuel)*	
	Short term	*Long term*	*Short term*	*Long term*		*Shorter term*	*Longer term*
Hydrogen: via biomass gasification and subsequent syngas processing. Combined fuel and power production possible; for production of liquid hydrogen additional electricity use should be taken into account.	60% (fuel only) (+ 0.19 GJe/GJ H2 for liquid hydrogen)	55% (fuel) 6% (power) (+ 0.19 GJe/GJ H2 for liquid hydrogen)	480 (+ 48 for liquefying)	360 (+ 33 for liquefying)	4	9–12	4–8
Methanol: via biomass gasification and subsequent syngas processing. Combined fuel and power production possible.	55% (fuel only)	48% (fuel) 12% (power)	690	530	4	10–15	6–8
Fischer-Tropsch liquids: via biomass gasification and subsequent syngas processing. Combined fuel and power production possible.	45% (fuel only)	45% (fuel) 10% (power)	720	540	4	12–17	7–9
Ethanol from wood: production takes place via hydrolysis techniques and subsequent fermentation and includes integrated electricity production of unprocessed components.	46% (fuel) 4% (power)	53% (fuel) 8% (power)	350	180	6	12–17	4–7

Ethanol from sugar: production via fermentation; some additional energy inputs are needed for distillation. As feedstock, sugar beet are assumed.	43% (fuel only) 0.065 GJe + 0.24 GJth/GJ EtOH	43% (fuel only) 0.035 GJe + 0.18 GJth/GJ EtOH	290	170	5	25–35	20–30
Biodiesel RME: takes places via extraction (pressing) and subsequent esterification. Methanol is an energy input. For the total system it is assumed that surpluses of straw are used for power production.	88%; 0.01 GJe + 0.04 GJ MeOH per GJ output Efficiency power generation on shorter term: 45%, on longer term: 55%		150 (+ 450 for power generation from straw)	110 (+ 250 for power generation from straw)	5 4	25–40	20–30

Source: Based on Faaij and Hamelinck, 2002; Hamelinck and Faaij, 2002; Tijmensen *et al.*, 2001; de Jager *et al.*, 1998; Ogden *et al.*, 1999; Wyman *et al.*, 1993; International Energy Agency, 1994; Williams *et al.*, 1995.

Notes:

Due to the variability of data in the various references and conditions assumed, all cost figures should be considered as indicative. Notes summarize assumptions, generally reflecting EU conditions.

1. Assumed biomass price of clean wood: 2 Euro/GJ. RME cost figures varied from 20 Euro/GJ (short term) to 12 Euro/GJ (longer term), for sugar beet a range of 12 to 8 Euro/GJ is assumed. All figures exclude distribution of the fuels to fueling stations.
2. For equipment costs, an interest rate of 10%, economic lifetime of 15 years is assumed. Capacities of conversion unit are normalized on 400 MWth input on shorter term and 1000 MWth input on lonsger term.
3. Diesel and gasoline production costs vary strongly depending on the oil prices, but for indication: recent cost ranges are between 4–7 Euro/GJ. Longer term projections give estimates of roughly 6–10 Euro/GJ. Note that the transportation fuel retail prices are usually dominated by taxation and can vary between 50–130 Euroct./litre depending on the country in question.

13.3 Economics of Biomass Energy Systems

Absolutely crucial for the success of bio-energy systems is their economic performance. Biomass is a competitive alternative in many situations, but currently competitive use is generally observed where cheap, or even negative costs biomass residues or wastes are available. In order to make bio-energy (especially from specially cultivated biomass) competitive with fossil fuels, the conversion technologies, biomass production, as well as the total bio-energy systems, require further optimization.

13.3.1 Power Generation

With biomass prices of about 2 U$/GJ state of the art combustion technology at a scale of 40–60 MWe can result in COE of around U$ct 5–6/kWh produced (Department of Energy, 1998; Dornburg and Faaij, 2001; Solantausta *et al.*, 1996). Co-combustion, particularly at efficient coal-fired power plants can obtain similar cost figures. If BIG/CC technology becomes available commercially, COE could drop further to about 3–4 U$ct/kWh, especially due to higher electrical efficiencies. For larger scales (i.e. over 100 MWe), cultivated biomass will be able to compete fully with fossil fuels in many situations. The benefits of lower specific capital costs and increased efficiency certainly outweigh the increase in costs and energy use for transport for considerable distances if a reasonably well-developed infrastructure is in place (Marrison and Larson, 1995; Dornburg and Faaij, 2001).

Localized power (and heat) production are generally more expensive, but better suited for off-grid applications. The costs that could ultimately be obtained with gasifier/diesel systems are still unclear and depend strongly on what emissions and fuel quality are considered acceptable. Combined heat and power generation is generally attractive when heat is required with high load factors. An overview of performance and cost date of different power and heat production technologies from biomass is given in Table 13.3.

13.3.2 Production of Liquid and Gaseous Fuels from Biomass

Table 13.4 gives a compact summary of estimates for costs of various fuels that can be produced from biomass. A difference is made between performance levels in the short and in the longer term. Generally speaking, the economy of 'traditional' fuels like RME in ethanol from starch and sugar crops in moderate climate zones is poor and unlikely to reach competitive price levels. Methanol, hydrogen and ethanol from ligno-cellulosic biomass offer much better prospects and offer competitive fuel prices in the longer term.

When the use of such fuels (especially hydrogen en methanol) in fuel cell vehicles (FCVs) or advanced hybrid cars is considered, the overall chain ('well – to – wheel') efficiency can drastically improve compared to current diesel/gasoline powered internal combustion engine vehicles (Faaij and Hamelinck, 2002). Furthermore, FCVs offer the additional and very relevant benefits of zero or near zero emission of compounds like NO_x, CO, hydrocarbons and small dust particulates, which are to a large extent responsible for large-scale problems with (urban) air quality in most urban zones in the world.

13.4 Heat, Power and Fuels from Biomass: Key Markets

As discussed in Section 13.1, the production of heat and electricity dominates current bio-energy use. At present, the main growth markets for bio-energy are the European Union, Central and Eastern Europe and South-East Asia (Thailand, Malaysia, Indonesia), especially with respect to efficient power generation from biomass wastes and residues. Two key industrial sectors for application of state-of-the-art biomass combustion (and potentially gasification) technology for power generation are the paper and pulp sector and cane-based sugar industry worldwide.

In most European countries heat production from biomass has stagnated somewhat (unless in CHP applications) (Harmelinck *et al.*, 2002). Bio-energy developed in North America has stalled in recent years, in particular due to political developments, the one exception being ethanol production linked to the corn production and processing industries. As a result, RD&D efforts in developing advanced ethanol production technology (including hydrolysis techniques) are significant and various demonstration projects are underway that may pave the way for large-scale commercial use of this technology before 2020 or so. Table 13.5 gives a rough overview of the prospects for various markets combined with main biomass resources in the short and the longer term. These prospects are:

- In traditional bio-energy use, heat for cooking and space heating are crucial. It is not expected that the traditional use of biomass will decrease in the coming decades, since traditional biomass use is often linked to poverty and underdevelopment, which prove difficult problems to solve. Nevertheless, modernizing bio-energy use for the poorer sector of the populations is an essential component of sustainable development schemes in many countries. This creates opportunities and major markets, for example, for improved stoves, production of high quality fuels for cooking (e.g. biofuel-based such as ethanol and Fischer-Tropsch liquids) with considerable efficiency and health advantages. Furthermore, biogas, e.g. produced with digestors on village level, proved very effective in various countries (such as China and India) in solving waste treatment problems and supplying high quality energy carriers (clean gas and power when used in gas engines).
- Commercial heat production from biomass stalls in most countries with significant biomass utilization. Reliable technology (e.g. gasification, advanced stoves, etc.) are commercially available for many applications (industrial, district and domestic heating), but profitability of power generation (or CHP) seems better in most current markets. In particular, for specific industrial applications heat production from biomass seems most attractive.
- Power generation from biomass by advanced combustion technology and co-firing schemes is the real growth market worldwide. Mature, efficient and reliable technology is available to turn biomass into power. In various markets the average scale of biomass combustion schemes rapidly increases due to improved availability of biomass resources and the economic advantages of economies of scale of conversion technology. It is also in this field that competitive performance compared to fossil fuels is possible once lower cost residues are available. This is in particular true for co-firing schemes, where investment costs can be minimal. Specific (national) policies (such as carbon taxes, renewable energy support, e.g. by direct investment subsidies or feed-in tariffs)

Table 13.5 *Integrated (and generic) overview of performance projections for different options and biomass markets on shorter (~5) and longer (> ~20) years*

Biomass feedstock category	*Heat*		*Electricity*		*Transport fuels*	
	Short term; roughly stabilizing market	*Longer term*	Short term; strong growth market worldwide	*Longer term; growth may stabilize due to competition of alternative options*	*Short term; growing market, but dependent on policies and financial support*	*Longer term; potential key market for (cultivated) biomass*
Organic wastes (i.e. MSW a.o.)	Undesirable for domestic purposes (emissions); industrial use attractive; in general competitive.	Especially attractive in industrial setting and CHP. (advanced combustion and gasification for fuel gas).	<3–5 U$ct for state-of-the art waste incineration and co-combustion. Economics strongly affected by tipping fees and emission standards.	Similar range; improvements in efficiency and environmental performance, in particular through IG/CC technology at large scale.	N.A.	In particular possible via gasification routes.
Residues: – forestry – agriculture	Major market in developing countries (<1–5 U$/kWhth); stabilizing market in industrialized countries.	Especially attractive in industrial setting and CHP. Advanced heating systems (domestic) possible but not on global scale.	4–12 U$ct/kWh (see below; major variable is supply costs of biomass); lower costs also in CHP operation and industrial setting depending on heat demand.	2–8 U$ct/kWh (see below; major variable is supply costs of biomass).	N.A.; only for surpluses of food crops.	See below.
Energy crops: – oil seeds – sugar/starch – sugar cane – perennials	N.A.	Unlikely market due to high costs feedstock for lower value energy carrier; possible niches for pellet or charcoal production in specific contexts.	6–15 U$ct/kWh High costs for small scale power generation with high quality feedstock (wood) lower costs for large scale (i.e. >100 MWth) state-of-the art combustion (wood, grasses) and co-combustion.	3–9 U$ct/kWh Low costs especially possible with advanced co-firing schemes and BIG/CC technology over 100–200 MWe.	8–25 U$/GJ; lower figures for ethanol from sugar cane; higher for biodiesel (RME) and sugar and starch crops in Europe and North America.	5–10 U$/GJ; low costs obtainable with lignocellulosic biomass (<2 U$/GJ) advanced hydrolysis techniques and large scale gasification (i.e. <1000 MWth) for MeOH/H_2/FT.

accelerate this development. Gasification technology (integrated with gas turbines/combined cycles) offers even better perspectives for power generation from biomass on medium term and can make power generation from energy crops competitive in many areas in the world once this technology would be proved on a commercial scale. Gasification (in particular larger-scale CFB concepts) also offers excellent possibilities for co-firing schemes.

- Biomass-derived transportation fuels currently represent a modest 0.5 EJ (or less than 1%) of total bio-energy use worldwide (largely covered by ethanol production from sugar and starch crops). But it is especially in this field that global interest is growing, again particularly in Europe, but also Brazil, the USA and Japan. Four main drivers can be distinguished for this growing interest:
    - the transport sector is particularly difficult to tackle in terms of GHG emission reductions;
    - the strategic importance to reduce the dependency on imported oil is growing;
    - technological developments offer perspectives of much cheaper and efficient production of biofuels from biomass, most notably ethanol via hydrolysis and fermentation techniques and fuels as Fischer-Tropsch, methanol, DMM/DME and hydrogen via gasification;
    - in addition, in the medium term (e.g. after 2020), biomass use for transport fuels may prove a more effective way to reduce GHG emissions using biomass than power generation.

For countries where sugar cane production is feasible, commercially available technology allows production of relatively low cost ethanol (Rosillo-Calle and Cortez, 1998). The Brazilian experience shows that almost competitive ethanol production with gasoline is possible. Ethanol production capacity based on sugar cane is, slowly, increasing in some African and several Latin American countries. Blending biofuels with conventional gasoline and diesel are popular utilization routes (Thuijl *et al.*, 2003). In Europe, policy targets are likely to create a major push for biofuels and demonstration of Fischer-Tropsch liquid production via gasification of biomass is likely in the foreseeable future (e.g. before 2010) (EC, 2001). Various countries (the Netherlands, Germany, Sweden) have shown an interest in move in this direction. Crucial for the economic feasibility of such schemes is their application on the large scale (i.e. over 1000 MWth). Related development and investment risks (also concerning a secure supply of biomass) are therefore considerable. Ethanol production from ligno-cellulosic biomass offers similar perspectives as well as technological and development challenges.

13.5 Summary

From a regional to a national focus in the 1980s and 1990s, biomass and bio-energy are increasingly becoming an international matter. Biomass markets are developing in international markets and the international trade in biomass and biomass-derived energy carriers is on the rise. Furthermore, certificate and emission trading as well as projects realized under the intended Clean Development Mechanism or as Joint Implementation activity make it more and more difficult to maintain very specific national policies. The recent EC biofuel directive is another interesting example of pan-European targets that potentially has important consequences for a European bio-energy market, both for raw materials and high quality transport fuels (Commission of European Communities, 2001). Similar arguments hold for technology developments and the RDD&D trajectories needed to commercialize more advanced, competitive and efficiency conversion capacity in particular for the production of electricity and fuels.

Biomass is one of the renewable energy sources capable of making a great contribution to the future world's energy supply. Although the actual role of bio-energy will depend on its competitiveness versus (and thus price levels of) fossil fuels and agricultural policies, it seems realistic to expect that the current contribution of bio-energy of 40–55 EJ will increase considerably. A range from 200 to 300 EJ may be observed looking well into this century, making biomass a more important energy supply option than mineral oil today.

A key issue for bio-energy is that its use should be modernized to fit in with sustainable development. In particular, the production of electricity via advanced conversion concepts (i.e. gasification and state-of-the-art combustion and co-firing) and modern biomass-derived fuels such as methanol, hydrogen and ethanol from ligno-cellulosic biomass are very promising and can reach competitive cost levels within a few decades. Flexible energy systems in which biomass and fossil fuels can be used in combination are likely to be the backbone of a low-risk, low-cost energy supply system. Prolonged RD&D efforts and biomass market development are essential prerequisites to achieve this. At the end of the day, policy support and international collaboration are essential.

References

Biewinga, E. E. and van der Bijl, G. (1996) *Sustainability of Energy Crops in Europe: A Methodology Developed and Applied*, Centre for Agriculture and Environment, CLM 234-1996, Utrecht.

Braber, K. (1995) Anaerobic digestion of municipal solid waste: a modern waste disposal option on the verge of breakthrough, *Biomass and Bioenergy*, **9**, 365–376.

Bridgewater, A. V. (1998) *The Status of Fast Pyrolysis of Biomass in Europe*, Proceedings of the 10th European Biomass Conference and Technology Exhibition, Wurzburg, Germany, pp. 268–271.

Consonni, S. and Larson, E. D. (1994a) Biomass-gasifier/aeroderivative gas turbine combined cycles, part A: technologies and performance modelling, paper prepared for Cogen Turbo Power '94, The American Society of Mechanical Engineers' 8th Congress and Exposition on gas turbines in cogeneration and utility, industrial and independent power generation, Portland, OR, 25–27 October 1994.

Consonni, S. and Larson, E. D. (1994b) Biomass-gasifier/aeroderivative gas turbine combined cycles, part B: performance calculations and economic assessment, paper prepared for Cogen Turbo Power '94, The American Society of Mechanical Engineers' 8th Congress and Exposition on gas turbines in cogeneration and utility, industrial and independent power generation, Portland, OR, 25–27 October 1994.

De Jager, D., Faaij, A. and Troelstra, W. P. (1998) *Cost-effectiveness of Transportation Fuels from Biomass*, ECOFYS, Dept. Of Science, Technology and Society, Utrecht University, Innas B. V., Report prepared for NOVEM (EWAB report 9830).

Department of Energy (1998) *Renewable Energy Technology Characterizations*, Office of Utility Technologies, Washington, DC.

Dornburg, V. and Faaij, A. (2001) Efficiency and economy of wood-fired biomass energy systems in relation to scale regarding heat and power generation using combustion and gasification technologies, *Biomass and Bioenergy*, **21**, 91–108.

EC Commission of the European Communities (2001) *Communication from the Commission to the European Parliament, the Council, the Economic and Social Committee and the Committee of the Regions on Alternative Fuels for Road Transportation and on a Set of Measures to Promote the Use of Biofuels*, COM (2001) 547, Brussels, European Communities, November 2001.

Elliott, P. and Booth, R. (1993) *Brazilian Biomass Power Demonstration Project*, Special project brief, Shell, London, September 1993.

Faaij, A. (2006) Bio-energy in Europe: changing technology choices, *Energy Policy*, in press.

Faaij, A. and Hamelinck, C. (2002) Long term perspectives for production of fuels from biomass; integrated Assessment and RD&D priorities, paper prepared for the 12th European Conference on Biomass for Energy, Industry and Climate Protection, Amsterdam, the Netherlands.

Faaij, A., Hekkert, M., Worrell, E. and van Wijk A. (1998a) Optimization of the final waste treatment system in the Netherlands, *Resources, Conservation and Recycling*, **22**, 47–82.

Faaij, A., Meuleman, B. and van Ree, R. (1998b) *Long Term Perspectives of BIG/CC Technology, Performance and Costs*, Department of Science, Technology and Society, Utrecht University and the Netherlands Energy Research Foundation (ECN), report prepared for NOVEM (EWAB 9840).

Faaij, A., van Ree, R., Waldheim, L., Olsson, E., Oudhuis, A., van Wijk, A., Daey Ouwens, C. and Turkenburg, W. (1997) Gasification of biomass wastes and residues for electricity production, *Biomass and Bioenergy*, **12**, 387–407.

Faaij, A., van Wijk, A. van Doorn, A., Curvers, A., Waldheim, L. Olsson, E. and Daey-Ouwens, C. (1997) Characteristics and availability of biomass waste and residues in the Netherlands for gasification, *Biomass and Bioenergy*, **12**, 225–240.

Hamelinck, C. N. and Faaij, A. (2002) Future prospects for production of methanol and hydrogen from biomass, *Journal of Power Sources*, **111**, 1–22.

Hamelinck, C., Faaij, A., den Uil, H. and Boerrigter, H. (2004) Production of FT transportation fuels from biomass; technical options, process analysis and optimisation and development potential, *Energy, the International Journal*, **29**, 1743–1771.

Hamelinck, C., van Hooijdonk, G. and Faaij, A. (2004) Future prospects for the production of ethanol from ligno-cellulosic biomass, *Biomass and Bioenergy*, in press.

Hamelinck, M., Voogt, M., Joosen, S., de Jager, D., Palmers, G., Shaw, S., Cremer, C. (2002) *PRETIR, Implementation of Renewable Energy in the European Union until 2010*, Report executed within the framework of the ALTENER programme of the European Commission, DG-TREN. ECOFYS BV, 3E, Fraunhofer-ISI, Utrecht, the Netherlands.

Hillring, B. (2002) Rural development and bioenergy – experiences from 20 years of development in Sweden, *Biomass and Bioenergy*, **23**, 443–451.

Hoogwijk, M., Faaij, A., van den Broek, R., Berndes, G., Gielen, D. and Turkenburg, W. (2003) Exploration of the ranges of the global potential of biomass for energy, *Biomass and Bioenergy*, **25**, 119–133.

International Energy Agency (1994) *Biofuels*, Energy and Environment Policy Analysis Series, OECD/IEA, Paris.

Kaltschmitt, M., Reinhardt, G. A. and Stelzer, T. (1996) *LCA of Biofuels under Different Environmental Aspects*, Institut für Energiewirtschaft und Rationelle Energieanwendung (IER) Universität Stuttgart, Stuttgart.

Kaltschmitt, M., Rosch, C. and Dinkelbach, L. (eds) (1998) *Biomass Gasification in Europe*, Institute of Energy Economics and the Rational Use of Energy (IER), University of Stuttgart. Report prepared for the European Commission, DG XII, EUR 18224.

Lynd, L. R. (1996) Overview and evaluation of fuel ethanol from lignocellulosic biomass: technology, economics, the environment and policy, *Annual Review, Energy Environment*, **21**, 403–465.

Marrison, C. I. and Larson, E. D. (1995) *Cost Versus Scale for Advanced Plantation-based Biomass Energy Systems in the US*, EOA Symposium on Greenhouse Gas Emissions and Mitigation Research, Washington, D.C.

Meuleman, B. and Faaij, A. (1999) *Overview of Co-combustion Options for Coal-Fired Power Plants*, report prepared within the framework of the JOULEIII COBIOCOWA project, Dept. of Science, Technology and Society, Utrecht University, Utrecht.

Morris, M., Waldheim, L., Faaij, A. and Stahl, K. (2005) Status of large-scale biomass gasification and prospects, in *Handbook Biomass Gasification*, Kurkela, *et al.* (eds), in press.

Naber, J. E., Goudriaan, F. and Louter, A. S. (1997) Further development and commercialisation of the small scale Hydro-Thermal Upgrading Process for Biomass liquefaction, in *Proceedings of the Third Biomass Conference of the Americas*, Montreal.

Nikolaisen, L. (ed.), Nielsen, C., Larsen, M. G., Nielsen, V., Zielke, U., Kristensen, J. K. and Holm-Christensen, B. (1998) *Straw for Energy Production*, Centre for Biomass Technology, Denmark. Available at: http://www.videncenter.dk

Ogden, J. M., Steinbugler, M. M. and Kreutz, T. G. (1999) A comparison of hydrogen, methanol and gasoline as fuels for fuel cell vehicles: implications for vehicle design and infrastructure development, *Journal of Power Sources*, **79**, 143–168.

Rodrigues, M., Faaij, A. and Walter, A. (2003) Techno-economic analysis of co-fired biomass integrated gasification/combined cycle systems with inclusion of economies of scale, *Energy, The International Journal*, **28**, 1229–1258.

Rosillo-Calle, F. and Cortez, L. A. B. (1998) Towards Pro-Alcool II – a review of the Brazilian bioethanol programme, *Biomass and Bioenergy*, **14**, 115–124.

Serup, H. (ed.) (1999) *Wood for Energy Production*, Centre for Biomass Technology, Denmark. Available at: http://www.videncenter.dk

Solantausta, Y., Bridgewater, T. and Beckman, D. (1996) *Electricity Production by Advanced Biomass Power Systems*, VTT Technical Research Centre of Finland, Espoo, Finland.

Stassen, H. E. (1995) *Small-Scale Biomass Gasification for Heat and Power Production: A Global Review*, World Bank Technical Paper No. 296, Energy Series, World Bank, Washington, DC.

Thuijl van, E., Roos, C. J. and Beurskens, L. W. M. (2003) *An Overview of Biofuel Technologies, Markets and Policies in Europe*, Energy Research Foundation of the Netherlands, Petter.

Tijmensen, M. J. A., Faaij, A. P. C., Hamelinck, C. N. and van Hardeveld, M. R. M. (2002) Exploration of the possibilities for production of Fischer-Tropsch liquids via biomass gasification, *Biomass and Bioenergy*, **23**, 129–152.

Turkenburg, W. C. (2000) *Renewable Energy Technologies*. Chapter 7 of the *World Energy Assessment of the United Nations, UNDP, UNDESA/WEC*, UNDP, New York.

Van den Broek, R., Faaij, A. and van Wijk, A. (1996) Biomass combustion power generation technologies, *Biomass and Bioenergy*, **11**, 271–281.

Van Loo, S. and Koppejan, J. (eds) (2002) *Handbook Biomass Combustion and Co-Firing*, Twente University Press, Enschede, the Netherlands.

Van Thuijl, E., Roos, C. J. and Beurskens, L. W. M. (2003) *An Overview of Biofuel Technologies, Markets and Policies in Europe*, Energy Research Foundation of the Netherlands, report ECN-C—03-008, Petten, the Netherlands.

Williams, R. H. and Larson, E. D. (1996) Biomass gasifier gas turbine power generating technology, *Biomass and Bioenergy*, **10**, 149–166.

Williams, R. H., Larson, E. D., Katofsky, R. E and Chen, J. (1995) *Methanol and Hydrogen from Biomass for Transportation, with Comparisons to Methanol and Hydrogen from Natural Gas and Coal*, Center for Energy and Environmental Studies, Princeton University, NJ. report no. 292.

Wyman, C. E., Bain, R. L., Hinman, N. D. and Stevens, D. J. (1993) Ethanol and methanol from cellulosic biomass, in *Renewable Energy, Source for Fuels and Electricity*, Island Press, Washington, DC.

14

Assessment of Biofuels

James A. Duffield, Hosein Shapouri and Michael Wang

14.1 Introduction

Countries throughout the world depend almost entirely on petroleum liquid fuels for public and private transportation (US Department of Energy, 2005a). Over time, as world oil demand increased rapidly, the dwindling supply of oil reserves has become more and more concentrated in a few oil-producing countries. Consequently, global economic growth is becoming increasingly dependent on oil from the Organization of the Petroleum Exporting Countries and a few other major oil producers. This growing dependency has encouraged many countries to adopt programs to increase alternative liquid-fuel production from domestic resources. The desire to replace petroleum fuels with a more reliable and sustainable energy source has created much interest in biofuels. In addition, replacing petroleum fuels with biofuels can help reduce air pollution and greenhouse gases. The aim of this chapter is to assess the potential supply and sustainability of biofuels.

14.2 Background

While biofuel technologies have been available since the advent of motorized vehicles, petroleum-based gasoline and diesel fuel have dominated the transportation fuel market. Interest in biofuels was initiated in the 1970s when the Arab oil embargo triggered world oil shortages. About this same time, environmental and health concerns began to surface over the overwhelming amount of air pollution being generated by auto transportation in urban areas. This created additional interest in biofuels because they often generate less

Renewables-Based Technology Edited by J. Dewulf and H. Van Langenhove

harmful emissions than petroleum fuels. More recently, energy security concerns have been heightened with increasing market volatility and record oil prices. Another major recent development is the threat of global climate change that many scientists attribute to an increase in greenhouse gases related to fossil energy use. The increasing concerns over energy dependency and environmental sustainability are responsible for a recent rise in biofuel use.

There are number of technologies that can be used to produce biofuels. Ethanol, which is the most popular bio-transportation fuel today, is generally made from the fermentation of sugar or starch into alcohol. Biodiesel is generally made from oilseeds or animal fats. The most common technology used to make biodiesel is called transesterification, which separates the glycerin from the oil or fat, resulting in a diesel fuel substitute. In addition, other biofuels, such as methanol, Fischer Tropsch (FT) diesel, dimethyl ether (DME), and biocrude could be made from biomass via gasification. The term biomass generally refers to energy feedstocks that are derived from any plant-derived organic matter and animal waste (US Department of Energy, 2004a). With the gasification process, biomass is gasified at high temperatures to produce a synthetic gas called syngas – a mixture of carbon monoxide (CO), hydrogen (H_2), methane (CH_4), and carbon dioxide (CO_2). The syngas then goes through a process that synthesizes the gas into a transportation fuel. In addition, biomass can be put through a pyrolysis process and converted to biocrude (US Department of Energy, 2004b). The biocrude can further be refined into diesel fuel, gasoline, and other petroleum substitutes. This chapter will concentrate mostly on ethanol and biodiesel, the only biofuels to move beyond the market development phase and achieve commercial success.

14.3 Biofuel Feedstocks

Various energy sources, commonly called feedstocks, are used to produce biofuels. With current technologies, starch and sugar have become the common feedstocks used for ethanol production, and oil crops and animal fats are presently being used to produce biodiesel. The choice of feedstock for a given fuel product varies around the world, mostly depending on a country's ability to produce a particular feedstock. For example, the United States, which is the world's largest corn producer, makes 98% of its ethanol from corn starch. Sugar-producing countries in South and Central America, and the Caribbean tend to make ethanol from sugar. For example, Brazil, the world's largest sugar producer, makes ethanol from sugar cane. India and Australia, also large sugar producers, use sugar cane for ethanol production. Ethanol from the European Union (EU) is primarily made from sugar beets, wheat, and barley.

Oil-bearing crops, such as soybeans, rapeseed, canola, mustard seed, palm oil, coconut oil, peanuts, and sunflower are used to produce biodiesel. Rapeseed, which is the dominant oil seed grown in Europe, is the primary feedstock used for EU biodiesel production. On the other hand, the majority of vegetable oil produced in the United States comes from soybeans, so soybean oil has become the dominant feedstock used in biodiesel production (Duffield *et al.*, 1998). Used cooking oil that is refined into yellow grease has also become a popular feedstock in the United States – approximately 2 million gallons of biodiesel were made from yellow grease in 2004. The primary supplier of yellow grease is the rendering industry that is positioned to become a leader in biodiesel production due to its access and control of large volumes of low-value feedstocks (Talley, 2004). Animal fats from meat packing and waste from poultry, turkey and food processing can also be used to produce biodiesel. Producing biodiesel with animal wastes, fats, and rendered products can reduce feedstock cost significantly compared to soybean oil and other oil crops. Recycled fats and oils are currently used in the production of animal feed supplements and fatty acids that are used in plastics, lubricants, paints, soaps, cosmetics, and many other industrial products. However, using some animal products in food and feed markets is now prohibited in many countries because of health concerns over bovine spongiform encephalopathy, commonly known as mad cow disease. Food processors and renderers are currently searching for new uses for these products and biodiesel has the potential of providing them with a significant market outlet.

In addition to the common feedstocks currently used, there is the potential of producing biofuels from low-valued biomass, such as trees, grass, forestry residues, wood waste, industrial waste, animal waste, and municipal solid waste (MSW). With advances in cellulosic conversion technologies, biomass may be used to produce cellulosic ethanol, methanol, FT diesel, DME, and hydrogen. In the past decade, there has been a considerable amount of research conducted on advancing the technology of fractionating organic materials into cellulose, hemi-cellulose, and lignin. Cellulose and hemi-cellulose can be further broken down into sugars for ethanol production (US Department of Energy, 2004c). The lignin portion of the biomass can be converted into chemicals or burned to produce steam and electricity for biomass ethanol plants. A small volume of cellulosic ethanol has been produced in laboratories and small pilot plants in the United States and Canada using grass and wood as feedstocks.

If cellulosic conversion technology were to become commercially viable, the size of the ethanol industry would grow significantly. The first feedstocks to enter this market would be agricultural and forestry residues and other low cost biomass materials. Crop residues include corn stover, straw, cotton trash, cotton gin, nut shells, residue from processing sugar cane to sugar (bagasse), and orchard trimmings. Forestry residues are primarily branches left over from tree harvesting in forests. Also, small diameter and dead trees that are removed from the forest to promote healthier trees and prevent forest fires can be used as a feedstock. Scrap wood leftover from lumber processing could provide a cheap feedstock for biofuel production.

Another inexpensive source of biomass could come from MSW. Currently most MSW is landfilled in the United States, however, the organic materials, including paper, cardboard, wood, grass, fiber, and waste food, account for 40% of the total weight. This organic material could be sorted out and used for biofuel production. Furthermore, as MSW decomposes, it produces landfill gas, primarily methane gas. There are a growing number of landfills throughout the United States that are using various methods to recover landfill gas for energy production (Energy Information Administration, 2004a). Landfill gas can be used to produce direct heating, chemical feedstocks, pipeline-quality gas, or to generate electricity. It can also be converted to liquid fuel such as methanol, or diesel fuel using the Fischer and Tropsch process. Manure could also be a source of methane gas – every year, millions of tons of manure are produced in livestock confinement operations, such as dairy farms, cattle feedlots, and hog operations. In addition, large amounts of chicken and poultry manure are produced annually. Recovering methane gas from MSW and manure is also good for the environment, since methane is a potent greenhouse gas that would otherwise be emitted into the atmosphere. Converting wastes into energy also can help reduce water pollution from animal confinement operations and reduce odor.

14.4 Bio-Transportation Fuels and Fuel Additives

To overcome the limitations of fuel delivery, storage, and distribution infrastructure, biofuels are generally blended with petroleum fuels or used as petroleum fuel additives. For example, most ethanol used in the United States is blended with gasoline at levels between 5 and 10% to increase the oxygen content and/or octane value of the fuel. Oxygenates, such as ethanol, are added to gasoline to meet Federal-reformulated gasoline requirements. Biodiesel blends are used in the EU; Germany, Austria, and Sweden have promoted a 100% biodiesel called B100; France encourages the use of B5, and B30 is commonly used in fleet applications; and the most popular biodiesel blend in Italy is B5. The most common biodiesel blend in the United States is B20, but lower blends, such as B5, are becoming more popular. Also, there is much interest in developing a biodiesel lubricity additive – adding 1 to 2% biodiesel to diesel fuel can increase lubricity (Schumacher, 2004). The demand for lubricity additives is expected to increase, because, beginning in 2006, US environmental regulations will require refiners to remove most of the sulfur from the diesel fuel they produce for transportation. The desulfurization process removes lubricity from diesel fuel, so petroleum companies will have to replace the loss in lubricity with an additive, which could spur a new market for biodiesel.

Biofuel additives or blends can easily be integrated into current petroleum distribution and marketing systems. However, use of higher levels of biofuels in vehicles may require minor engine modifications and special handling and storage systems are sometimes necessary to accommodate these fuels. For example, flexible fuel vehicles (FFVs) manufactured in Brazil are specifically engineered to run on gasoline and ethanol, or any combination of ethanol and gasoline. Hydrous ethanol and various anhydrous ethanol-blends are sold throughout the country and all gasoline sold in Brazil must contain 22 to 26% ethanol by volume. Since ethanol is currently cheaper than gasoline in Brazil, ethanol blends have become a very popular choice for consumers and the demand for FFVs is on the rise. In the United States, several auto manufacturers have designed FFVs that run on gasoline or any combination of gasoline and ethanol up to a blend of 85% ethanol and 15% gasoline (E85). There are about four million FFVs in the United States, however, unlike Brazil, ethanol blends are not widely available (National Ethanol Vehicle Coalition, 2005).

In recent years, hybrid electric vehicles have been introduced into the automobile market. About 150 000 hybrids were sold in the United States, as of the end of 2004. Relative to conventional internal combustion engine vehicles, hybrids can offer an improvement in vehicle fuel economy up to 100%. Current hybrid vehicle models in the US market are gasoline engine based. These hybrids, with minor modifications, can be powered with E85. Automakers are also developing hybrid vehicles powered with diesel engines, especially for the European market. Diesel engine-based hybrids could be run with biodiesel, FT diesel, and DME.

There is also much interest in developing fuel-cell vehicles (FCVs) that run on hydrogen. At present, the major US auto companies are conducting intensive research programs and have developed several FCV prototypes for demonstration. Similar efforts are on-going in Europe and Asia. Fuel-cell vehicles could be far more energy-efficient than internal combustion engine vehicles and FCVs have zero vehicle operation emissions. Hydrogen can be produced from fossil fuels or biofuels. The US Department of Energy (DOE) has developed a major research and development program to address hydrogen production and distribution issues (US Department of Energy, 2005b). One hydrogen production pathway that is being actively pursued by DOE is to produce hydrogen from ethanol at retail refueling stations to avoid the expense of building a hydrogen distribution system. It is much easier to transport liquid fuels, such as ethanol, than to transport hydrogen.

14.5 Current Supply of Biofuels

In 2004, the world produced 40.7 billion liters of ethanol. Brazil is the largest ethanol producer with about 37% of world production, followed closely by the United States that produced 32% (Figure 14.1). China is a distant third, producing almost 3.785 billion liters or about 9% of the world market. Following China is the EU and India, with 6 and 4% of the world market respectively. Available data show that about 20 other countries also produced ethanol in 2004 (F.O. Licht, 2004).

World biodiesel production has not reached the same level as ethanol, but biodiesel industries are beginning to emerge in a few countries, most notable in the EU, where tax incentives have stimulated new demand for biofuels. The EU countries produced 2.2 billion liters of biodiesel in 2004 (European Biodiesel Board, 2005). Germany, the

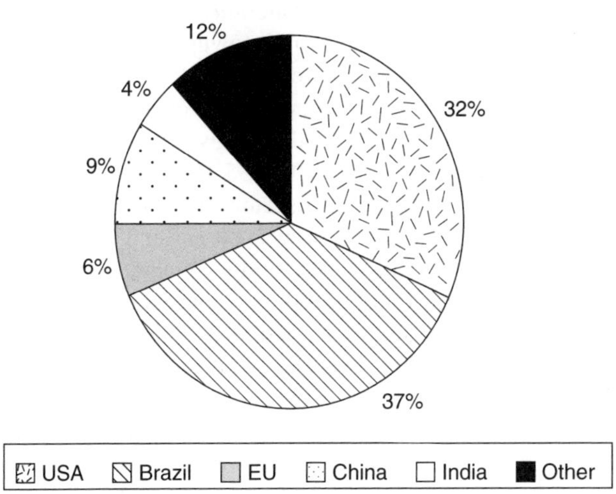

Figure 14.1 *Share of world fuel ethanol production, 2004*
Source: Licht (2004); Energy Information Administration (2005).

largest biodiesel producer in the world, produced about 1.15 billion liters in 2004 – about half of the EU's biodiesel production. The next largest producer is France with about 387 million liters, followed by Italy that produced about 356 million liters in 2004. Biodiesel production in the United States remained dormant for several years, but with recent government incentives, production increased from about 1.89 million liters in 1999 to over 250 million liters in 2005. Many other countries around the world are developing biodiesel industries and are expected to significantly increase their biodiesel production in the near future (Foreign Agriculture Service, 2004).

14.6 Future Supply of Biofuels

Although the world supply of biofuels is growing at a rapid pace, biofuels provide only a fraction of the world's transportation fuel needs. For example, the 12.87 billion liters of ethanol produced in the United States in 2004 was only about 2% of total annual gasoline consumption (Federal Highway Administration, 2004). US ethanol production required about 38.36 million metric tons of corn or about 12% of the 2004 corn crop (US Department of Agriculture, 2005). Farmers and agricultural policy-makers support biofuel incentive programs that increase the demand for commodities because agricultural markets have a tendency to generate surpluses and low market prices. The ethanol market has had a positive effect on corn prices, but so far ethanol production has not been large enough to have a major effect on corn supply. US corn demand is still dominated by animal feed and export markets. These markets, along with global market conditions, are the primary forces behind grain prices and grain production. However, as ethanol production continues to rise, the supply of corn may not be able to meet increasing demand. With higher corn prices, farmers would have the incentive to increase production by growing additional corn on idled land and substituting corn for other crops, i.e., replacing soybeans with corn. In addition, other crops could be used for ethanol production, but eventually land constraints and competition from other uses of crops would drive ethanol feedstock prices up, raising production costs and slowing ethanol growth.

It is difficult to predict how much grain-ethanol will be produced in the future. To address this question, it is best to take an economic approach, because the growth rates of agricultural products are generally determined more by economic factors than agronomic factors. An economic analysis conducted by the US Department of agriculture (USDA) estimated the economic effects on the farm sector from increasing ethanol production to 18.92 billion liters by 2012. This scenario was based on a legislative proposal called the Renewable Fuels Standard (RFS). The study concluded that increasing ethanol production, primarily from corn, to 18.92 billion liters would not put major stress on agricultural markets (US Department of Agriculture, 2000). The study concluded that farm income would increase by $2.9 billion and estimated that the price of corn would be $12.60 per metric ton higher in 2012 than in the absence of the RFS. Furthermore, results from the study showed that as more land is brought into corn production and less land is used for other crops, the prices of most other crops also rise. Thus, the production of ethanol from corn can benefit the entire agricultural sector.

Five billion gallons of ethanol in 2012 would require almost 16% of the US corn crop and provide just over 3% of US gasoline consumption (US Department of Agriculture,

2005). How much more ethanol can be produced from grain? The exceptional growth rate experienced in ethanol production will likely slow down over the next few years, as the grain-ethanol industry matures (Figure 14.2). Assuming annual growth rates will level off to around 5% ethanol production would approach 8 billion gallons in 2015, about 5% of the US gasoline consumption. This would require almost 25% of annual corn production (ibid). At this point, it seems that resource constraints and economic forces would limit further use of grain for ethanol production.

It's even more difficult to predict a growth rate for biodiesel in the United States, since the biodiesel industry is at the beginning stage of its development. In the United States, the total supply of oil crops is about 10.66 million metric tons annually (Economic Research Service, 2003). Other potential sources of biodiesel are animal fats and greases with an annual production of about 4.54 million metric tons (Economic Research Service, 2003; US Department of Commerce, 2004). Together these potential feedstocks could produce about 15 billion liters of biodiesel or about 10% of the diesel fuel used in US motor vehicles today (Energy Information Administration, 2004b). How much of these potential feedstocks will be used for biodiesel is unknown, but it seems unlikely that there would be a dramatic shift away from the traditional uses of these commodities. Currently, crop oils and animal fats are used primarily for food products, animal feed and for industrial purposes. As the demand for these commodities grow, biodiesel producers may find it too costly to take these commodities away from their traditional uses. Thus, without a major structural change in agricultural markets, the use of oil crops and animal fats for biodiesel use will be constrained by competing uses. Thus, it may be unrealistic to expect much more than 20% of the potential feedstock supply to be used for biodiesel. Looking at current annual production of oils and fats, 20% would provide enough feedstock for about 3 billion liters of biodiesel.

Countries in the EU have a goal of producing enough biofuels to provide 5.75% of their fuel consumption by 2010 (Foreign Agricultural Service, 2004). Currently in Brazil, about 25% of its gasoline pool is ethanol. Moreover, Brazil's agricultural sector has the capacity to expand sugar cane production significantly and ethanol production is

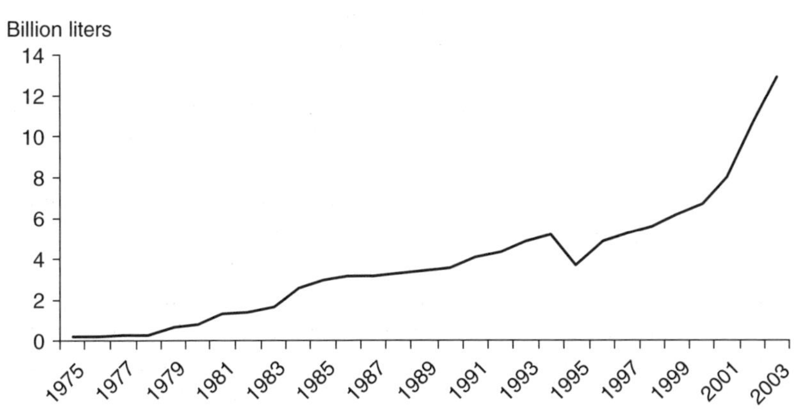

Figure 14.2 *US corn-ethanol production*
Source: National Corn Growers Assoc. (1992); Energy Information Administration (2005).

expected to increase almost 50% from current production to 22.71 billion liters by 2010 (F.O. Licht, 2004). Brazil has also recently launched a major biodiesel program that will increase domestic biodiesel production and reduce diesel fuel imports. Biodiesel blends will be phased in, starting with a 2% blend by 2008 (Renewable Fuel News, 2005).

14.6.1 Potential Production of Cellulosic Ethanol

Clearly, the production of biofuels made from traditional crops and animal fats is relatively small compared to transportation fuel demand. The desire to extend biofuel production to a much larger scale has created much interest in cellulosic ethanol. Cellulosic ethanol has not yet been developed on a commercial scale, however, with technological advances, it could play a significant role in the transportation fuel market in the future. Hopefully, research will lead to an economical process of converting large quantities of low-value plant materials and waste into fuel ethanol. As the industry expands, dedicated crops like switchgrass and other fast-growing perennial crops, such as hybrid poplar and willow could become economical feedstocks. In the United States, cellulosic feedstocks from agriculture, forestry, and municipal solid waste could potentially produce over one billion dry metric tons of biomass annually, resulting in 189 to 265 billion liters of cellulosic ethanol (Oakridge National Laboratory, 2005). However, without a major increase in government support, moving from the current pilot plant production to commercial operation could be years away.

14.7 Measuring the Sustainability of Biofuels

The supply of fossil fuels is finite and inevitably world oil supplies will be depleted. How long will it take before the world runs out of oil is subject to much debate and it may be years before world oil production cannot satisfy world demand. However, the majority of industrialized countries around the world have limited oil reserves and have already reached the point where domestic oil supply cannot keep up with domestic demand. These countries are looking internally for new sources of energy such as biofuels to reduce their dependence on oil imports. Biofuels can provide a sustainable source of energy because biomass feedstocks used for biofuel production are renewable. The energy in biomass comes from plant matter which derives energy from the sun through the process of photosynthesis.

Like all types of transportation fuels, it requires energy to produce biofuels. The amount of energy required to produce a biofuel, relative to its energy content, can be evaluated using net energy balance (NEB) methods. Net energy balance is defined as the energy content of a fuel, minus the energy content of the petroleum and other fossil energy sources used over the fuel's entire production cycle. All petroleum fuels have negative NEBs, because when converting one form of energy into a more useful form of energy, the transformation process uses energy from the initial state. As stated by the second law of thermodynamics, 'in all energy exchanges, if no energy enters or leaves the system, the potential energy of the state will always be less than that of the initial state'. For example, when converting crude oil into gasoline, the net energy ratio (NER), i.e., energy output of gasoline divided by energy input is less than 1 (Figure 14.3). The average NER of the total US electricity system is only about 0.45. Biofuels, on the other hand, can have a positive NEB because the solar energy captured by biomass is considered to be energetically free (Stout, 1990). When solar energy is not included in the NEB estimation, it is possible for a biomass fuel to have a NER greater than 1, i.e., converting biomass to energy results in a net energy gain (Figure 14.3). Net energy balance calculations for biofuels are very comprehensive, including the energy used for energy extraction, energy transportation, feedstock production, and feedstock conversion. The energy content of materials that are made from energy resources, such as fertilizers, pesticides, and other petrochemicals are also included in NEB estimations. If the conversion process results in multiple products, the total input energy should be allocated among the various co-products. For example, in a wet-milling plant, the starch is extracted from the corn kernels to produce ethanol. The remaining non-starch components are used to produce corn oil, corn gluten meal, and corn gluten feed. Only the energy specifically used to produce ethanol should be credited to ethanol, with the remaining energy credited to the other coproducts. Methods for allocating co-product credits are discussed in Chapter 5.

There have been many studies estimating the NEB of ethanol (International Energy Agency, 2004). While there is some disagreement over how much energy is required to produce ethanol, most recent studies conclude that the NEB of ethanol is positive (Shapouri *et al.*, 2003; Macedo *et al.*, 2004). As reported in Chapter 5, we found that depending on the estimation method used, the NEB of corn ethanol is positive and the NER ranges between 1.35 and 1.67. The NER is much higher when sugar cane is converted into ethanol, ranging between 8.3 and 10.2 (Macedo *et al.*, 2004). The fossil energy input used to convert sugar cane into ethanol is largely reduced because the ethanol plant uses bagasse to produce process steam. Bagasse is a cellulose fibre residue that is a by-product of sugar cane processing. The conversion of cellulosic feedstocks into ethanol also has a higher NEB than corn ethanol because the energy demands for cellulosic biomass are much less than for growing corn. The left-over lignin from converting the cellulose can be used to generate steam and electricity, reducing the

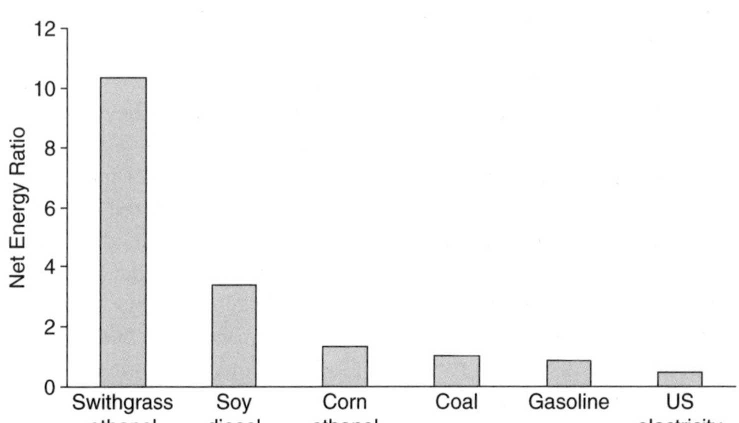

Figure 14.3 *Net energy ratios of renewable and fossil energy sources*
Source: Wang (2005).

need for fossil energy in a cellulosic ethanol plant. An analysis conducted by Argonne National Laboratory on ethanol made from switchgrass found an NER of 10.31 (Wang, 2005). Biodiesel has not been studied nearly as much as ethanol in the United States, however, a comprehensive analysis was conducted by USDA and DOE in 1998, which reported a 3.2 NER for biodiesel made from soybean oil (Sheehan *et al.*, 1998). There is also a number of studies from Europe and Canada that show the net energy savings and GHG benefits of using biodiesel made from rapeseed (International Energy Agency, 2004).

14.7.1 Environmental Benefits of Biofuels

Biofuels not only provide a sustainable source of energy, their use can also improve the environment. Using biofuels in conventional engines can lower some tailpipe emissions and are less toxic to handle relative to petroleum fuels. For example, about 7.5% by volume of ethanol is added to the gasoline in some US cities during the winter months to control carbon monoxide (CO), as mandated under the Clean Air Act of 1990 (CAA). Carbon monoxide is a poisonous gas produced by the incomplete combustion of gasoline in conventional engines and is harmful to human health. Using oxygenates, such as ethanol, has reduced ambient CO emissions in some US cities by as much as 10% (National Science and Technology Council, 1997). Ethanol is also used in reformulated gasoline (RFG) that is required in US cities with the worst air pollution to reduce harmful emissions of ozone, which can cause respiratory and other human health problems. RFG reduces cancer risk from gasoline by 12–19% (US Environmental Protection Agency, 2004). The CAA requires that RFG contain 2.3% by weight oxygen to meet the emission performance standards. The two most common oxygenates used in RFG are ethanol and methyl tertiary butyl ether (MTBE). Lately, ethanol has become the additive of choice for RFG, because MTBE in gasoline causes contamination of drinking water. Low levels of MTBE found in drinking water have made the water undrinkable because of its offensive taste and odour (Blue Ribbon Panel, 1999). Due to public concerns, many states have banned MTBE, and ethanol is rapidly becoming the dominate oxygenate used in RFG throughout the United States.

Adding ethanol to gasoline also reduces total hydrocarbon exhaust emissions and sulfur through dilution (National Science and Technology Council, 1997). However, blending ethanol-gasoline blends can increase evaporative emissions of volatile organic compounds (VOCs). VOC emissions are problematic because they contribute to the formation of ground-level ozone, which can contribute to serious respiratory problems. Evaporative emissions are particularly a problem during the summer months when ozone formation is the highest. Due to this problem, ethanol is often restricted during the summer months or refineries lower the volatility of RFG by adjusting the blending components.

Biodiesel has received much attention for its ability to reduce total particulate emissions (PM) from diesel engines, which are significant contributors to air pollution. Emissions of PM contribute to a wide variety of environmental and health problems, including heart and lung disease. A review of biodiesel emission studies conducted by the US Environmental Protection Agency (EPA) showed that biodiesel and biodiesel blends can significantly reduce PM, as well as emissions of CO and total hydrocarbons (THC) (US Environmental Protection Agency, 2002). Adding 20% biodiesel to diesel fuel reduces PM emissions of heavy-duty engines by 10%. CO is reduced by 11%, and THC drops by 21%, relative to using conventional diesel fuel. One disadvantage of using B20 is that NO_x emissions are about 2% higher compared to diesel fuel. Although biodiesel's elevated effect on NO_x is worrisome, diesel fuel additives are available to reduce NO_x to a tolerable level (McCormick *et al.*, 2002). Moreover, starting in 2006, EPA's low-sulfur diesel fuel rule will require new diesel engines to be equipped with after-treatment devices that control NO_x emissions.

14.7.2 Greenhouse Gas Reduction Effects

Perhaps the most important environmental benefit of using biofuels is their ability to neutralize or reduce greenhouse gases (GHGs). The observed global increase in carbon dioxide (CO_2) and other GHGs has caused great concern around the world. There is much uncertainty over the environmental, economic, and social consequences of anthropogenic GHG emissions, however, most countries recognize the seriousness of global climate change and are developing policies to reduce GHG emissions. One strategy for reducing GHG emissions, which some countries plan to use to help meet their Kyoto Protocol requirements is to replace fossil fuels with biofuels. The ability of biofuel to reduce GHGs is related to feedstock growth, i.e., the carbon in biofuels is derived from the air during feedstock growth via the photosynthesis process. The carbon from biofuel combustion is then recycled by carbon uptake during feedstock growth. On the other hand, the combustion of fossil fuels releases carbon that has been stored underground for millions of years. Therefore, replacing fossil fuels with biofuels can help stabilize GHG emissions.

Estimating GHG emissions for biofuels is a complex task that involves identifying and measuring fuel-cycle or 'well-to-wheels' GHG emissions, including emissions from energy extraction, feedstock production, feedstock conversion, transportation, and fuel consumption. See Chapter 5 for more details on the methodology of fuel-cycle analysis. According to a recent survey of studies on GHG emissions by the International Energy Agency (IEA), most researchers conclude that replacing petroleum fuels with biofuels significantly lower GHG emissions. These studies generally include emissions of CO_2, N_2O, and CH_4, which are aggregated and reported on a CO_2 equivalent basis. There have been many ethanol studies conducted over the past ten years, which generally agree that, compared to gasoline, ethanol derived from grains reduces GHG emissions by 20–40% (International Energy Agency, 2004). Studies have also shown that replacing diesel fuel with biodiesel reduces GHG emissions significantly. Most of the biodiesel studies surveyed by IEA were from Europe and Canada where rapeseed and canola are the primary feedstocks used for biodiesel production. The estimates for net GHG emissions reductions from biodiesel ranged between 40 and 60% compared to conventional diesel fuel, as reported by IEA. The 1998 USDA/DOE study mentioned above reported that biodiesel made from soybean oil reduced CO_2 by about 78% and CH_4 declined by about 2.5%. Results for N_2O were not reported (Sheehan *et al.*, 1998). Figure 14.4 shows results from this study indicating that biodiesel generates about 136 grams of CO_2 per-brake-horsepower-hour (g-CO_2bhp-h), compared to 633 g-CO_2/bhp-h for petroleum diesel. B20 generated about 534 g-CO_2/bhp-h over the entire fuel cycle, about a 16% reduction compared to petroleum diesel. The study concluded that fuel-cycle emission reductions for biodiesel varied linearly from about 78% for B100 to almost 4% for B5. The CO_2 reduction achieved by biodiesel is a direct result of carbon recycling from soybean plants.

The effects of biofuels on GHG emissions depend on many factors, including type of biofuel, type of feedstock, conversion technology, type of motor vehicle used for fuel consumption, and the petroleum fuel being replaced. This is illustrated by Figure 14.5 that shows results from research conducted by Argonne National Laboratory estimating GHG emission reductions of ethanol made from corn compared to cellulosic ethanol made from switchgrass. For further comparison, both ethanol fuels are blended with gasoline at the E10 (10% ethanol and 90% gasoline) and E85 (85% ethanol and 15% gasoline) levels. The E10 is used in a conventional gasoline vehicle (CGV), E85 is used in a flexible fuel vehicle (FFV), and both corn ethanol and switchgrass ethanol are used as the stock fuel to produce hydrogen for a fuel-cell vehicle (FCV). The base fuel used for comparison is reformulated gasoline (RFG) containing no oxygenates. All GHG emissions are measured on a CO_2 equivalent basis. Results are reported for a gallon of ethanol, not per gallon of ethanol blend. Figure 14.5 shows that replacing RFG with corn ethanol

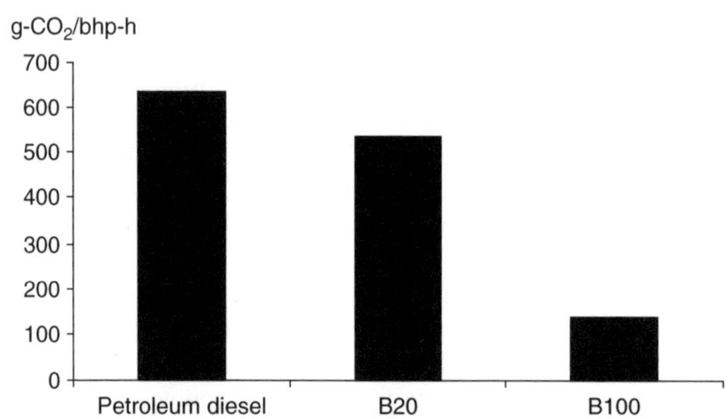

Figure 14.4 *Fuel-cycle CO_2 emissions for petroleum diesel, B20, and B100*
Source: Sheehan *et al.* (1998).

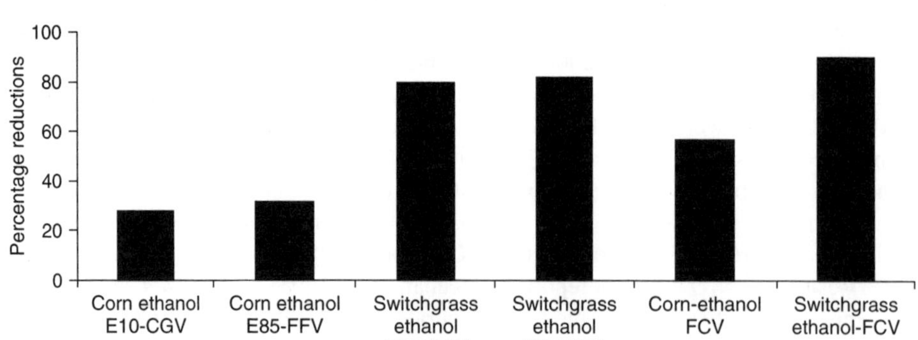

Figure 14.5 *GHG emission reductions for ethanol relative to RFG on a per gallon of ethanol basis*
Source: Argonne National Laboraotry, GREET model (2004).

reduces GHG emissions by about 30% when the fuel is used in a CGV or a FFV. In this case, vehicle type had only a small effect on the results, i.e., the FFV achieved only about 4% greater emission reduction compared to the CGV. However, when using a fuel-cell vehicle, the fuel-cycle GHG emission reduction of corn ethanol is 57% compared to conventional vehicles using RFG, almost two times greater than the emission reductions from a CGV or a FFV. Figure 14.5 also shows that type of feedstock and conversion technology can have a major effect on the results. Using ethanol made from switchgrass in a CGV or a FFV reduces GHG emissions by about 80% more than twice the reduction as in the case of corn ethanol. GHG emissions are lower for ethanol made from switchgrass primarily because cellulosic conversion technology can utilize the left-over lignin coming from the fractionation process for steam and electricity generation in ethanol plants. In addition, switchgrass growth requires less fertilizer and pesticides compared to growing corn.

14.8 Summary

The desire of many countries to increase their energy independence and reduce air pollution from motor vehicles has created a great deal of interest in developing bio-transportation fuels. While there are a number of biofuels available, only ethanol and biodiesel have gained commercial success. Brazil, the world's largest producer of ethanol, makes ethanol from sugar cane and the United States makes grain ethanol. Together, these two countries produced the majority of the world's 40.7 billion liters of ethanol in 2004. The EU is by far the world's largest biodiesel producer. However, a biodiesel industry is beginning to emerge in the United States and other parts of the world. Although, the world supply of biofuels is growing at a rapid pace, biofuels currently provide only a small fraction of our transportation fuel needs. In the future, some countries may be able to replace a significant amount of their oil imports with a sustainable supply of domestic biofuels made from traditional crops. However, countries with large energy demands, such as the United States, are looking to advance cellulosic conversion technologies in order to extend domestic biofuel production to a much larger scale. If economical conversion technologies for cellulosic materials were to become available, the United States could produce 189 to 265 billion liters of ethanol per year, on a renewable basis. Biofuels not only provide a sustainable source of energy, they also can improve environmental quality. Because biofuel feedstocks are able to utilize energy from the sun, they generally have greater energy balances than petroleum fuels. In addition, the carbon from biofuel combustion is recycled by carbon uptake during feedstock growth, therefore replacing fossil fuels with biofuels can help stabilize global GHG emissions.

References

Argonne National Laboratory (2004) *The Greenhouse Gases, Regulated Emissions, and Energy Use in Transportation (GREET) Model*, Transportation Technology R&D Center, US Department of Energy, Argonne, IL. (Accessed 2004.) Available at http://www.transportation.anl.gov/software/GREET/index.html

Blue Ribbon Panel (1999) *Achieving Clean Air and Clean Water: The Report of the Blue Ribbon Panel on Oxygenates in Gasoline*, EPA420-R-99-021, commissioned by the US Environmental

Protection Agency, September 1999. (Accessed 2005.) Available at: http://www.epa.gov/oms/consumer/fuels/oxypanel/blueribb.htm
Duffield, J., Shapouri, H., Graboski, M., McCormick, R. and Wilson, R. (1998) *Biodiesel Development: New Markets for Conventional and Genetically Modified Agricultural Products*, ERS Report No. 770, US Department of Agriculture, Economic Research Service, Washington, DC.
Economic Research Service (2003) *Oil Crops Situation and Outlook Yearbook*, OCS-2003, US Department of Agriculture, Economic Research Service, Washington, DC.
Energy Information Administration (2004a) *Landfill Gas*. (Accessed 2005.) Available at: http://www.eia.doe.gov/cneaf/solar.renewables/page/landfillgas/landfillgas.html
Energy Information Administration (2004b) *Fuel Oil and Kerosene Sales 2003*, DOE/EIA-0535(03). US Department of Energy, Energy Information Administration, Office of Oil and Gas, Washington, DC.
Energy Information Administration (2005) EIA-819 *Monthly Oxygenate Report: Petroleum Supply Monthly*, February, 2005, US Department of Energy, Energy Information Administration, Office of Oil and Gas. (Accessed 2005.) Available at: http://www.eia.doe.gov/oil_gas/petroleum/data_publications/monthly_oxygenate_telephone_report/motr.html
European Biodiesel Board (2005) *Statistics: The EU Biodiesel Industry*. (Accessed 2005.) Available at: http://www.ebb-eu.org/stats.php
Federal Highway Administration (2004) *Motor Fuel Use 2003: Highway Statistics, 2003*. Federal Highway Administration, US Department of Transportation. (Accessed 2005.) Available at: http://www.fhwa.dot.gov/policy/ohim/hs03/mf.htm
F.O. Licht (2004) *F.O. Licht's World Ethanol and Biofuels Report*. Volume 3, No 7, December 3, 2004. Available at: http://www.agra-net.com
Foreign Agriculture Service (2004) *More Oil Seed Production Going toward Biodiesel*, Production Estimates and Crop Assessment Division, Foreign Agricultural Service, US Department of Agriculture, July 2004. (Accessed 2004.) Available from http://www.fas.usda.gov/pecad2/highlights/2004/07/WorldBiodiesel/index.htm
International Energy Agency (2004) *Biofuels for Transport: An International Perspective*, The Organization of Economic Cooperation and Development, International Energy Agency, Paris.
Macedo, I., Leal, M. and Silva, J. (2004) *Assessment of Greenhouse Gas Emissions in the Production and Use of Fuel Ethanol in Brazil*, Report prepared for the Secretary of the Environment, Government of the State of São Paulo, April 2004.
McCormick, R., Alvarez, J., Graboski, M., Tyson, K. and Vertin, K, (2002) *Fuel Additive and Blending Approaches to Reducing NO_x Emissions from Biodiesel*, Society of Automotive Engineers Technical Paper No. 2002-01-1658.
National Corn Growers Association. (1992) *The World of Corn: A Comprehensive Look*, National Corn Growers Association and National Corn Development Foundation, Washington, DC.
National Ethanol Vehicle Coalition (2005) *For All the Right Reasons*. National Ethanol Vehicle Coalition, Jefferson City, Missouri. (Accessed 2005.) Available at: http://www.E85Fuel.com
National Science and Technology Council (1997) *Interagency Assessment of Oxygenated Fuels*. Committee on Environmental and Natural Resources, June 1997. (Accessed 2005.) Available at: http://www.epa.gov/otaq/regs/fuels/ostpexec.pdf
Oakridge National Laboratory (2005) *Billion Ton Feedstock Supply for a Bioenergy and Bioproducts Industry*, DOE/USDA Joint Feedstock Portfolio Review Through the Stage Gate Process, 14–15 March 2005. Oakridge National Laboratory, US Department of Energy, 2005. (Accessed 2005.) Available at: http://feedstockreview.ornl.gov
Renewable Fuel News (2005) Volume XVII, No 8, 28 February 2005. Hart Energy Publishing, Potomac, MD.
Shapouri, H., Duffield, J. A. and Wang, M. (2003) The energy balance corn ethanol revisited, *American Society of Agricultural Engineers*, 46 (4), 959–968.
Sheehan, J., Camobreco, V., Duffield, J., Graboski, M. and Shapouri, H. (1998) *Life Cycle Inventory of Biodiesel and Petroleum Diesel for Use in an Urban Bus*, NREL/SR-580-24089. US Department of Energy, Office of Fuels Development and US Department of Agriculture, Office of Energy, Washington, DC.
Schumacher, L. (2004) Biodiesel Lubricity, paper presented at Biodiesel utilization workshop, September 9–10, 2004, Boise, Idaho, University of Idaho.
Stout, B. A. (1990) *Handbook of Energy for World Agriculture*, Elsevier Applied Science, London and New York.

Talley, D. (2004) Biodiesel: a compelling business for the rendering industry, *Render, The National Magazine for Rendering*, February. *Render Magazine*, Camino, California.
US Department of Agriculture (2000) *Analysis of Ethanol Production Under a Renewable Fuels Requirement*, Staff Analysis, US Department of Agriculture, Office of the Chief Economist, Washington, DC.
US Department of Agriculture (2005) *USDA Agricultural Baseline Projection to 2014*, Baseline Report OCE-2005-1, February 2005, US Department of Agriculture, World Agricultural Outlook Board. (Accessed 2005.) Available at: http://www.usda.gov/agency/oce/waob/index.htm
US Department of Commerce (2004) *M311 Series for Fats and Oils, Production, Consumption and Stocks, Annual Summaries 2000–2002*, US Department of Commerce, Bureau of Census, Washington, DC.
US Department of Energy (2005a) Transportation Energy Basics: Energy Efficiency and Renewable Energy, US Department of Energy, 2005. (Accessed 2005.) Available at: http://www.eere.energy.gov/EE/trans_basics.html
US Department of Energy (2005b) *Multi-Year Research, Development and Demonstration Plan: Planned Program Activities for 2003–2010*, US Department of Energy, Energy Efficiency and Renewable Energy, Hydrogen, Fuel Cells, and Infrastructure Technologies Program, Washington, DC. (Accessed 2005.) Available at: http://www.eere.energy.gov/hydrogen and fuelcells/mypp
US Department of Energy (2004a) *Biomass Program*, Energy Efficiency and Renewable Energy, US Department of Energy. (Accessed 2004.) Available at: http://www.eere.energy.gov/biomass/biomass_today.html
US Department of Energy (2004b) *Pyrolysis and Other Thermal Processing*, Energy Efficiency and Renewable Energy, US Department of Energy. (Accessed 2005.) Available at: http://www.eere.energy.gov/biomass/pyrolysis.html
US Department of Energy (2004c) *Sugar Platform*, Energy Efficiency and Renewable Energy, US Department of Energy. (Accessed 2005.) Available at: http://www.eere.energy.gov/biomass/sugar_platform.html
US Environmental Protection Agency (2002) *A Comprehensive Analysis of Biodiesel Impacts on Exhaust Emissions*, EPA420-P-02-001, October 2002. Office of Transportation and Air Quality, US Environmental Protection Agency. (Assessed 2005.) Available at: http://www.epa.gov/otaq/models/biodsl.htm
US Environmental Protection Agency (2004) *Reformulated Gasoline: Basic Information*, US Environmental Protection Agency, Air and Radiation. (Accessed 2005.) Available at: http://www.epa.gov/otaq/rfg/information.htm
Wang, M. (2005) An Update of Energy and Greenhouse Emission Impacts of Fuel Ethanol, paper presented at the 10th Annual National Ethanol Conference, Scottsdale, AZ, 7–9 February.

15

Assessment of Organic Waste Treatment

Jan-Olov Sundqvist

15.1 Introduction

Waste management in modern countries is developing rapidly. Previously the normal method of treatment for organic degradable waste was landfilling. According to the EU Directive on landfilling of waste (1999/31/EC), the amounts of landfilled biodegradable municipal waste were to be successively decreased by 25% by 2006, by 50% by 2009, and by 75% by 2016. Several EU countries have decreed stronger rules. For example, in Sweden there is ban on landfilling of 'organic waste' from 2005. These requirements have focused on alternative methods to treat organic waste. Organic waste is mainly of renewable origin and has the potential to substitute other energy sources, non-renewable as well as renewable. Recovery of energy from organic waste is thus an interesting option from an energy perspective. At the same time most organic wastes also contain plant fertilizers, e.g. nitrogen and phosphorus, which make recovery of fertilizers also interesting.

This chapter describes the results of a system study where the recovery of energy and plant nutrients from waste was assessed with respect to the environment, energy and the economy. The full study was published in Swedish (Sundqvist *et al.*, 2002). Shorter reviews in English have also been presented (Sundqvist, 2001; 2002; 2004).

15.2 General Description of Options for Organic Waste Treatment

By organic waste we mean the organic biodegradable wastes of biological origin. Biodegradable waste is defined in the EU's landfill directive as follows: 'Biodegradable

Renewables-Based Technology Edited by J. Dewulf and H. Van Langenhove

waste means any waste that is capable of undergoing anaerobic or aerobic decomposition, such as food and garden waste, and paper and paperboard.'

There are four main ways to treat organic biodegradable waste:

- incineration;
- landfilling;
- anaerobic digestion;
- composting.

Anaerobic digestion and composting require the organic degradable waste to be sorted out before treatment, either in a central mechanical separation plant, or by pre-sorting at source. When incinerated or landfilled, the organic degradable waste can be mixed and co-handled with other municipal waste.

15.2.1 Incineration

Incineration or combustion is a thermal treatment with an excess of air. The organic matter is oxidized to mainly CO_2 and H_2O, which is discharged into the air recipient as flue gas after cleaning. Non-combustible, inorganic material is discharged slag and ash. The ash and slag are usually disposed of in a landfill. The energy released during the combustion can be used for the production of steam, district heating and electricity.

15.2.2 Landfilling (of mixed waste)

A landfill is a waste disposal site for the deposit of waste onto or into land. When organic degradable waste is deposited in the landfill, it will undergo different degradation processes. Methane-producing processes are of major importance. The methane phase start usually within a few years after the deposit and can continue for 25–100 years. Modern landfills have gas recovery systems, where the landfill gas is extracted. The gas usually contains more than 50% methane. The recovered gas can be used for production of steam, heat and electricity.

15.2.3 Anaerobic digestion

Anaerobic digestion is a biological degradation process without access to air. The digestion results in a methane-rich biogas (usually 55–70% vol. methane) and a stabilized organic residue often called digestate, containing nutrients (nitrogen, phosphorus and potassium). There are several different types of anaerobic digestion, e.g. wet, dry and wet/dry. The process can be thermophilic (with a temperature of around 55 °C) or mesophilic (around 37 °C). The process occurs in closed reactor vessels, where the residence time is about 20–30 days.

The biogas can be used for energy production: fuel for buses and cars, heat generation or electricity generation. The organic residue is an excellent fertilizer on agricultural fields because of the nutrients content. The stabilized organic material also has positive effects as soil conditioner.

15.2.4 Composting

Composting is an aerobic biological degradation process with access to air. The degradation results in an oxidation to mainly carbon dioxide and water, and a stabilized organic residue, called compost. Ammonia is released during the degradation. The temperature often reaches 60–70 °C within a few days and then declines slowly. The compost can be used as fertilizer and soil conditioner. Since the composting process gives rise to a release of ammonia, the compost will have a lower nitrogen content than the digestate from anaerobic digestion. The phosphorus content will be similar. Today, no energy is recovered from compost sites. However, theoretically low-grade heat can be recovered from the off-gases (50–70 °C) by heat pumps, but this process is of no importance today.

There is a large variety of composting technologies, from small simple compost containers in private gardens to advanced, highly technological centralized plants. Smaller composts are often open, e.g. as heaps or windrows. Larger compost plants are usually closed, with separate collection and cleaning of the off-gases.

15.3 Environmental Characteristics of Organic Waste Treatment

15.3.1 Incineration

Incineration results in discharges of emissions into the environment of the following:

- *Flue gases*: the flue gases are treated in advanced flue gas cleaning facilities. The discharged flue gases can contain minor amounts of hydrochloric acid, sulfur dioxide, heavy metals, polycyclic aromatic hydrocarbons (PAH), chlorinated organic compounds (for example, TCDD and other 'dioxins'), etc.
- *Wastewater*: the flue gas cleaning process usually give rise to contaminated water, which is treated before discharge. The major contaminants in the wastewater are heavy metals.
- *Flue gas cleaning residue*: from the flue gas cleaning and the cleaning of water different kind of flue gas cleaning residues can be generated. This residue is usually disposed in a landfill, where it can give rise to emissions of metals and some organic compounds.
- *Slag or bottom ash* can be disposed of in a landfill, giving rise to emissions of e.g. heavy metals. Nowadays, the slag is often used as construction material for roads and similar.

15.3.2 Landfilling (of Mixed Waste)

The major emissions from MSW landfills are:

- *emissions of landfill gas*: the major emittant is methane gas (which is a greenhouse gas), but different volatile constituents in the waste, as well as volatile degradation products may occur in the emitted gas.
- *emission of leachate water*, polluted by both organic compounds and metals. The leachate water is produced from excess precipitation water.

As mentioned above, the landfill gas can be recovered and used as a fuel. Unfortunately, this is a rather an inefficient energy recovery method. The losses to the environment are large and usually only about 50% of the produced gas can be recovered.

15.3.3 Anaerobic Digestion

The major emissions connected with anaerobic digestion are as follows:

- the discharge from the digestion reactor and after-storage of the digestate can give rise to emissions of methane;
- if the gas is processed to vehicle fuel (e.g. removing the carbon dioxide by wet scrubbing), there is usually a small portion of methane emitted to the ambient air;
- the spreading of the digestate on arable land gives rise to emissions of released ammonia. New spreading technologies have been developed, where the digestate during spreading is simultaneously covered by tilting the soil.
- the use of the digestate can also give rise to emissions of nitrogen and phosphorus. If the digestate is applied in excess, emissions of nitrogen and phosphorus can occur in larger amounts than from ordinary chemical fertilizers.

15.3.4 Composting

During the composting process ammonia is released. In a closed process (reactor composting) the ammonia can be collected. In open composts, e.g. windrow composting and home composts, the released ammonia will be emitted to the air. Ammonia contributes to both acidification and eutrophication.

Compost can be used as fertilizer, replacing chemical fertilizers. The use of compost, if applied in excess, can also give rise to eutrophicating emissions of e.g. phosphorous.

15.4 Results of a Life Cycle Assessment of Organic Waste

15.4.1 General

The ORWARE Model

The simulation model ORWARE has been used to assess waste management schemes for biodegradable waste and other wastes. The framework of the model has been developed during the past ten years in several research projects. Descriptions of ORWARE can be found in several reports and articles, e.g. (Björklund, 1998; Carlsson 1997; Dalemo *et al.*, 1997; 1998; Eriksson *et al.*, 2000; 2005; Soneson *et al.*, 1997, Sundqvist *et al.*, 1999; 2002).

ORWARE consists of a number of separate submodels, which may be combined into a virtual waste management system. Each submodel describes a real process in the waste management system, e.g. waste collection, waste transport, different waste treatment processes. All submodels calculate the turnover of materials (including the emissions), energy and financial resources in the process. Materials turnover is characterized by: (1) the supply of waste materials and process chemicals; (2) the output of products and secondary wastes; and (3) emissions to air, water and land. Energy turnover is the consumption of energy raw materials (primary energy carriers). The financial turnover is defined as life cycle costs (LCC) of individual processes.

The submodels are combined to describe the waste management system in e.g. a city or municipality. A conceptual model of the waste management system in ORWARE is shown in Figure 15.1. At the top in the conceptual model there are different waste sources, followed by different transport and treatment processes. The dashed line defines the boundaries of the waste management core system, where wastes are treated and different products are formed.

Another important topic is the 'from-cradle-to-grave' perspective. Both upstream processes ('cradle') and downstream processes ('grave') are considered. Examples of upstream and downstream processes are:

- If electricity is consumed in a process, the environmental impact and resources consumption from the production are included as an upstream process. If the electricity is produced from coal condense power, also the landfilling of coal ash is included as a downstream process.
- If oil is consumed in a process, the environmental impact and resource consumption related to the production and distribution of the oil are included.

The analysis carried out in ORWARE generates data on emissions from the system. The individual emissions are aggregated into different environmental impact categories, according to LCA practice as defined by the international standard ISO 14042 (ISO, 1999). The following impact categories are assessed: global warming, acidification, eutrophication and photo-oxidant formation. Also ecotoxicity and human toxicity can be

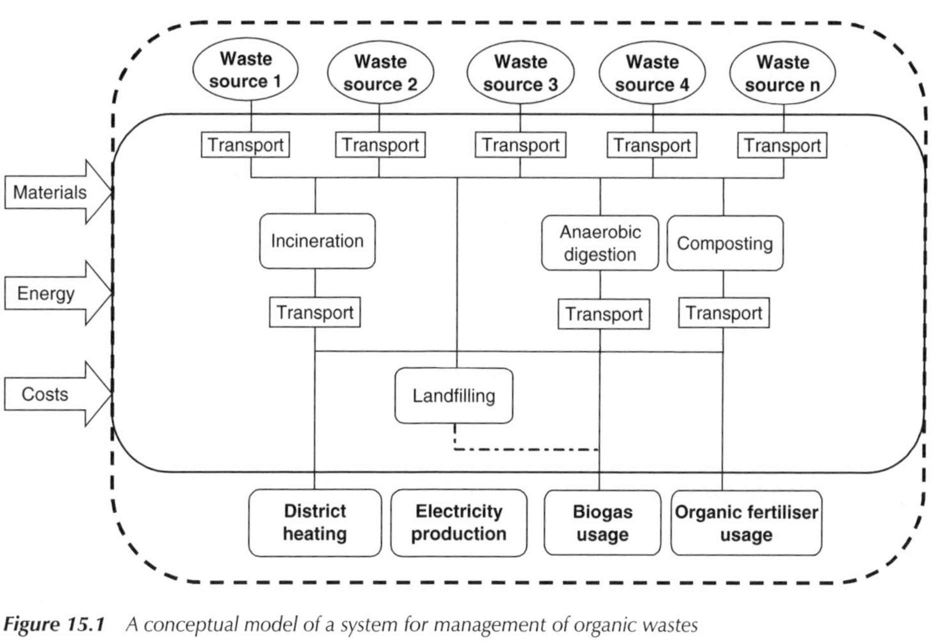

Figure 15.1 A conceptual model of a system for management of organic wastes

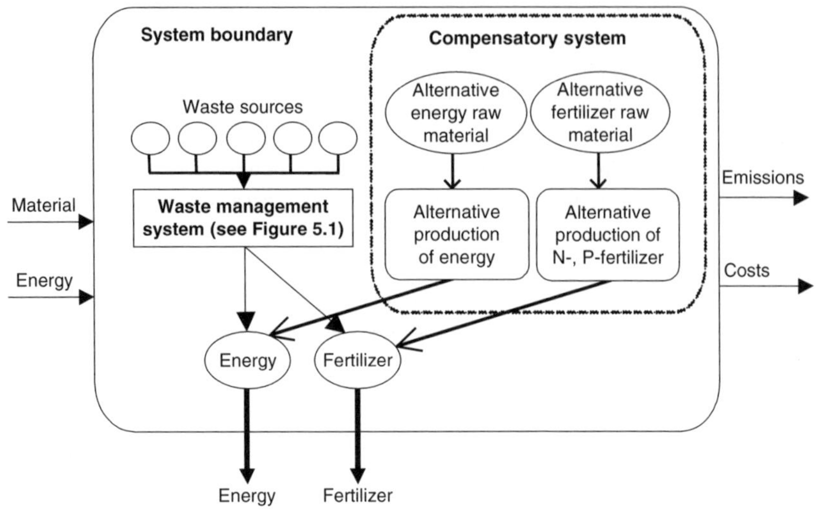

Figure 15.2 *The studied system consists of the waste management system (according to Figure 15.1) and the compensatory system where alternative energy and fertilizer are produced*
Source: Sundqvist *et al*. (1999, 2002).

studied, but the models for weighting different toxic emissions are not fully developed so the results have to be considered carefully.

The system studied works with several functional units. The major function of the waste management system is to treat the waste from a specified region. The waste management system can also produce different products (functions) from the waste:

- energy: district heating, electricity and biogas (the biogas can be used as vehicle fuel or as an energy source for electricity and district heating);
- fertilizers: a digestion residue or compost containing e.g. nitrogen (N) and phosphorus (P).

When a product is produced from waste, it replaces a product from a virgin source. Each of the waste products has an alternative virgin raw material source with a production process that has been included in the studied system. In the study, all studied systems (all scenarios) produce the same amount of energy (district heating, electricity, and vehicle fuel) and fertilizers (N- and P-fertilizers), either from waste or from an alternative virgin source, see Figure 15.2. This alternative system is called the compensatory system. The use of the compensatory system enables a quantitative comparison of environmental, economic and energy parameters between the use of waste as raw material and the use of virgin raw materials. The compensatory systems also have upstream and downstream processes.

Overview of Some Sub-Models in ORWARE

- *Collection and transport*. The transport submodel was described in Dalemo *et al.* (1997). The transport sub-model includes three different kind of transports: 'garbage truck', 'ordinary truck', and 'truck and trailer'. Energy consumption is calculated from the transport distances and the kind of truck. The emissions are calculated from the fuel consumption by emission factors.

- *Incineration*. The original incineration model was described in Dalemo *et al.* (1997) and Björklund (1998). Since then the model has been updated (Sundqvist *et al.*, 1999; Sundqvist *et al.*, 2002). The modelled incineration plant is a modern plant that fulfils the requirements of the EU waste incineration directive 2000/76/EC by a good margin. The plant has an advanced flue gas cleaning system, including a flue gas condensation step where energy is recovered by heat pumps. The incineration plant produces district heat. The flue gas cleaning residue, ash and slag are disposed in a landfill.
- *Landfill*. A conventional municipal solid waste landfill has been modelled. The landfill model was described in detail in earlier reports (Björklund, 1998; Fliedner, 1999). The time aspects in our landfill model for organic materials are based on the two time horizons:

  - the surveyable time period, which is the time until some kind of pseudo-steady state is obtained. For a conventional municipal solid waste landfill, this corresponds to the end of the methane production phase, which is estimated to be of the magnitude of one century.
  - the hypothetical, infinite time period, which is the period until the landfilled material is completely released into the environment.

  In this chapter, only the emissions during the surveyable time period are presented. During the surveyable time close to 100% of sugar, starch, hemicellulose, fats, and proteins are degraded and about 70% of the cellulose. Plastics are degraded to 3% and humus and lignin are assumed to be nondegradable. We have assumed that 50% of the theoretically produced landfill gas is recovered (during the surveyable time period) (Sundqvist, 1999; Björklund, 1998) and can be used as fuel in a gas engine producing electricity and heat. The rest of the landfill gas will migrate to the ambient air. During the passage through the landfill cover about 15% of the methane gas is oxidized into carbon dioxide.
  The leachate water is treated in a local treatment plant, reducing COD-, N- and P-emissions to water. Flue gas cleaning residue, ash and slag from incineration are also landfilled. The emissions from the ash and slag landfill are calculated with the same time horizons as the organic waste. In the ash landfill, no degradation occurs. The major emissions are from leaching of metals.
- *Anaerobic digestion*. The original anaerobic digestion model was described in earlier reports (Dalemo, 1996; Dalemo *et al.*, 1997). Since then the model has been adapted to a thermophilic process (Sundqvist *et al.*, 1999).

  The incoming domestic food wastes are packed in plastic bags. In a bag separator the bags are cut and emptied and separated from the degradable material. The separated bags are sent to the incineration plant. Slaughterhouse wastes are hygienized at 70 or 130 °C before digestion. The water content is adjusted to about 85% and the waste is fed into the digestion reactor. The hydraulic residence time in the reactor is 20 days.

  The study addresses two ways to use the biogas:

  - The gas is upgraded to 97% CH_4 and compressed to 250 bars. The gas is then used as fuel for buses.
  - The gas is directly combusted in a gas engine, where both electricity and heat (for district heating) are produced.

The digestion residue or digestate is spread on agricultural land, substituting chemical P- and N-fertilizer. During the spreading ammonia can be released. A new, improved spreading technology was modelled where the digestate is worked down in the soil simultaneously with the spreading.

- *Composting*. In the study, open windrow composting is modelled. About 50–75% of the organic material is degraded into carbon dioxide, water and compost. The degradation rate is different for different organic compounds. Ammonia is released into the air in the process. In a separate study also closed composting or reactor composting was studied. The exhaust gases are treated (e.g. by cooling) to decrease the ammonia release to environment and increase the amount of nitrogen fertilizer that can be recovered. The compost is spread on agricultural land, thus substituting chemical nitrogen and phosphorous fertilizers.

Scenarios

Six different scenarios are studied:

- Incineration: all waste is treated by incineration (no source separation of bio-waste).
- Landfilling: all waste is disposed at a landfill (no source separation of bio-waste).
- Anaerobic digestion – bus fuel: source-separated biodegradable waste are treated by anaerobic digestion. The biogas is used as fuel for buses. The rest of the waste is incinerated.
- Anaerobic digestion – electricity and heat: source-separated biodegradable waste is treated by anaerobic digestion. The biogas is used for production of district heat and electricity. The rest of the waste is incinerated.
- Open composting: source-separated biodegradable waste is composted in open windrows. The rest of the waste is incinerated.
- Reactor composting: source-separated biodegradable waste is composted in a closed reactor where the off-gases are cleaned. The rest of the waste is incinerated.

Important Assumptions

The study is based on a hypothetical municipality. The chosen hypothetical municipality has 186 000 inhabitants and consisted of a city area, a 'suburban' area, and a rural area. The municipality has a waste incineration facility and a district heating system. All transport distances and similar demographic data is taken from a real Swedish city, and is varied in a sensitivity analysis.

The waste studied is domestic waste and similar waste from business and industry. The total amount of waste is approximately 69 000 tonnes/year. About 18 000 tonnes/year of organic waste is sorted out in the anaerobic digestion and composting scenarios. In all scenarios newsprint paper (75%), glass packages (70%) and metal packages (50%) are sorted out and recycled 'outside' the studied system (not included in the 69 000 tons of waste). The remaining paper, glass and metal are present in the waste studied.

Important assumptions are the choices of upstream and compensatory energy sources. In this study, the electricity is assumed to be produced by combined heat and power plants (CHP). CHP is judged to be the so-called base load marginal technology, in which new investments are made. A change in electricity generation or consumption in the studied system (including the compensatory system) is supposed to affect the base load technology.

The design of the district heating systems is special for each municipality. The fuel that competes with waste can be peat, wood chips, oil or coal, depending on the heat demand, when the demand is raised, the prices for different fuels, existing combustion facilities, etc. In this chapter it is assumed that fossil oil is the primary compensatory heat source. This assumption was made to illustrate the potential of organic waste as a renewable energy source. However, in the original study (Sundqvist *et al.*, 2002), we used bio-energy as primary compensatory heat, since bio-energy is the most important energy source in district heating systems in Sweden today. In this chapter the bio-energy case is presented in the sensitivity analysis.

The studied parameters are: global warming, acidification, eutrophication, photo-oxidant formation, environmental costs, consumption of primary energy carriers, financial life cycle costs, and welfare costs (which are the sum of environmental costs and financial life cycle costs, exclusive environmental taxes and fees).

Sensitivity Analyses

Several important choices is made in the goal definition and scoping stage, e.g. when choosing system boundaries and when using process data. The importance of different 'municipality-specific' and 'site-specific' parameters is studied in a sensitivity analysis. The results is thoroughly analysed to identify parameters that can have an influence on the results.

15.4.2 Results

The most important results are presented in Figures 15.3–15.5. In the figures the waste system is displayed as one data category (the lower black part in the bars). This category includes all processes in the waste managements system as collection, transports, and the type of treatment used in the specific scenario including downstream processes as landfilling of residues and spreading of compost and digestate. The upper parts of the bars represent different processes in the compensatory system: district heat generation, electricity generation, production of bus fuel, and production of chemical fertilizers. In the figures the results has been normalized to environmental impact per capita by dividing the calculated emissions for the model municipality with the number of inhabitants.

Environmental Impact

Figure 15.3 presents the results for global warming (emissions of greenhouse gases). Greenhouse gases are fossil carbon dioxide (CO_2) from combustion of fossil fuels and plastic, methane gas (CH_4) from landfilling of organic waste and laughing gas (N_2O). These substances has different global warming potential: 1 kg of methane is equal to

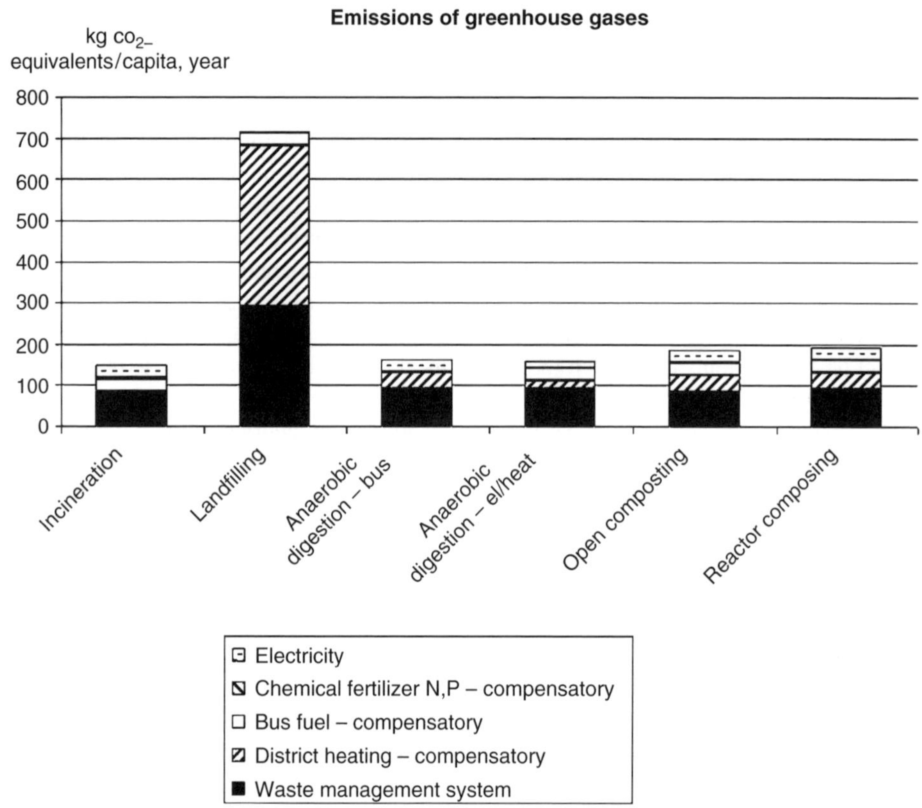

Figure 15.3 *Emissions of greenhouse gases from the studied system (waste system and compensatory system)*

21 kg of fossil carbon dioxide, and 1 kg of laughing gas is equal to 310 kg of fossil carbon dioxide. The landfilling scenario gives the worst impact due to methane emissions from the landfill. Incineration gives the lowest impact, but only slightly lower than anaerobic digestion.

Acidifying substances are mainly gases such as sulphur dioxide SO_2 (from e.g. fossil fuels), hydrochloric acid HCl (from waste incineration), nitrous oxides NO_X (from all combustion processes: incineration, district heating production, engines, etc.), and ammonia NH_3 from composting and spreading of compost and digestate. These substances have different acidification potential: 1 kg NO_X is equal to 0.7 kg SO_2, 1 kg of NH_3 is equal to 1.88 kg SO_2, and 1 kg of HCl is equal to 0.88 kg of SO_2. The ranking between the scenarios is (from lowest impact to highest):

1–2 Incineration, and anaerobic digestion – bus fuel
3 Reactor composting
4 Anaerobic digestion with generation of electricity and heat
5 Windrow composting
6 Landfilling.

Landfilling gives the highest emissions of acidifying substances, due to NO_X emissions from the landfill gas combustion and from district heat production in the compensatory system. Windrow composting gives a high emission due to ammonia releases from the compost process. Anaerobic digestion with production of heat and electricity gives high NO_x emissions from the combustion engine.

Eutrophicating substances are nitrogen- and phosphorus-compounds (N- and P-compounds) in water, COD (chemical oxygen demand) in water, NO_X in combustion gases, and ammonia (NH_3) releases from spreading of anaerobic digestion residue and compost. Different substances have different eutrophicating potential: 1 kg of NO_X is equal to 6 kg of COD, 1 kg of NH_3 is equal to 16 kg COD, 1 kg of NO_3^- is equal to 4.4 kg COD and 1 kg of phosphorus is equal to 140 kg COD. The emissions of eutrophicating substances have a similar pattern as acidification with the same ranking between the scenarios.

Photo-oxidant formers are divided into VOC (volatile organic compounds) and NO_X. Methane is included in the VOC but with a relatively low factor: 1 kg of methane is equal to 0.006 kg ethene, CO is equal to 0.03 kg ethene and NMVOC (non-methane volatile organic compounds, used as a summary parameter) is equal to 0.416 kg of ethene. The ranking between the scenarios considering VOC emissions is (from lowest to highest impact):

1 Incineration
2 Open composting
3 Reactor composting
4 Anaerobic digestion – electricity and heat
5 Anaerobic digestion – bus fuel
6 Landfilling.

Landfilling gives the highest emissions due to the methane emissions. Methane emissions also put anaerobic digestion on a lower ranking than composting.

The ranking between the scenarios considering the NO_X emissions is:

1 Anaerobic digestion – bus fuel
2 Incineration
3–4 Windrow composting, and reactor composting
5 Anaerobic digestion – electricity and heat
6 Landfilling.

Different environmental impact categories can be weighted together by different methods. One common approach is to express all impacts in monetary terms, which gives the environmental costs. However, there are several methods and philosophies regarding how to place values on different emissions. We have used three methods:

- ORWARE, which was developed in a previous ORWARE project (Carlsson, 1997);
- EPS 2000 (Steen, 1999);
- ECOTAX 99 (Johansson, 1999).

The economic weighting factors are shown in Table 15.1. The major difference between the three methods is how consumption of non-renewable resources is valued. There are also differences between the value on specific emissions.

Table 15.1 *Economic weighting factors used in the study*

	Economic weighing ORWARE €/kg	*Economic weighing EPS 2000 $/kg*	*Economic weighing EcoTax '99 €/kg*
CO_2 (fossil) (to air)	0.044	0.10	0.044
Particles and dust (to air)		34	3.5
N-NO_X (to air)	5.9	2.0	3.8
N-N_2O (to air)	14	36	9 700
S-SO_2 (to air)	3.7	3.1	5.9
CH_4 (to air)	0.92	2.5	0.37
N-NH_3 (to air)			5.1
HCl (to air)	7.5		
N-NH_4 (to water)	5.2	−0.40	6.0
N-NO_3 (to water)			1.70
COD (to water)	0.33		0.42
P (to water)	48.2	0.05	
VOC (to air)	0.16	2.0	13
CO (to air)	0.012	0.31	0.07
Pb (to air)	34 000	2 700	860 000
Pb (to water)	34 000		11 000
Pb (to land)	34 000		410
Cd (to air)	123 000	10	410 000
Cd (to water)	123 000		68 000
Cd (to land)	123 000	5.1	3 300
Hg (to air)	25 000	57	430 000
Hg (to water)	25 000		2.20
Hg (to land)	25 000	181	143
Cu (to air)			430 000
Cu (to water)			2,20
Cu (to land)			143
Cr (to air)		19	66 000
Cr (to water)			63
Ni (to air)			61 000
Ni (to water)			370
Ni (to land)			370
Zn (to air)		40	13 000
Zn (to water)			0.74
Zn (to land)			340
Consumption of biomass		0.04	
Consumption of crude oil		0.47	1.29
Consumption of coal		0.05	0.02
Consumption of natural gas		1.0	0.97

All three methods give the same ranking between the scenarios (from lowest to highest costs):

1 Incineration
2–3 Anaerobic digestion (only minor differences between the two scenarios)
4–5 Composting (only minor differences between the two scenarios)
6 Landfilling.

However, the total costs differ between the methods. For example, the environmental cost for the incineration scenario is 30 €/capita, year with the ORWARE method, 53 €/capita, year with EPS 2000 and 55 €/capita, ton in ECOTAX.

Energy – Consumption of Primary Energy Carriers

In Figure 15.4 the energy consumption for the whole system (including the compensatory system) is shown. There is a net consumption of energy for the whole system. Incineration gives the lowest energy consumption. The two anaerobic digestion scenarios have slightly higher energy consumption than incineration. The landfill scenario has the highest energy consumption, because of the production of district heating, fuels and fertilizers in the compensatory system.

Figure 15.4 also illustrates the possibility of using organic waste as a renewable energy source. If the landfilling scenario represents a kind of state-of-art scenario with low recovery of energy, we can see that the different energy recovery methods (incineration and anaerobic digestion) mainly replaces non-renewable energy. Another important result from the study, not shown in Figure 15.4, is that the energy consumption for collection and transport of waste is small compared to the energy consumption of the other processes in the studied system.

Financial Life Cycle Costs

The incineration scenario gives the lowest financial life cycle costs. In Figure 15.5 the life cycle costs correspond to the two lowest parts of the bars (life cycle costs for the waste management system and the compensatory system). Incineration has the lowest life cycle costs. Biological treatment (anaerobic digestion and composting) has higher costs than

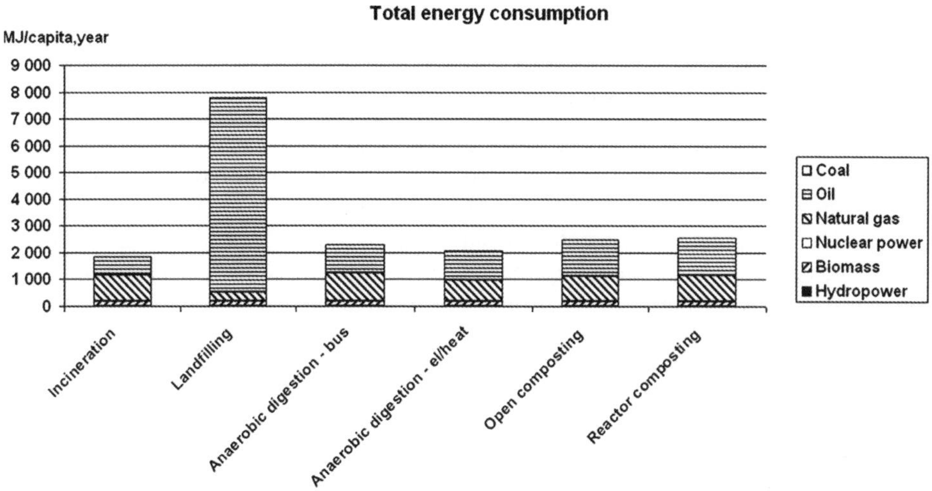

Figure 15.4 *Consumption of primary energy carriers in the studied system (waste system plus compensatory system)*

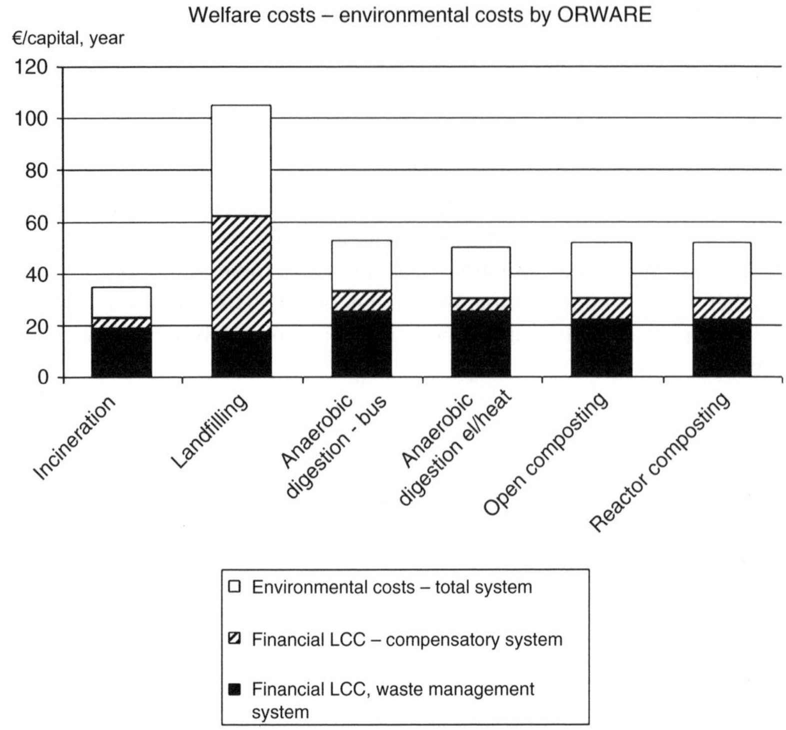

Figure 15.5 Welfare costs for total system, with environmental costs according to ORWARE environmental cost model

incineration. Landfilling is the most expensive waste treatment, mainly depending on the costs for producing new district heating and vehicle fuel in the compensatory system.

Welfare Economy (Financial Plus Environmental Costs)

Welfare costs are calculated as the financial life cycle costs (excluding environmental taxes and fees) plus the environmental costs (with three different methods, see above). The welfare cost with environmental cost calculation according to ORWARE is shown in Figure 15.5. All three environmental valuation methods show that biological treatment (anaerobic digestion and composting) has higher welfare costs than incineration. Landfilling is definitely the most expensive option in all three methods.

Sensitivity Analysis

The results from the sensitivity are shown in Table 15.2. As shown in the table, the result is relatively robust. This indicates that the qualitative result (the ranking between different treatment options) should be valid for a large variety of municipalities.

Table 15.2 *Summary of sensitivity analysis*

Changes	*Changes compared to base scenario*
Production of electricity	
Coal condense	Anaerobic digestion with production of electricity and heat gives slightly lower total energy consumption than incineration. Ranking is not changed.
Swedish average electricity*	No change in ranking
District heating production	
Biomass	The two anaerobic digestion alternatives give lower emissions of greenhouse gases than incineration. Other impact categories show the same ranking as base assumptions.
Thermal efficiency	No change in ranking
NO_X-emissions (high NO_X-emission for waste incineration, low NO_X-emission for biofuel or oil)	No change in ranking
NO_X-emissions (low NO_X-emission for waste incineration, high NO_X-emission for alternative biofuel or oil)	Incineration gives lower NO_X-emission, lower emissions of acidifying substances, lower emissions of eutrophicating substances than anaerobic digestion.
Power production	
Combined power and heat production	No change in ranking
Transport distance	
500 km distance to incineration	No change in ranking for environmental impacts categories and energy consumption. However, one of the three environmental cost valuation methods anaerobic digestion gives lower welfare costs than incineration.
Spreading of anaerobic digestion residue and compost	
Distance to arable land 50 km	No change in ranking
The compost and the digestion residue can not be spread but have to be incinerated	The welfare costs for composting becomes lower than for anaerobic digestion, but still higher than for incineration.
Valuation of resources	
Doubled price for energy and doubled environmental costs for greenhouse gases	No change in ranking, but all financial life cycle costs, and welfare costs increases.
Energy price and valuation of greenhouse gases increases by a factor = 5	No change in ranking, but life cycle costs and welfare costs for all scenarios increases.
The phosphorus price increases by a factor = 10	No change in ranking. The phosphorus price has to increase by a factor 100 to make the welfare costs for anaerobic digestion and composting equal to the costs for incineration.

Note: *The Swedish average electricity production is based on mainly hydropower and nuclear power, which gives very small environmental impact with the impact categories that have been studied.

15.5 Discussion

The study is based on typical Swedish conditions. The most 'country-specific' parameter is probably the performance of the incineration plant. The incineration plant has a flue gas condensation system, which gives a very efficient flue gas cleaning, but also high energy recovery rates. The net energy recovery is about 100% (relative to the lower heating value) and the energy is used for district heat, which can be sold during almost the whole year. The high energy recovery rate results in relatively favourable results for incineration. In several other countries the opportunities for selling or using energy are more limited.

The modelled municipality has a city area, a suburban area and a rural area. The majority of the waste is collected from the city and suburban region. The structures of the city and suburban area are estimated as no different from other European or North American municipalities and cities. The rural part in the study was typical Swedish, but has a relatively low impact on the total result.

15.6 Summary

This chapter describes the results from a system study where recovery of energy and plant nutrients from waste was assessed with respect to environment, energy and economy. The most obvious conclusion from the study is that landfilling should be avoided. Wastes that can be incinerated (combusted), anaerobically digested or composted should not be landfilled. From the system perspective there are small differences between incineration and aerobic digestion of easy degradable organic waste. Both incineration and anaerobic digestion are preferable to composting and landfilling. An important conclusion of this study is that energy recovery from organic waste has several advantages, especially if the energy from waste replaces non-renewable energy sources, e.g. oil in district heating (both incineration and aerobic digestion) or diesel oil as bus fuel (biogas from anaerobic digestion).

Anaerobic digestion and incineration should not be seen as competing options, but as complementary options. Since it is impossible to obtain 100% biological treatment, there will always be some organic waste that has to be incinerated.

Transport of waste, when the waste has been collected, is of very low importance considering the consumption of energy resources, environmental impact and costs, if the transport is performed in an efficient manner. The type of collection system has very small influence on the total consumption of energy resources, the total environmental impact and the total welfare costs.

The results show that organic waste has the potential as a renewable energy source with respect to environment, energy and economy.

References

Björklund, A. (1998) Environmental systems analysis of waste management with emphasis on substance flows and environmental impact, PhD thesis, the Royal Institute of Technology, Department of Chemical Engineering and Technology, Sweden.

Carlsson, M. (1997) *Economics in ORWARE: A Welfare Analysis of Organic Waste Management*, Report No. 117, Department of Economy, Swedish University of Agricultural Sciences.

Dalemo, M. (1996) *The Modelling of an Anaerobic Digestion Plant and Sewage Plant in the ORWARE Simulation Model*, Report No. 213, Department of Agricultural Engineering, Swedish University of Agricultural Sciences.

Dalemo, M., Sonesson, U., Björklund, A., Mingarinai, K., Frostell, B., Jönsson, H., Nybrant, T., Sundqvist, J-O. and Thyselius, L. (1997) ORWARE – A Simulation Model for Organic Waste Handling Systems, Part 1: Model Description, *Resources, Conservation and Recycling*, **21**, 17–37.

Dalemo, M., Sonesson, U., Jönsson, H., and Björklund, A. (1998) Effects of including nitrogen emissions from soil in environmental systems analysis of waste management strategies, *Resources, Conservation and Recycling*, **24**, 363–381.

Eriksson, O., Frostell, B., Björklund, A., Assefa, G., Sundqvist, J.-O., Granath, J., Carlsson, M., Baky, A. and Thyselius, L. (2000) ORWARE: A simulation tool for waste management, *Resources, Conservation and Recycling*, **36**, 287–307.

Eriksson, O., Reich, M., Frostell, B., Björklund, A., Assefa, G., Sundqvist, J. -O., Granath, J., Baky, A. and Thyselius, L. (2005) Municipal solid waste management from a systems perspective, *Journal of Cleaner Production*, **13**, 241–252.

Fliedner, A. (1999) Organic waste treatment in biocells: a computer-based modelling approach in the context of environmental systems analysis, Master's thesis, Department of Industrial Ecology, Royal Institute of Technology Stockholm, Sweden.

ISO (1999) *Life Cycle Impact Assessment*, ISO 14042, ISO, Geneva.

Johansson, J. (1999) A monetary valuation weighting method for life cycle assessment based on environmental taxes and fees, Master's thesis, Department of Systems Ecology, Stockholm University, Stockholm.

Soneson, U., Dalemo, M., Mingarinai, K. and Jönsson, H. (1997) ORWARE – a simulation model for organic waste handling systems, Part 2: Case study and simulation results, *Resources, Conservation and Recycling*, **21**, 39–54.

Steen, B. (1999) *A Systematic Approach to Environmental Priority Strategies in Product Development (EPS)*, Version 2000, CPM report 1999:4, Chalmers University of Technology, Technical Environmental Planning, Göteburg.

Sundqvist, J.-O. (1999) *Life Cycle Assessments of Solid Waste*, AFR Report 279 (Swedish Environmental Protection Board 1999, Available at: *www.naturvardsverket.se*).

Sundqvist, J.-O. (2001) Material and nutrient recycling and energy recovery from solid waste: a systems perspective, in Lens, P. *et al.*, *Water Recycling and Resource Recovery in Industries: Analysis, Technologies and Implementation*. IWA Publishing, London.

Sundqvist, J.-O. (2002) LCA for treatment and disposal of urban waste, in *Proceedings from International Conference on Life Cycle Analysis, Efficiency and Cost-Benefit*, 7 November 2002, Madrid.

Sundqvist, J.-O. (2004) System Analysis of organic waste management schemes – experiences of the ORWARE model, in Lens, P. *et al.*, *Resource Recovery and Reuse in Organic Solid Waste Management*, IWA Publishing, London.

Sundqvist, J.-O., Baky, A., Björklund, A., Carlsson, M., Eriksson, O., Frostell, B., Granath, J. and Thyselius, L. (1999) Systemanalys av energiutnyttjande från avfall – utvärdering av energi, miljö och ekonomi. Översiktsrapport (*System analysis of energy from waste – assessment of energy, ecology and economy. Summary Report*) IVL, Stockholm (In Swedish, with English summary.) Available at: www.ivl.se/reports

Sundqvist, J.-O., Baky, A., Carlsson, M., Eriksson, O., and Granath, J. (2002) Hur skall hushållsavfallet tas omhand – Utvärdering av olika behandlingsmetoder (*How should municipal municipal solid waste be disposed of – assessment of different methods of treatment*) IVL, Stockholm. (In Swedish, with English summary.) Available at: *www.ivl.se/reports*

16

Oleochemical and Petrochemical Surfactants

An Overall Assessment

Erwan Saouter, Gert Van Hoof, Mark Stalmans and Alan Brunskill

16.1 Introduction

Cleaning products such as laundry detergents, dishwashing liquids, shampoos, etc., are all dependent on surfactants for their cleaning ability. Surfactant is a shorthand word for a surface-active agent which removes the dirt and grease from fabrics, dishes, skin, etc., and holds them in solution/suspension in water. Ideally, manufacturers have access to a broad range of surfactants, which gives them the flexibility to achieve optimum cleaning performance under a broad range of conditions. For specific purposes, additional active ingredients such as builders (water softeners), enzymes and bleaches (stain removers) can be added according to the end-use needs of the product formulation.

All surfactants currently available can be separated into two source groups: those from feedstocks based on fats and oils from plant and animal sources, often called 'natural'; and those that are derived from petrochemicals based on crude oil, natural gas or coal, often called 'synthetic'. In the past decade, there has been a lot of debate about the pros and cons of these two types of sourcing. 'Natural', more correctly termed 'Oleochemical' based surfactants, are often assumed to be better for the environment because of their origin, and people assume that they should, therefore, be selected first. However, their raw material sources sometimes are questioned on environmental (deforestation) grounds. Conversely, petrochemicals are questioned because of the non-renewable nature of their feedstock and get unfairly labelled because of a general antipathy to the chemical industry in general. All these are subjective assessments but there are many facts that can guide us to a sound conclusion on surfactant usage.

Renewables-Based Technology Edited by J. Dewulf and H. Van Langenhove

The use of the terms 'natural' and 'synthetic' to describe the origin of surfactants has led to much confusion. Technically, these are not accurate descriptions for these materials. Petrochemical and oleochemical based surfactants both come from *natural* sources, but of different ages. Crude oil/gas/coal derived from vegetation that existed millions of years ago, became trapped in rocks, modified into their present forms by intense heat and pressure. Fats and oils, however, are extracted from living trees/fruits/plant seeds or from the carcasses of recently slaughtered animals. Both types of surfactants could also be termed 'synthetic' in that both oleochemical and petrochemical feedstocks require further substantial chemical processing before they become the surfactants we can use.

Historically, selection of chemicals for formulations was done simply on the basis of availability, efficiency, cost, and safety to the consumer. Environmental safety took its place in selection criteria only in the second half of the past century, while sustainability has only seriously entered the selection criteria in the past few years. For this reason, the concept of sustainability is relatively little understood by the public and tends to be thought of as being, in some way, 'how long something will last'. In our company (Procter and Gamble), managing 'sustainability' is considered as managing the 'economic', 'environmental' and 'social' objectives in a holistic manner. Some people refer to this as the 'triple bottom line' or 'people, planet and profit'.

Applying this thinking to the world of surfactants, we then have to consider, not only performance, price, etc., but the places, people and companies involved in the entire supply chain (Stalmans *et al.*, 2003). For example a palm kernel oil-based surfactant used in the USA or in Europe will also have involved countries and people from South-East Asia who cultivate the palm trees and the overall welfare of the rural people who work in the plantations, while a petrochemical-based surfactant also needs to be concerned with the environmental aspects of more exploitation of reserves in environmentally sensitive areas and the safety of complex chemical factories. Similarly, the fact that petroleum is a finite and non-renewable resource has to be balanced against increasing planted areas for agriculture at the expense of reduced rainforest area and the resulting reduced biodiversity.

In the rest of this chapter we will not attempt to cover all areas coming under the overall sustainability complex but will concentrate on those which are most pertinent to the natural/synthetic debate and which have become areas of public debate in recent years. These are:

- renewable versus finite resources;
- environmental profile;
- palm oil and deforestation.

We will attempt to show how the whole supply chain of the surfactant industry is working to minimize the long-term depletion of resources. This will include the area of formulation where trends to lower wash temperatures – to save energy, lower chemical usage, to save resources and lower water usage, are presenting new challenges to the manufacturers.

In terms of the overall economic and social aspects of the sustainability debate, we will simply note that the production and use of surfactants provide benefits to the consumer (cleaning/hygiene), to the producers (employment/return to shareholders), and to rural economies in developing countries (employment from growing oil and coconut palms), etc.

First, however, we need to clarify some of the terminology of our materials.

16.2 Main Chemical and Structural Differences

All surfactants have the same basic structure: a hydrophilic/lipophobic (water-loving/fat-hating) 'head' and a hydrophobic/lipophilic (water-hating/fat-loving) 'tail', comprising a long hydrocarbon chain. The tail binds to and mobilizes soil particles, and the head, which bonds to water, works to pull the soil-surfactant couplet to the water phase, to be removed with the wastewater.

The polar head groups tend to be common to both, i.e. the same headgroup can be attached to both oleochemical and petrochemical hydrocarbon chains. There are three main classes of polar groups – anionic, nonionic and cationic. Anionics include Alcohol Sulphates (SO_4^{-2}), Sulphonates (SO_3^{-2}), Carboxylates (CO_3^{-}), normally neutralized with sodium hydroxide. Nonionics are normally Ethoxylates, i.e., a number of (-O-CH_2-CH_2-) groups added to the hydroxyl group on a fatty alcohol. Cationics are normally Quaternary ammonium compounds where a nitrogen atom is attached to four hydrocarbon units, e.g. Distearyl, Dimethyl Ammonium Chloride. Most cationics are used as fabric conditioners rather than as cleaning agents.

As the subject matter of this chapter is primarily about the oleo–petro differences we will not delve further into the chemistry of the polar part of the surfactant molecule and concentrate on the organic portion of surfactants. More information on surfactants and their roles in detergent and cleaning products can be found in the recent overview by Hauthal and Wagner (2004).

The 'tail' is made of a long hydrocarbon chain. In oleochemical surfactants, also referred to as 'natural', it is mainly derived from natural oils and fats from plant oils such as tallow (from cattle), palm oil, palm kernel oil or coconut oils. Soap, an anionic surfactant (carboxylate), was the original surfactant and was produced by boiling this material with sodium hydroxide. Tallow and coconut oil were the normal feedstocks. Soap, however, has a major disadvantage in that its effectiveness is hampered by water hardness and its usage has been largely restricted to personal washing or hand laundry in recent times. For most cleaning purposes other anionics or nonionics are now preferred because of their lower sensitivity to water hardness and in recent years tallow has decreased in usage as a surfactant raw material due to lower laundry washing temperatures. In contrast, fruit or tree-based oils like palm kernel or coconut oils (also wrongly referred to as vegetable oils) have been gaining ground because of their shorter hydrocarbon chains and increased availability. It is pertinent to mention here that other plant/vegetable oils such as soya bean oil or rapeseed/canola oil are largely unsuitable as feedstocks for surfactant production due to the composition of their longer and more unsaturated hydrocarbon chains. Petrochemical surfactants are derived from crude oil, natural gas or coal and are also known as 'synthetic' surfactants.

There are, however, some minor chemical differences between the two surfactants. Hydrocarbon chains from oleochemical feedstocks are always linear and even-numbered, while synthetic feedstocks may have slightly branched hydrocarbon chains and contain both even and odd numbers of carbon atoms. These differences may seem subtle, but they can have a significant impact on cleaning performance, especially in mixed surfactant systems. Some petrochemical based surfactants have a more complex hydrocarbon, e.g. in alkyl benzene sulphonate the hydrocarbon group is comprised of an alkyl chain attached to a benzene ring.

16.3 Resource and Usage

Part of the natural/synthetic debate revolves around the availability of viable raw materials on a long-term basis. In the case of surfactants, fats and oils come mainly from perennial trees with fruiting lifetimes of 25 years or more and which can be replanted. In contrast, there is a continuous debate about how long the world's reserves of oil and gas will last. These points lead to the simplistic conclusion that natural is best for the long term but a look at the overall material balance shows the simplistic conclusion to be misleading. To add perspective, we need to look at relative usage and raw material availability.

16.3.1 Relative Usage

The surfactant industry currently uses both 'natural' oleochemicals and 'synthetic' petrochemicals (Figure 16.1). Overall, if soap is included, there is a rough balance between 'natural' and 'synthetic' sourcing. Of the major surfactants, soap is completely based on fats and oils while alkyl benzene sulphonates are completely based on petrochemicals. Surfactants based on fatty alcohols can be either of oleochemical or petrochemical origin and so are an instructive comparison for the relative merits of the two sourcing groups.

Presently, about 50% of alcohol-based surfactants are derived for petrochemicals and 50% from oleochemicals. The most important categories are anionic (alcohol sulphates and ethoxy- sulphates) and nonionic surfactants (ethoxylates), the hydrocarbon chains used are linear or only slightly branched to ensure biodegradability. Highly branched chains adversely affect biodegradation and so have been eliminated from cleaning product formulation for many years in most parts of the world.

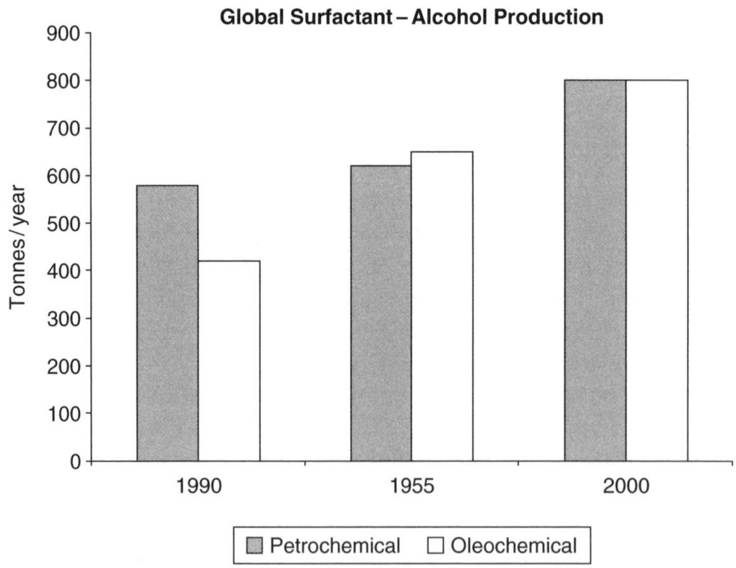

Figure 16.1 *World wide production of alcohols used in surfactant production in 2000 in tonnes/yr*
Source: (CESIO, 2001).

16.3.2 Petrochemicals

Approximately 90% of the annual usage of almost 4 billion tonnes of crude oil used globally is for the energy and transportation sector. About 8% of petroleum is used by the chemical industry (CESIO, 2001; Stalmans *et al.*, 2003). A fraction of this (1.5% of the total petrochemical use) is used for the production of surfactants. This corresponds to about 0.1% of the worldwide crude oil consumption.

In addition, the petrochemical industry is often based on the by-products of the oil refining industry and is therefore not the primary cause of reserve depletion. Also as we will see, the use of petrochemical-based surfactants enables manufacturers to produce detergents that work at lower temperatures, thus offsetting some of the resources involved in their production and use.

16.3.3 Oleochemicals

Looking at oleochemical raw material from the same perspective, we note that the technical, or non-edible uses of fats and oils only account for 10–15% of an annual global production of ca. 130 million tones (author's estimation based on CESIO 2001). Oils and fats are produced primarily for food usage and the majority of them, e.g. soya bean oil, rapeseed/canola oil, sunflower oil etc., are unsuitable for surfactant production due to the length and degree of unsaturation of their hydrocarbon chains.

The oils which are used for most oleochemical-based surfactants are palm kernel oil and coconut oil. Although used for edible purposes, these are not mainstream food oils and, in the case of palm kernel oil, it is a by-product of palm oil production. Coconut oil and palm kernel oil are often referred to as 'lauric oils' because of their high levels of C-12 and C-14 chain lengths, in contrast to other oils and fats which are predominantly C-18 and C-16. This C-12/14 content makes these oils particularly suitable for surfactant production. Tallow, which is widely used as a soap and softener raw material, is a by-product of the beef and dairy industries.

Palm kernel oil and coconut oil production totals 6–6.5 million tonnes per year and only about 50% of this goes into surfactants, i.e., about 2.5% of overall oils and fats production. As in petrochemicals, the raw material need is driven by another and bigger user, no one is going to increase the planted area for palm because of surfactant needs. Conversely, more palm kernel oil will become available as the rapidly increasing world demand for edible oils will push for increases in palm oil production, palm is the most efficient crop in terms of oil yield per hectare, see below. World demand for oils and fats has been increasing by ca. 4 million tonnes per year over the past 10 years. To meet this demand, palm oil production has increased by 14 million tonnes (100%) and this has made ca. 1.5 million tonnes of palm kernel oil available for oleochemicals. In contrast, coconut oil production was almost static over the same period.

However, if oleochemical-based surfactants were to completely replace those from petrochemicals sources, then either all non-surfactant uses of coconut oil and palm kernel oil would have to cease or the area planted would need to be increased by at least 50% (author's calculation). Given that these trees are confined to tropical regions because of their need for continuous high rainfall and temperature, such large increases in planted areas could lead to further encroachment onto virgin forest. This will be looked at later under the issue of palm oil sustainability.

On the plus side for oleochemicals, the growth of the feedstock is a carbon sink, i.e., the growth of organic matter is a major consumer of atmospheric carbon dioxide.

To sum up the raw material position, while neither type can dominate for reasons mentioned, there is a place for both without undue stress on raw material availability as long as a balance is maintained.

16.4 Environmental Profile

A considerable number of environmental assessment have been done on oleo- and petro-chemical based surfactants (Zoller, 2004). In this chapter we will first review the overall environmental impact of natural versus synthetic surfactants using the life cycle assessment (LCA) approach. We will then consider two specific aspects: biodegradability and aquatic toxicity.

The desire to make better, fully informed decisions about different options (i.e. which one has the lowest environmental impact?) has resulted in the emergence of a relatively young scientific discipline called environmental life cycle assessment. LCA is a tool used to evaluate the potential environmental impact of a product, process or activity throughout its entire life cycle by quantifying the use of resources ('inputs' such as energy, raw materials, water) and environmental emissions ('outputs' to air, water and soil) associated with the system that is being evaluated. LCA seeks to encourage the efficient use of energy and material resources so that emissions and wastes will be reduced. Since the publication of the Society of Environmental Toxicology and Chemistry code-of-conduct (SETAC, 1993), its standardization has taken place in ISO with the 14040 series (ISO 14040, 2000).

One of the first steps before starting an LCA is to define the 'functional unit', which is related to the function that a product or service will deliver. When conducting an LCA on laundry detergent or chemicals used in detergent, we often report the results on the basis of 1000 wash cycles. The definition of a functional unit is actually very much linked to the question asked. There is not one functional unit, but many, depending on the type of questions we want to answer, but it is important that two products or two services are compared on the same basis. Energy and raw materials consumption as well as associated environmental emissions are calculated on the basis of this functional unit.

Procter & Gamble (P&G) participated in a life cycle inventory study group composed of more than 13 surfactant producers and detergent formulators (Stalmans *et al.*, 1995). The group's goal was to quantitatively assess resource requirements and environmental releases associated with the production of surfactants sourced from oleochemical versus petrochemical feedstocks. This life cycle inventory showed that each type of surfactants has an impact on the environment via the consumption of broad range of resources as well environmental releases during the production and transportation of these chemicals. No environmental superiority could be claimed for one category of surfactant versus the other. Taking into account energy consumption and emissions, all have some positive and some negative but more importantly, the study revealed opportunity for improvement for all (ibid.).

For illustration purposes, we report here a few selected inputs/outputs that relate to well-known environmental themes: energy consumption relates to resource depletion; CO_2 emission relates to greenhouse effect, and biological oxygen demand (BOD) and chemical oxygen demand (COD) relate to surface water pollution. In a complete LCA study, a life cycle inventory (as conducted by Stalmans *et al.*, 1995) is usually followed by a life cycle assessment that converts all inputs and outputs into environmental themes or categories relative to resource use, human health and ecological areas.

The results, shown in Figure 16.2 for an oleochemically and a petrochemically derived alcohol ethoxylate (AE7), make it clear that neither surfactant can be supported as environmentally superior. Rather, there are trade-offs: lower environmental resource requirements were offset by higher emissions. The total energy consumption to produce 1000 kg for those surfactants (i.e. the functional unit for this study) which includes the energy contained in the material produced (EMR: energy of material resource; Jansen, 1995) is only ~ 10% higher for the petrochemical based surfactant (Alcohol Ethoxylate – AE7 Petro) than for its oleo equivalent (AE7 PKO – palm kernel oil – or AE7 CNO – coconut oil). The petro chemical based surfactant also shows a higher CO_2 emission (direct link to higher energy requirement). However, if we take into consideration few waterborne emissions, oleo-based surfactants show higher emissions, while total solid waste are similar. Of course, to conclude on overall environmental superiority of one system versus the other, all emissions must be analyzed as was done by the author of the study (Stalmans *et al.*, 1995), but the main facts presented here remain valid.

Since then there have been several improvements to the environmental performance of the plantation business. At the time of the LCI many palm plantations burned the waste from the trees (palm fronds, processed fruit bunches, etc.). Now, with a ‘zero burn’ policy, these materials are returned to the land as mulch, returning organic matter to the soil and bacterial sludge from wastewater processing/clean-up is returned to the soil as fertilizer.

One major plantation group in Malaysia, United Plantations Berhad, have published data that show an overall level of 93% utilization of oil palm biomass residues and waste (Table 16.1). It was calculated that the materials recycled on their plantations were equivalent to a fertilizer value (or cost reduction) of US$3.1 million. In addition, much research is going into higher value outlets for biomass waste by bodies such as the Malaysia Palm Oil Board.

In recent years, a number of studies have been conducted to assess the potential CO_2 saving via the total replacement of petrochemical surfactants by oleo-based surfactants (Patel *et al.*, 1998; ECCP, 2001). Assuming a full replacement of all petrochemical surfactants by oleochemical surfactants in laundry detergents, for Germany only, resulted in 0.8% saving of the total industrial CO_2 emissions (Patel *et al.*, 1998). Extrapolating this approach across all European detergents, this results in a possible reduction of fossil CO_2 emissions of around 1.5 to 2 million tonnes of CO_2 (author’s own estimation). Compared to the total fossil CO_2 emitted in the atmosphere in Europe, which was 3148 million tones in 2001 (European Environment Agency, 2001), the maximal potential saving is therefore 0.07%, which is a very low value as compared to the savings potential from other measures that are presently under consideration (e.g. in the transport or energy sector). The European Climate Change Program also concluded that saving from replacement of petrochemical surfactants is rather limited (ECCP, 2001).

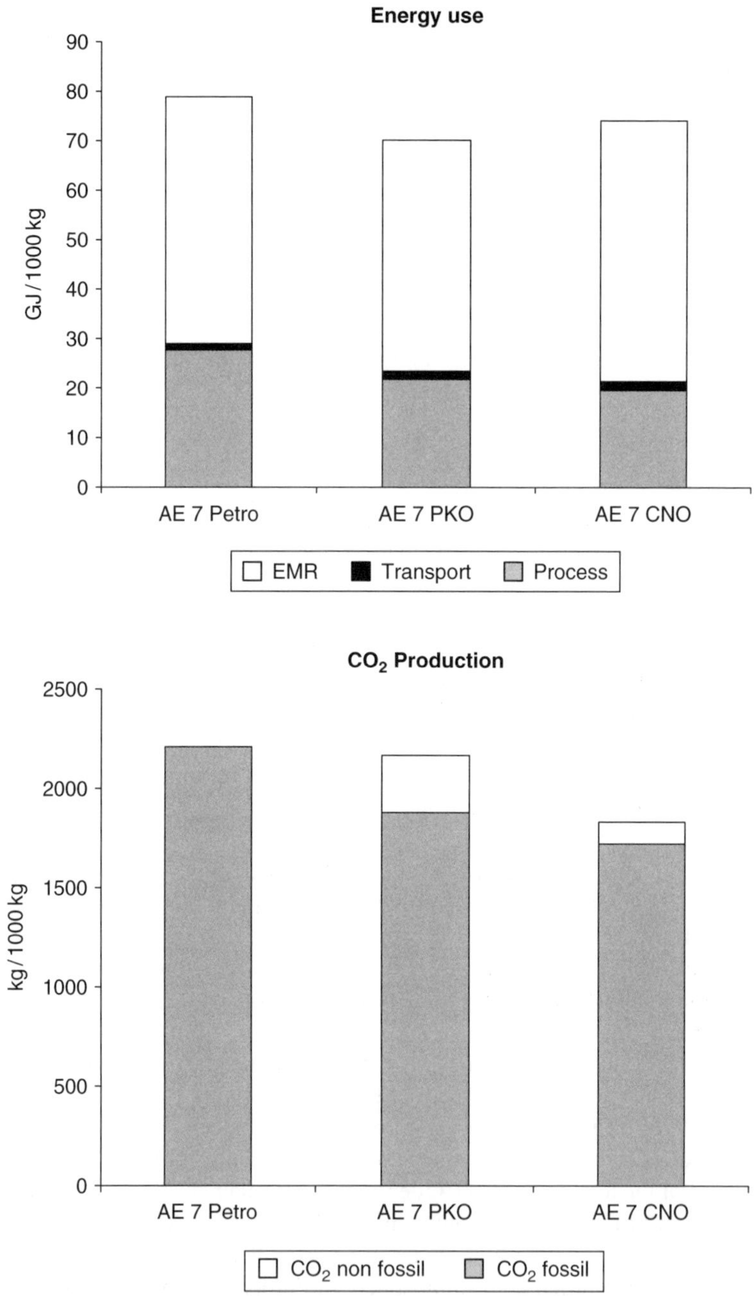

Figure 16.2 *Selected life cycle inventory results of petrochemicals versus oleo chemical surfactants: Energy use (EMR: Energy of Material Resource), CO_2 emissions, BOD and COD waterborne emissions and solid waste (AE7 – Alcohol Ethoxylates, Petro – Petrochemical; PKO – Palm Kernel Oil, CNO – Coconut Oil, oleo-chemicals)*
Source: Stalmans et *al*. (1995).

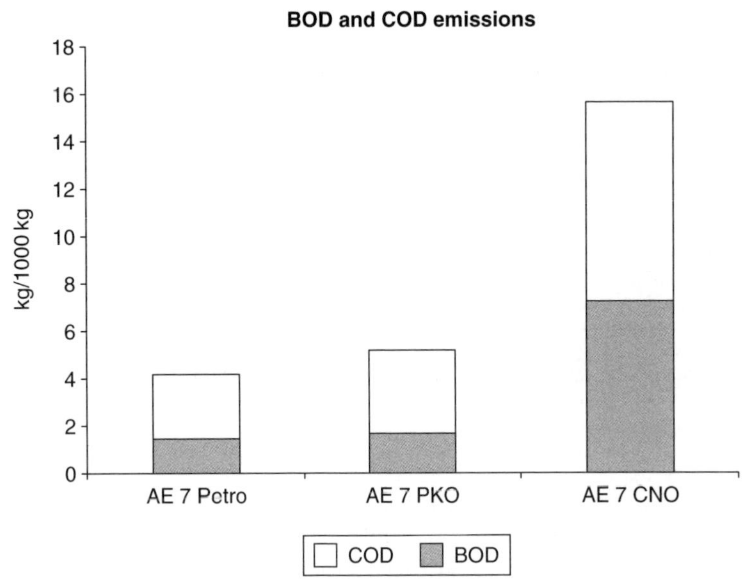

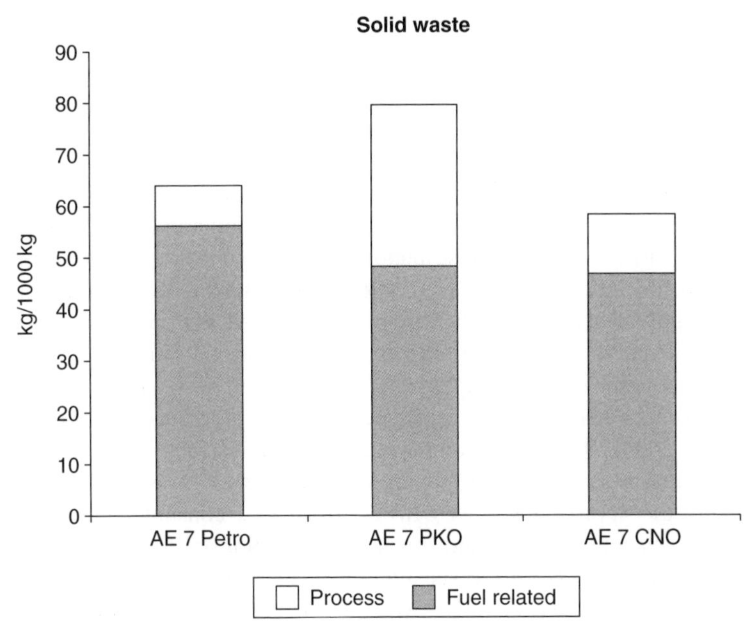

Figure 16.2 *(Continued)*

However, looking at the overall laundry process, significant saving can be achieved via lowering wash temperature. Recent studies (Saouter *et al.*, 2002, 2004), evaluated to what extent the washing process in general, and detergent ingredients specifically, contribute to the emission of global warming gases such as CO_2. This life cycle study holistically

Table 16.1 Percentage of biomass residues and waste re-utilization from palm oil production from United Plantations Berhad, Malaysia

Biomass	*Utilization (%)*	*Method*
Pruned fronds[1]	100	Mulch[a]
Trees at replanting[2]	80	Mulch[a]
Spent male flowers[3]	100	Organic matter[b]
Fibre[4]	95	Fuel[c]
Kernel shells[5]	95	Fuel[c]
Mill effluent[6]	90	Nutrient/fertilizer[d]
Empty fruit bunches[7]	90	Mulch and bunch ash[e]

Notes:
[1] Dead/dying palm leaves removed from tree to harvest fruit
[2] Old trees cut down and chipped
[3] Dead flowers after pollination
[4] Mesocarp fibre removed during pressing to obtain oil
[5] Casing of palm kernels
[6] Organic residue from wastewater treatment process at oil mill
[7] Palm fruit bunches after removal of palm fruits for processing

[a] Material returned to ground to rot down into soil
[b] Material allowed to fall and rot down
[c] Burnt in oil mill boilers to provide steam and power
[d] Dried material has high nutrient content and can replace bought fertilizer
[e] As (a) or burnt and ash used as fertilizer

Source: www.unitedplantations.com.

evaluated the entire life cycle of the 'wash process', starting from the production of detergent raw materials, and then continuing with the formulation of these materials into consumer detergents, as well as the distribution activities and the 'household consumer use' phase, and then concluded with the post-consumer disposal and waste treatment phase.

In these studies, it was shown that around 60–80% of the consumed energy of the washing process (and thus 60–80% of the total CO_2 emissions) was related to the heating of the wash water to the desired temperature (e.g. 40, 60 or 90 °C). Modern detergents are now capable of achieving a high wash performance at reduced wash temperatures (Saouter *et al.*, 2004). The more common practice of washing at lower temperatures would thus represent a considerable saving of energy and CO_2 emissions. Overall, this study supports the approach to focus R&D efforts and consumer behaviour towards lowering the wash temperatures and towards a reduction of total product dosage, as opposed to focusing efforts on the substitution of one chemical with another within a given detergent formulation. A lower wash temperature can considerably reduce the wash-related emissions of global warming gases like CO_2, whereas even a complete substitution of all petrochemical surfactants by oleochemical surfactants would result in relatively small changes in the CO_2 emission pattern.

In addition to the overall environmental impact as judged by the LCA which concentrates on overall resource consumption and environmental emissions, we need to assess what happens at the end of the wash processes when the wastewater goes into the environment. In many countries this is via the municipal sewage systems which clean up water before discharging it into rivers, lakes, the sea, etc., but direct discharge is still practised in some areas of the world and in these areas, the direct impact of chemicals on the water resources is more apparent.

Two aspects of a surfactant are relevant to water pollution. First, how quickly does the material biodegrade, the long hydrocarbon chain being converted by bacterial action into water and carbon dioxide, and second, what is the direct effect of materials on aquatic organisms before biodegradation takes place?

The assumption is frequently made that natural products are more biodegradable than synthetic. Is this true? The answer is clearly 'no'. The biodegradability of a material is solely related to its chemical structure and its solubility, not its origin. Surfactants that share the same structure will perform and biodegrade equally well, regardless of whether they were derived from oleochemical or petrochemical alcohol feedstocks. Today, all hydrocarbon chains used are linear or only slightly branched and are all easily biodegradable.

This false assumption relates back to the 1950s and 1960s when intense foaming and adverse environmental effects occurred in many countries, in sewage treatment plants as well as surface waters. These foaming phenomena were caused by the increased usage of washing machines, the increased consumption of laundry detergents, the general lack of adequate wastewater treatment, and the use of poorly biodegradable 'synthetic' surfactants. In this period, highly branched surfactants had become available and were used to replace the conventional soap-based detergents. The detergent industry responded and conducted detailed environmental research and introduced formal environmental criteria into the development process. The detergent and surfactant industry subsequently developed essentially linear, more rapidly biodegradable surfactants to replace the highly branched surfactants in the 1960s. Many of the linear surfactants developed in this period are still used today as the major surfactants in modern detergents (so-called 'workhorse' surfactants, i.e. Linear Alkylbenzene Sufonate, Alkyl Sulfate, etc.).

To show the true position, a series of side-by-side comparative biodegradation studies were conducted using exactly the same test system (OECD biodegradation test 301B) and test conditions (same concentrations, pre-exposure of bacterial inoculum) (Stalmans and Cavalli, 1993). The surfactants included Coconut/Palm Kernel Alcohol sulphate (C_{12}-C_{14} chain length), Palm Alcohol sulphate C_{16-18} chain length), Petrochemical alcohol sulphate C_{12-13}, C_{12-15}, C_{14-15}). The biodegradation curves of oleochemical and petrochemical surfactants were very similar, showing that both types of modern detergent surfactants (oleochemical and petrochemical) had similarly good biodegradation properties (ibid.). Overall, it is very important that detergent surfactants have a good biodegradation profile since they are used in high volume consumer products, which are discharged down the drain after use. Thus, a rapid loss via biodegradation ensures that they are present in the environment at concentrations well below levels of possible concern. With these examples, it has been shown how important it is to evaluate the environmental impact of detergents and their ingredients in a very holistic way, but also that the detergent industry uses a considerable amount of oleochemical products, which meet the most modern environmental criteria and contribute to the sustainability of modern detergents.

In addition to laboratory biodegradation testing, practical studies are done to ensure that surfactants in waste waters will be effectively removed by sewage treatment. One such study compared the removal efficiency of four different types of surfactants in seven sewage

Table 16.2 Removal efficiency of four different types of surfactants in seven sewage treatment plants

	Influent (measured in mg/L)	Effluent (measured in mg/L)	Removal (measured) (%)
LAS[1]	5.2	0.039	99.2
AES[2]	3.2	0.007	99.6
Soap	28.0	0.174	99.0
AE[3]	3.0	0.006	99.8
AS[4]	0.6	0.006	99.2

Notes: [1] linear alkylbenzene sulphonate; [2] alcohol ethoxy sulphates; [3] alcohol ethoxylates; [4] alcohol sulphates.
Source: Van de Plassche *et al.* (1999).

treatment plants (Table 16.2). No significant differences in removal were observed. Both types of surfactant were effectively removed, which means that very little surfactant reaches the rivers. Details are shown below with soap, the surfactant with the oldest history, as a reference point.

The toxicity of a material is also dependent on its chemical structure, not its origin. In addition to the study of the human toxicity of our surfactants for our employees and our customers, we also conduct tests on environmental toxicity. One aspect of environmental safety is biodegradability, as discussed previously. This ensures rapid removal of surfactant materials from the environment by conversion to carbon dioxide and water. However, biodegradation is not instantaneous and therefore we also need to ensure that the materials are safe for aquatic organisms and do not adversely affect the bio-systems in wastewater treatment plants as well as lakes, rivers, etc. Numerous researchers have done work on this area and many good summaries exist (Belanger, 1994; Madsen *et al.*, 2001; Talmage, 1994; Van de Plassche *et al.*, 1999; Zoller, 2004).

In general, the longer the fatty chain, the more toxic the surfactant is to aquatic organisms. Conversely, the shorter the fatty chain, the lower the toxicity. This phenomenon is well known in the scientific literature and is linked to hydrophobicity (the degree to which a surfactant dislikes water and dissolves in fats). More hydrophobicity correlates with higher toxicity, but this relationship only holds true as long as the surfactants are water-soluble. When the alkyl chain becomes so long ($> C_{16}$–C_{18}) that the surfactant is barely water-soluble, the surfactant will appear to be less available to the organisms and will be, practically speaking, less toxic (it is also useless as a surfactant).

16.5 Sustainability Aspects of Oleochemical Production

In Section 16.4 we reported on the improved plantation practices of the palm oil industry that took place in the past decade. Most framers are now returning most 'waste biomass' back to the land and this has improved yields and reduced fertilizer usage while biological pest and weed control have all but eliminated traditional herbicides and insecticides (Stalmans *et al.*, 2003). These are major steps forward and are becoming standard practice. However, the concern that is being addressed at the moment is one of plantation expansion versus forest/biodiversity losses. Oil palms are the most efficient plants in terms of oil yield per hectare. They produce two types of oil, palm oil from the mesocarp and palm kernel oil from the endocarp. An oil palm yields 3–4 tonnes of palm oil per hectare plus palm kernel oil at ca. 10% of palm oil levels. Most other oil crops yield 1 tonne or less per hectare. For this reason palm oil plantations have expanded rapidly in the past 30–40 years to meet the world's ever increasing demand for edible oil. In the past 40 years, Malaysia and then Indonesia, have put very large areas to planting oil palm trees. In many cases, either derelict land, old mining land, previously logged out land, or land converted from other crops, e.g., rubber, was used for oil palm but the advance of oil palm has caused a reduction in the rainforest area.

In the late 1990s there were extensive forest fires in Indonesia with consequent loss of forested area and increased pressure on biodiversity and endangered species. Much concern was raised that increasing areas under oil palm was leading to deforestation and loss of biodiversity in South-East Asia, particularly in Indonesia.

There were several interlinked reasons for the forest decline, including illegal logging and traditional 'slash and burn' cultivation by the indigenous population, exacerbated by very dry conditions due to an extreme El Niño. Although no palm planters were implicated in the origination of these fires, these issues raised the profile of the discussion on whether the rapid increases in palm acreage had led to detrimental loss of rainforest and biodiversity and what future pressures to produce more edible oil would do in this respect.

This issue was highlighted in Europe by certain NGOs and one Swiss supermarket group started buying only palm oil from older plantations which could not have contributed to recent deforestation. A similar campaign is ongoing against the large expansions of soy bean plantations in South America but as soya bean oil is not a surfactant raw material it is beyond our scope of this chapter.

It is also Procter and Gamble's position that forest resources worldwide must be responsibly managed to sustain them for both current and future generations, and in order to meet a wide range of societal needs, including economic, environmental, recreational value and preservation of natural beauty. P&G does not own or manage forests or plantations, but as a major purchaser of products derived from them, P&G has a social responsibility and economic interest in the long-term viability of the world's forest resources. It is our position that the sustainability of global forest resources is best achieved by a balance of three forestry approaches:

1 preservation of sufficient temperate, boreal and tropical forests to ensure that biological diversity, recreation and natural beauty are maintained;
2 mixed-use plans for maintaining forests through a combination of sustainable harvesting and management for environmental and recreational goals;
3 plantation forest management to maximize yields. This helps to reduce the pressure to commercially exploit forests that could be preserved, or those that are managed under mixed-use plans.

Arriving at the right balance of forest uses requires the involvement and cooperation of a wide variety of stakeholders and experts in academia, government, public organizations, and industry. The principles of sustainability include economically and environmentally responsible harvesting and reforestation management in order to minimize impacts on wildlife habitat, soil and water quality, with protection of special sites of unique geological, biological or historical significance.

Concern throughout the entire palm oil supply chain and consumer reaction in some European countries, combined with the position of the plantation companies and governments in South-East Asia that the expansion of palm had largely been done in a responsible manner led, in 2003, to the establishment of the 'Roundtable for Sustainable Palm Oil' (RSPO). This body, comprising key players from the palm oil industry includes palm plantation companies, edible oil refiners, food product companies, supermarkets, oleochemical producers, concerned NGOs, sustainable forestry experts, industry associations and government bodies. Their objectives are stated in detail on their website but the work of the RSPO will be based on developing criteria and best practices for planting and growing oil in order to increase yields on existing land, thus reducing the pressure for new land and to develop criteria for selection and development of new land without further unnecessary pressure on high conservation value forests and biodiversity. Another source of useful information on the Environmental/Sustainability direction of the oil palm industry is the Malaysian Palm Oil Association (MPOA).

16.6 Summary

Because of the limited range of alkyl structures available from fats and oils (i.e., only even-numbered chain lengths, no branching), formulating detergents with oleochemicals alone often restricts the manufacturers' flexibility and can require washing with warmer water. Petrochemical-based surfactants with their odd- and even-numbered and branched chains add additional formula flexibility. Using the total variety provided by both types, our formulators can develop detergents that will perform well at lower water temperatures and/or reduce the total amount of chemicals needed per wash load, thus reducing the overall environmental impact of the wash process.

The discussion above about the relative sustainability of natural and synthetic surfactants indicates that neither drives the depletion of resources, in comparison to the main uses of this same resource for energy and food production. The surfactant industry is small versus the food or energy/fuel industry and often uses the by-products or co-products resulting from the manufacture of the main product. Natural surfactants are similar in most environmental respects to their synthetic equivalents and their advantage appears to be mainly in their renewability and carbon recycling. However, they could not replace synthetic materials without putting undue pressure on existing tropical forest areas.

The bottom line is that no major environmental improvements would be expected from sole use of either type, whereas the increase in formulation flexibility may actually lead to reduced chemical usage and energy requirements, and decrease pollution.

References

Belanger, S. E. (1994) Review of experimental microcosm, mesocosm, and field tests used to evaluate the potential hazard of surfactants to aquatic life and the relation to single species data, in Hill, I. R., Heimbach, F., Leeuwangh, P. and Matthiessen, P. (eds) *Freshwater Field Tests for Hazard Assessment of Chemicals*. Lewis Publishers, Boca Raton, FL, pp. 287–313.

CESIO (2001) Surfactant statistics for Western Europe. Available at: www.cefic.org

ECCP (2001) European Climate Change Program. Available at: http://europa.eu.int/comm/environment/climat/eccpreport.htm

European Environment Agency (2001) Indicator carbon dioxide emission. Available at: http://themes.eea.eu.int/Environmental_issues/climate/indicators/Carbon_dioxide_emissions.

Hauthal, H. G. and Wagner, G. (2004) *Household Cleaning, Care and Maintenance Products*, Verlag für chemische Industrie, Augsburg, Germany.

ISO 14040 (2000) Environmental Management – Life Cycle Assessment – Principles and Framework. ISO/FDIS/TC207SC514040/1997(E), ISO, Geneva.

Jansen, D. (1995) Methodology of the European Life Cycle Inventory for detergent surfactant production, *Tenside Surfactants Detergents*, **2**, 110–121.

Madsen, T., Buchardt Boyd, H., Nylén, D., Rathmann Pedersen, A., Petersen, G. I. and Simonsen, F. (2001) *Environmental and Health Assessment of Substances in Household Detergents and Cosmetic Detergent Products*, Environmental Project No. 615 2001, Miljøstyrelsen, Miljø–og Energiministeriet.

Patel, M. K., Theib, A. and Worrell, E. (1998) Surfactant production and use in Germany: resource requirements and CO2 emissions, *Resources, Conservation and Recycling*, **25**, 61–78.

Saouter, E., Van Hoof, G., Feijtel, T. C. J. and Owens, J. W. (2002) The effects of compact formulations on the environmental profile of north European granular laundry detergents, Part II: Life cycle assessment, *International Journal of Life Cycle Assessment*, **7**, 27–38.

Saouter, E., Van Hoof, G. and White, P. (2004) A novel approach to the environmental of detergent consumer products, in Zoler, U. (ed.) *Handbook of Detergents, Part B: Environmental Impact*, Marcel Dekker, New York, pp. 195–209.

SETAC (1993) *Guidelines for Life Cycle Assessment: A Code of Practice*, Society of Environmental Toxicology and Chemistry, Pensacola, FL, Sesimbra, Portugal.

Stalmans, M., Berenbold, H., Berna, J. L, Cavalli, L., Dillarstone, A., Franke, M., Hirsinger, F., Janzen, D., Kosswig, K., Postlethwaite, D., Rappert, Th., Renta, C., Scharere, D., Schick, K.-P., Schul, W., Thomas, H., and Van Sloten, R. (1995) European life-cycle inventory for detergent surfactants production, *Tenside Surfactants Detergents*, **32**, 84–109.

Stalmans, M. and Cavalli, L. (1993) Biodegradation of commercial oleochemical and petrochemical alkyl sulphate surfactants, paper presented at workshop on 'The Environmental Fate and Effects of Surfactants', Veldhoven, The Netherlands, 13–15 September.

Stalmans, M., Saouter, E., Brunskill, A. and Walker, D. (2003) Oleochemicals and consumer detergents: a perspective on sustainability, paper presented at Proceeding of the Malaysian Palm Oil Board International Palm Oil Congress, Palm Oil: The Powerhouse for the Global Oils and Fats Economy, Malaysia, 24–28 August.

Talmage, S. S. (1994) *Environmental and Human Safety of Major Surfactants: Alcohol Ethoxylates and Alkylphenol Ethoxylates*, The Soap and Detergent Association and Lewis Publishers, Boca Raton, FL.

Van de Plassche, E. J., De Bruijn, J. H. M., Stephenson, R. R., Marshall, S. J., Feijtel, T. C. J., Scott, E. and Belanger, S. E. (1999) Predicted no-effect concentrations and risk characterization of four surfactants: Linear alkyl benzene sulfonate, Alcohol ethoxylates, Alcohol ethoxylated sulfates and Soap, *Environmental Toxicology and Chemistry*, **18**, 2653–2663.

Zoller, U. (2004) *Handbook of Detergents, Part B: Environmental Impact*. Marcel Dekker, New York.

17

Assessment of Bio-Based Packaging Materials

Andreas Detzel, Martina Krüger and Axel Ostermayer

17.1 Introduction

17.1.1 The Packaging Market

Traditionally, plastics packaging materials have been produced from non-renewable, petrochemical-based materials. In fact, packaging currently is the principal application of conventional plastics accounting for about 38% of all plastics consumed in Europe (APME, 2004). In the United States about 29% of the plastics produced are used for packaging (Stevens, 2002).

The packaging market is also projected to provide best opportunities for bioplastics, both, in Europe and the USA. This is also visible in the consumption of bioplastics in Europe which almost doubled from 2001 to 2003, showing a substantial contribution of biopackaging, i.e. packaging made from biopolymers, to this market growth (IBAW, 2005). US demand for degradable plastics is predicted to grow by 13.7% per annum over the next three years (2005–2008) with biopackaging as a driver (Plasticker, 2005).

Although the development of biopolymers has focused on biodegradability so far, the term biopolymers is increasingly being used to identify polymers that are fully or in part made from renewable raw materials (e.g. Crank *et al.*, 2004; Murphy and Bartle, 2004). This also applies here.

Polylactic acid (PLA), currently the only commercial biopolymer made completely from renewable feedstock, is one of the most versatile materials of the different types of biopolymers already available on the market. It is also suitable for more sophisticated packaging

Renewables-Based Technology Edited by J. Dewulf and H. Van Langenhove

applications such as beverage and food packaging (Vink *et al.*, 2004). Some 70% of current PLA production is used in packaging (Crank *et al.*, 2004). PLA has good mechanical properties similar to PET (polyethylene terephthalate) and PP (polypropylene) (Bastioli, 2000; Gruber and O'Brien, 2002; Crank *et al.*, 2004) and can be converted to (packaging) end products using the standard machinery for thermoplastics (Gruber and O'Brien, 2002). High value films and rigid thermoformed containers are the most promising bulk applications (Crank *et al.*, 2004). Given the growing interest, PLA will be in the focus of this chapter.

17.1.2 Waste Management Framework for Biopackaging

Plastics account for a substantial share of municipal solid waste (MSW). Related public concern has led to major activities in reducing the amount of packaging waste by stimulating recovery and recycling through a number of policy measures.

The European Packaging and Packaging Waste Directive (PPWD) (Council Directive 2004/12/EC) requires that no later than 31 December 2008 a minimum 22.5% of plastic packaging waste has to be recycled back into plastics in each EU Member State. Besides that, a minimum of 55% of recycling and 60% of recovery of all packaging waste have to be achieved.

In the directive, composting and anaerobic digestion of biodegradable packaging are considered as (organic) recycling and can thus contribute to the achievement of recycling and recovery targets. Interestingly, the PPWD also aims at introducing producer responsibility to minimize the environmental impact of packaging. The operators in the packaging chain in the future will have to ensure that the environmental impact of packaging throughout its life cycle is reduced as far as possible.

17.1.3 Life Cycle Assessment of Biopolymers

Environmental considerations have been an important motivation in the development of biopolymers as markets in Europe and the USA increasingly demand environmentally friendly packaging products. A favourable environmental performance of biopackaging as compared with their petrochemical counterparts will thus be a major factor in establishing biopolymers on the packaging market.

The method of choice when examining the environmental performance of a product is life cycle assessment (LCA). This is an approach used to realize life cycle thinking and has now developed into a widely recognized and standardized tool. The ISO 14040 series (ISO 1997–2000) provide a basis for a formalized analysis of environmental impacts of a product or service over its whole life cycle. LCAs with ISO conformity, testified by a third party review, have improved transparency and credibility and contribute to a public acceptance of results.

It is not uncommon that the market introduction of biopackaging is accompanied by LCA activities (see e.g. Matthews, 2004; Vink, 2004). During the past 15 years a number of biopolymer-related LCA-type studies have been performed (for an overview, see e.g. Patel *et al.*, 2003; Crank *et al.*, 2004). Despite the ongoing activities public LCA studies on PLA are quite scarce so far. Only two out of 20 LCA-related studies reviewed by Patel *et al.* (2003) dealt with PLA.

As a common result, biopolymer-related LCA studies often come to the conclusion that biopolymers are potentially beneficial in terms of savings of greenhouse gas (GHG) emissions and non-renewable energy resources as compared with conventional polymers (Ehrenberg, 2002). However, these findings still are of a preliminary character and require substantiation by comprehensive LCAs addressing a larger set of scientifically accepted environmental indicators.

17.1.4 Focus on Polymer Production and Waste Management

It has been shown in several packaging LCAs, covering both traditional polymers and biopolymers, that raw material production as well as the end-of-life disposal settings have an important role in the overall ecobalance of packaging applications and can strongly affect the final LCA results (Würdinger *et al.*, 2002; Detzel *et al.*, 2004). A better understanding of the related data and assumptions can help a proper design of (bio)packaging LCAs.

Using PLA and PET as examples, in the next two sections we examine the environmental performance of raw material production and packaging disposal using non-renewable resource consumption, global warming, acidification, eutrophication and ground-level ozone formation as environmental categories. Furthermore, we discuss improvement potentials, data issues, as well as less tangible environmental items that form part of an overall sustainability assessment but might not easily be covered within the scope of LCA methodology.

17.2 Environmental Aspects of Polymer Production

The analysis discussed in this chapter makes use of data sets on traditional polymers published by Plastics Europe, short name 'PET-1' (APME, 2002), as well as our own data described in Detzel *et al.* (2004), short name 'PET-2'. For PLA we refer to publicly available data (Vink *et al.*, 2003; Crank *et al.*, 2004) and in-house LCA work for several clients from public administrations and the packaging industry (e.g., Gärtner *et al.*, 2002; Würdinger *et al.*, 2002).

17.2.1 Ecoprofile Data for PLA and PET

The term ecoprofile is often used for inventories ideally covering all the relevant processes from cradle to polymer factory gate. Ecoprofiles can be used to derive environmental information on the overall polymer production chain. Given sufficiently detailed data, they can also provide insight into the relevance of individual process steps and the related improvement potential.

The PET production chain covers: (1) crude oil and gas extraction and delivery; (2) refinery and cracker processes; (3) purified terephthalic acid production; and (4) polycondensation (see also Detzel *et al.*, 2004). The process steps in current commercial PLA production are: (1) corn production; (2) corn wet milling; (3) fermentation; and (4) polymerization (Vink *et al.*, 2003). This suggests that, although using a fundamentally different feedstock, in both the PLA and the PET route, a substantial production effort is required.

And, indeed, the non-renewable process energy demand (Figure 17.1, black bars in left graph) is approx. 40 GJ/Mg for PET and 55 GJ/Mg for PLA. However, it should be pointed out that the overall non-renewable resource demand is higher in the PET production due to the fossil feedstock.

Regarding greenhouse gas emissions (GHG), right graph of Figure 17.1, there is a considerable difference between PET-1 and PET-2 which can be attributed to variations in the underlying fuel mix and energy efficiency. PLA polymer production shows higher specific GHG emissions originating from process energy use than PET-2 (see overall bar in right graph of Figure 17.1).

On the other hand, the bio-based PLA contains carbon taken up from the atmosphere during plant growth. As we will see, waste management is decisive as to how carbon dioxide (CO_2) fixed in the material can influence the overall GHG balance of a biopackaging product.

17.2.2 PLA: Process Chain Analysis

Process breakdown helps gain further insight into the environmental performance of PLA production. Figure 17.2 indicates that non-renewable process energy and GHG emissions are mainly caused in the fermentation and polymerization steps. On the other hand, acidification and eutrophication results show a major contribution from agricultural and corn wet mill operations. This highlights the importance of agricultural production for the

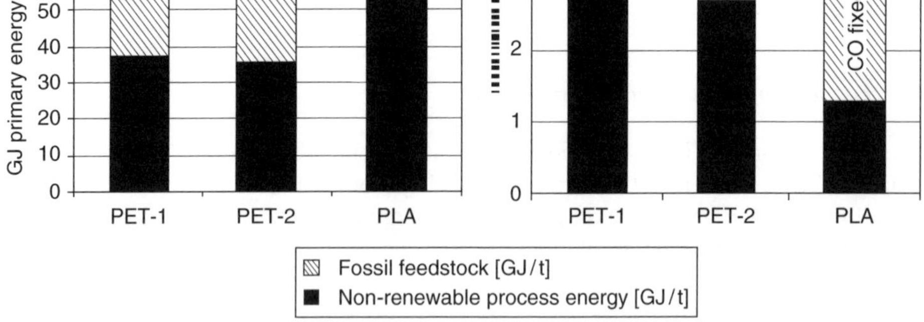

Figure 17.1 *Non-renewable energy demand and greenhouse gas emissions associated with the production of PET and PLA*

Sources: PET-1 source: APME (2002); PET-2 source: Detzel *et al.* (2004); PLA sources: Vink *et al.* (2003); Crank *et al.* (2004); assumptions of the authors.

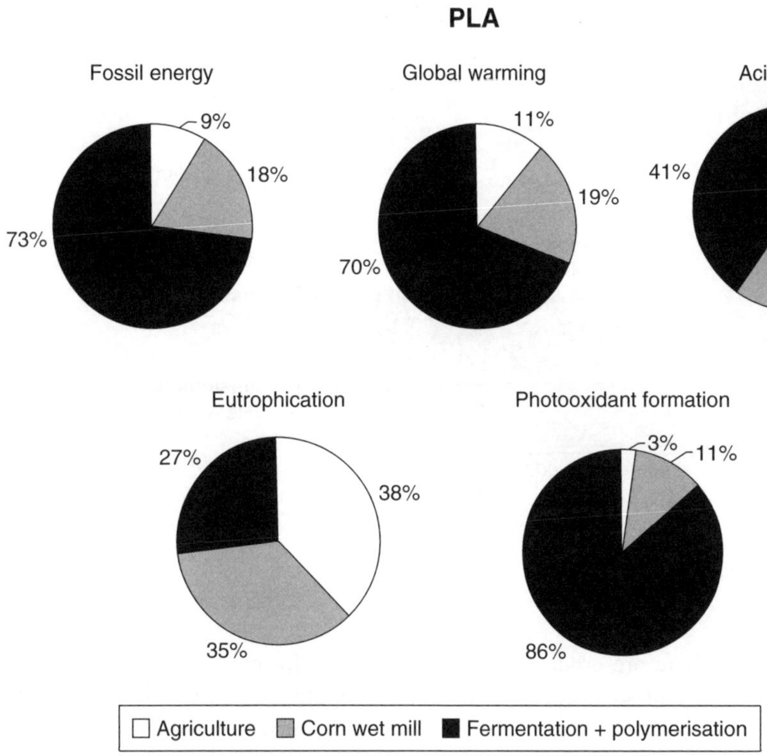

Figure 17.2 *Contribution of individual process steps to the overall environmental profile of PLA production*
Source: Data derived from Vink *et al.*, (2003) and Gärtner *et al.* (2002).

overall environmental performance of PLA production. As for photo-oxidant formation, related emissions originate mainly from the fermentation and polymerization steps but are caused by production pre-chains of process chemicals used.

It is clear that potential improvements must be located in different areas according to the individual environmental problem of interest. According to Vink *et al.* (2003), a reduction in fossil energy demand by about 10% is feasible through the use of enhanced fermentation technology. An additional 36% of reduction may be achieved by a production sequence using bio-refinery concepts. As shown in Figure 17.2, fossil fuel consumption and GHG emissions patterns are closely related, therefore considerable improvements in the GHG emissions can be expected for the future too.

Agriculture, given its relevance for impacts concerning acidification and eutrophication, deserves a deeper consideration in the following section.

17.2.3 Production of Agricultural Crops: Example Corn Growing

The production of corn, currently the most important feedstock for commercial PLA, not only includes processes related to corn growing but also pre-chains of material and energy supply such as fertilizer and fuel production (see Figure 17.3). Typical materials

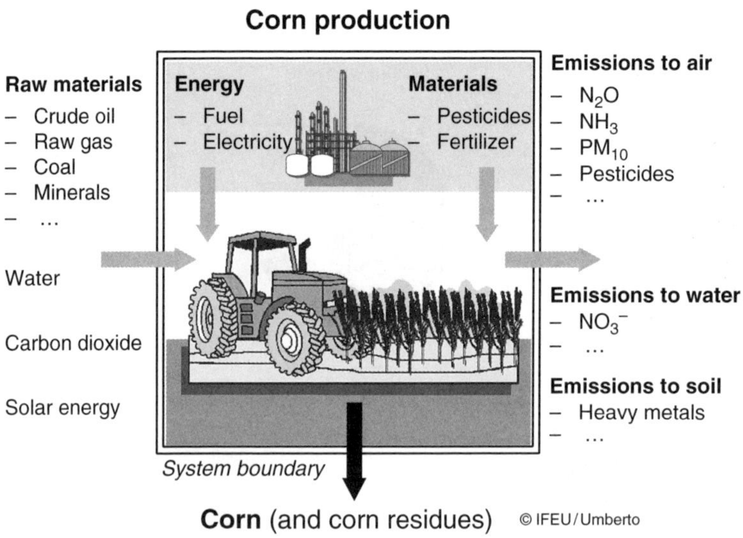

Figure 17.3 *Material flows in the corn production system*

taken from the ecosphere are crude oil, raw gas and coal for energy supply as well as water, carbon dioxide and solar energy for the photosynthesis process, the latter being characteristic of agricultural systems. On the other hand, agriculture-specific emissions are released into the air (such as nitrous oxide (N_2O) or ammonia (NH_3), particulate matter (PM_{10}) and pesticides), into water (such as nitrate, NO_3^-) and on soils (such as heavy metals from fertilizers and pesticides).

Unlike the rather 'closed' industrial system with generally clearly distinguishable emission pathways to the environment (e.g. chimneys, wastewater discharges), the agricultural system shows a rather 'open' character due to direct exchanges with the ecosphere via air, soil and water. Consequently, the addressed emissions vary largely as they are strongly influenced by surrounding conditions such as soil properties, weather or management practice of the farmer. For this reason, it is difficult to accurately measure emissions related to agriculture and available emission factors show rather wide ranges. A similar situation can be found in nutrient and soil fertility balances which depend on the preceding crop (e.g. soy bean in case of corn) as well as on environmental conditions.

Despite these methodological problems it is obvious that agriculture is a relevant source for a number of environmental issues related to biopolymer production. In this context implementation of biorefinery concepts is of particular interest as it offers potential improvements for the overall PLA ecoprofile. Notable research activities are under way concerning e.g. the use of food waste as fermentation feed material (Sakai *et al.*, 2004) or corn stover as raw material for sugar production (Vink, 2004).

17.2.4 Other Issues Related to Environmental Assessment

Environmental evaluations of product systems are typically based on life cycle inventories subsequently translated into potential impacts with the help of appropriate methods.

While for most emissions from technical processes, life cycle impact assessment methods have been developed and validated, the corresponding methods for a series of agricultural-specific emissions are missing. This is especially true concerning environmental topics such as biodiversity, human toxicity, ecotoxicity and soil fertility being affected by pesticides, nitrates, heavy metals, fertilizers and genetically modified crop plants among others. Methods developed so far (e.g. Giegrich and Fehrenbach, 1999; Klöpffer *et al.*, 2001; Guinée *et al.*, 2002) are not fully satisfying as yet.

Therefore, it is usually necessary for the assessment of agricultural products to discuss the environmental risks mentioned above only qualitatively, which in turn puts some limitations on comparisons with non-agricultural products. However, also in the case of 'conventional' fossil-based polymers, critical issues concerning the assessment of environmental impact can be found.

One example is the refinery system regarding fugitive emissions. Although the total amount of fugitive organic emissions (e.g. losses via leakages of on-site installations) can be derived from overall input–output balances the exact chemical composition and the emission pathways remain unknown in most cases. Consequently, it is often not possible to link environmental impacts with these emissions. Again, this imposes limitations to the comparative assessment of fossil and bio-based polymers.

As with agricultural systems, there are also environmental impacts associated with crude oil extraction that are not covered by the 'classical' environmental impact categories typically used for product LCAs. Examples include accidental release of crude oil into the marine, freshwater or soil environment caused by oil tanker accidents or offshore and onshore pipeline leakages as well as non-accidental but nevertheless unavoidable crude oil losses at drilling platforms. A basic problem here arises – even if released amounts are known – when the reference to a functional unit (such as 1 kg of crude oil) is not clearly given in the data source or the amount refers to a product mixture (such as the transport of non-specified mineral oil products). Related research is in progress and first efforts to include risk aspects concerning oil tanker and pipeline accidents among others in life cycle assessments have been made by Kurth *et al.* (2004).

17.3 Environmental Aspects of Packaging Disposal

The analysis done in this section makes use of in-house waste treatment process data.

17.3.1 Packaging Waste Disposal Pathways

The standard waste treatment pathways of plastics packaging waste can be: (1) landfill; (2) municipal solid waste incineration (MSWI); (3) mechanical biological treatment (MBT); (4) recycling; and (5) recovery (see also Figure 17.4). Scenarios (4) and (5) require the source-separated collection of plastics packaging waste, while the other options will be the fate of those packages ending up in the household rest waste. In practice, packaging waste will always occur in several of these pathways depending on the existence and organisation of source-separated collection programmes and the type of plastics packaging scheduled for collection.

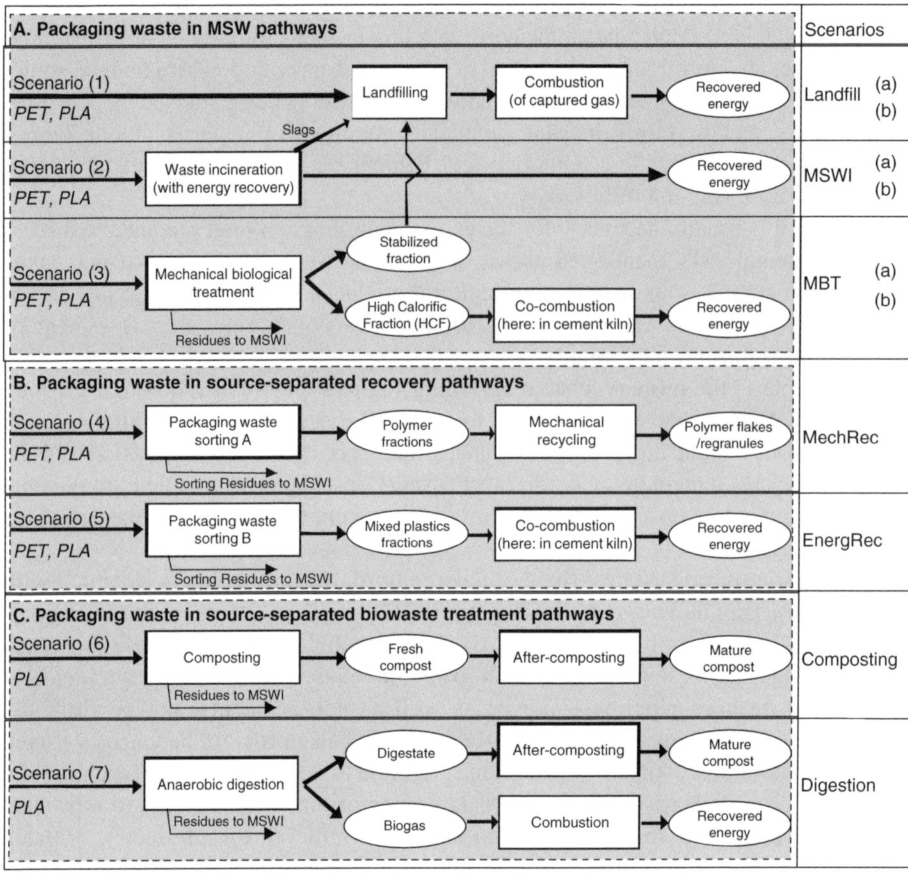

Figure 17.4 *Individual pathways for treatment of packaging wastes (left) and waste disposal scenarios investigated (right)*

Notes: Variants within scenarios: Landfill (a) 50% degradation of PLA, Landfill (b) 0.1% degradation of PLA, MSWI (a) 50% energy recovery, MSWI (b) 10% energy recovery, MBT(a) 90% sorted to HCF, MBT(b) 10% sorted to HCF.

If no additional measures are taken, packages such as, e.g. transparent trays made of PLA and PET, will be handled indiscriminately by the consumer. To a certain point, they will share a common fate in the mentioned waste pathways. In contrast to PET, PLA waste streams can also be processed using industrial composting and anaerobic digestion. For that purpose PLA packages have to be included within organic household collection schemes or special collection activities.

In principal, chemical recycling is technically possible for both PET and PLA. However, it is currently of minor practical relevance and therefore has not been included in the assessment presented here.

Given the availability of different options and the variety of their combination in waste management practice, it is important to understand how the individual scenarios can influence the environmental sustainability of PLA and PET. A set of appropriate scenarios has been examined to improve the existent knowledge in this regard. The considered scenarios along with their short names used in the text are summarized in Figure 17.4. Variations within the waste management scenarios are indicated by (a) and (b) extensions.

17.3.2 Waste Scenario Description

Scenario 1: Packaging Wastes to Landfill

In the landfill model a technical standard is assumed that complies with the requirements given by the EU Landfill Directive (Council Directive 1999/31/EC). The module accounts for the emissions and the consumption of resources for the deposition of domestic wastes on a sanitary landfill site.

Environmental loads considered here are methane emissions emitted in the landfill gas resulting from anaerobic degradation of PLA and emissions generated (1) during the combustion of captured landfill gas in a gas engine and (2) during fuel combustion in order to supply energy for the landfill system. Carbon dioxide emissions originating from PLA are not accounted for as they are of regenerative origin. In the calculation model, the electricity produced is assumed to replace average European grid electricity.

In the case of PET wastes, no degradation has been assumed. In the case of PLA it is not known to what extent degradation is likely to occur in landfills as respective studies are missing. For this reason, two variations have been examined. In variation (a) a degradation of 50% and in variation (b) a degradation of 0.1% has been assumed. The latter has been found in tests on the degradation behaviour of amorphous PLA in backyard composting systems (Klauss, 2004). The PLA material that has not been degraded is accounted for as a carbon sink.

Scenario 2: Packaging Wastes to MSWI

In the incineration model a technical standard is assumed that complies with the requirements given by the EU Incineration Directive (Council Directive 2000/76/EC). The model calculation considers a grid-firing with boiler system with steam turbine and flue gas cleaning. Environmental loads considered here are those linked to the flue gas emissions. Waste-specific emissions (e.g. CO_2, SO_2) reflect the elementary contents of the packaging material. Again, the CO_2 emissions originating from PLA are not accounted for.

The energy recovered during MSWI is assumed to be heat used for district heating purposes. In the model it replaces heat generated by natural gas and light fuel oil boilers in private homes.

The settings concerning rates and type of energy recovery in MSWI differ between plants and geographic regions. For the purpose of the present assessment, simplifications were necessary. For this reason, again, two variations have been examined. In variation (a) 50% and in variation (b) 10% recovery of the heat content of waste input has been assumed.

Scenario 3: Packaging Wastes to MBT

MBT is a major alternative to direct deposition in sanitary landfills and to the direct incineration of packaging wastes. The mechanical-biological step serves as a pre-treatment that separates the waste material into a high calorific fraction (also known as refuse-derived fuel, RDF) and a fraction directed to biological treatment generating a stabilized residue fraction.

In the MBT model the RDF fraction is assumed to be burned in a co-combustion unit (here a cement kiln) replacing hard coal as a primary fuel. The stabilized fraction is assumed to be sent to the landfill. The model reflects a closed plant with integrated cleaning system (scrapping of ammonia and regenerative thermal oxidation).

Different MBT plant concepts can be found in practice. For this reason, here too, two variations have been examined. In variation (a) 90% and in variation (b) 50% of the plastics packaging waste (i.e. both PLA and PET) is sorted into the high calorific fraction. The remaining fraction is subjected to biological treatment. A degradation of 80% of PLA ending up there has been assumed. The PLA material that has not been degraded is counted as a carbon sink.

Scenario 4: Packaging Wastes to Mechanical Recycling

During mechanical recycling plastic waste is ground, cleaned and eventually recycled into flakes or pellets. This scenario is used for biopolymers too (Degli-Innocenti, 2003). Pre-conditions for a valuable mechanical recycling of plastics are source-separated waste collection and sorting operations providing polymer-specific fractions. Current studies suggest that sorting of PLA fractions from mixed plastics waste streams is technically feasible (Food, 2005).

In the model used here, energy demand for and material losses due to sorting and recycling processes have been accounted for. The model is based on PET recycling data. For PLA the same effort per mass input as for PET has been calculated. The recycled polymers are assumed to displace the production of the corresponding virgin polymers.

Recycled PLA maintains the CO_2 fixed from the atmosphere during plant growth within the technical material cycle. This could be accounted for as a type of CO_2 sequestration (Gerngross and Slater, 2003) which has been implemented in the calculation model applied here.

Scenario 5: Packaging Wastes to Energy Recovery

Source-separated plastics packaging wastes which, for some reason, cannot be separated into polymer fractions, often end up in a so-called mixed plastics fraction (MPF). Co-combustion is the typical fate of this kind of high energy content waste material. In the calculation model the MPF is assumed to be burned in a cement kiln substituting hard coal as a primary fuel.

Scenario 6: PLA Packaging Waste to Composting

Composting of PLA shows a two-step degradation process: at first, hydrolysis breaks down the polymer into its basic building blocks (lactic acid), afterwards, the lactic acid is metabolized by micro-organisms into CO_2 (Klauss, 2004).

Biogenic waste from households is increasingly treated in plants with improved techniques, especially to minimize the odour pollution and also to reduce the area demand.

The composting model considered here refers to a medium commercial composting standard having an encapsulated system (container composting) for the main degrading step.

The compost is assumed to serve as a soil amendment and to replace mineral fertilizers and peat. Relevant emissions are CO_2, methane and other volatile organic compounds being generated in the degradation steps. Again, CO_2 emissions from degraded PLA are not accounted for. An overall degradation rate of 95% has been assumed for PLA.

The composting model used here is not applicable for home or backyard composting, where PLA showed almost no degradation (Klauss, 2004).

Scenario 7: PLA Packaging Waste to Anaerobic Digestion

Little is known about the behaviour of PLA during anaerobic digestion. The basic principle of degradation of organic compounds via hydrolysis and acidification and eventual conversion to mainly carbon dioxide and methane by micro-organisms also applies to PLA. However, the extent of degradation is still under discussion.

In the model, the same degradation rate as for composting, i.e. 95%, has been assumed. This corresponds to a generation rate of 0.84 m^3 biogas/kg PLA degraded. The biogas is combusted in a gas motor producing a certain amount of net electricity. In the calculation model the replacement of the average European grid electricity is assumed.

Waste-specific Input Data

In order to properly differentiate the behaviour of materials varying in elemental composition with respect to waste treatment, some basic data have to be used. In the context of the exercise here, a lower heating value of 18 MJ/kg and 26 MJ/kg has been applied for PLA and PET, respectively. Correspondingly there is a difference in carbon content between PLA and PET (500 g/kg and 625 g/kg) with the first being of regenerative origin. As for nitrogen content, 0.05 g/kg has been used for PLA and 6.5 g/kg for PET.

17.3.3 Findings

The results of waste scenarios are shown in Figure 17.5 (PET) and Figure 17.6 (PLA) using the environmental indicators global warming, acidification, eutrophication and photo-oxidant formation. In each graph, the individual scenarios are separated by dashed lines and their results are expressed by three bars. The left bar (black) represents the impact of the considered waste treatment itself. The medium bar (white) represents the environmental loads being avoided by the use of products obtained through waste recycling and recovery respectively. The right bar (grey) shows the net values. The net values are the reference for the overall evaluation. They are either positive, which means an overall environmental load caused by the waste management operation, or negative, which corresponds to a net reduction of environmental loads by the waste management operation. All indicator results refer to the disposal of 1000 kg of polymer.

Figures 17.5 and 17.6 show favourable results for mechanical recycling, not only for PET but also for PLA. In both cases the avoided burden is substantial while the loads caused by the sorting and recycling chain are relatively small.

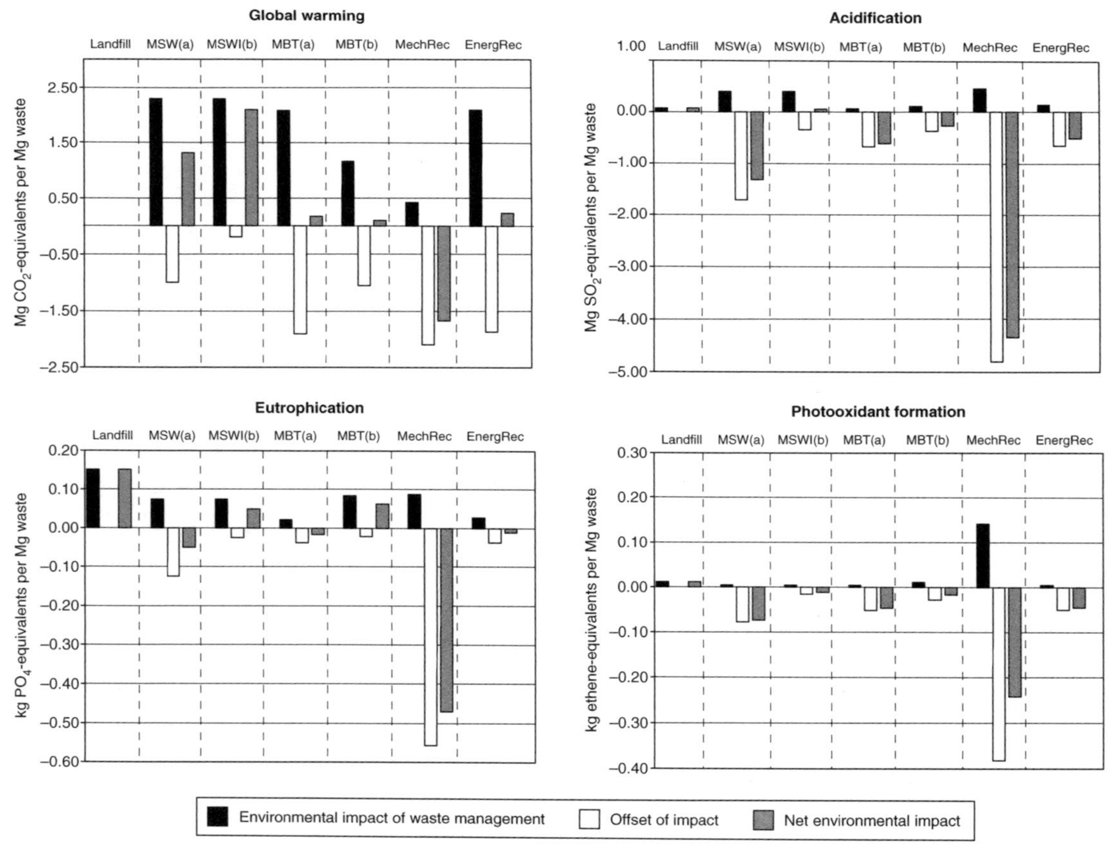

Figure 17.5 *Environmental performance of PET in individual waste disposal scenarios. Positive values on the y-axis indicate environmental loads whereas negative values represent avoided environmental loads*

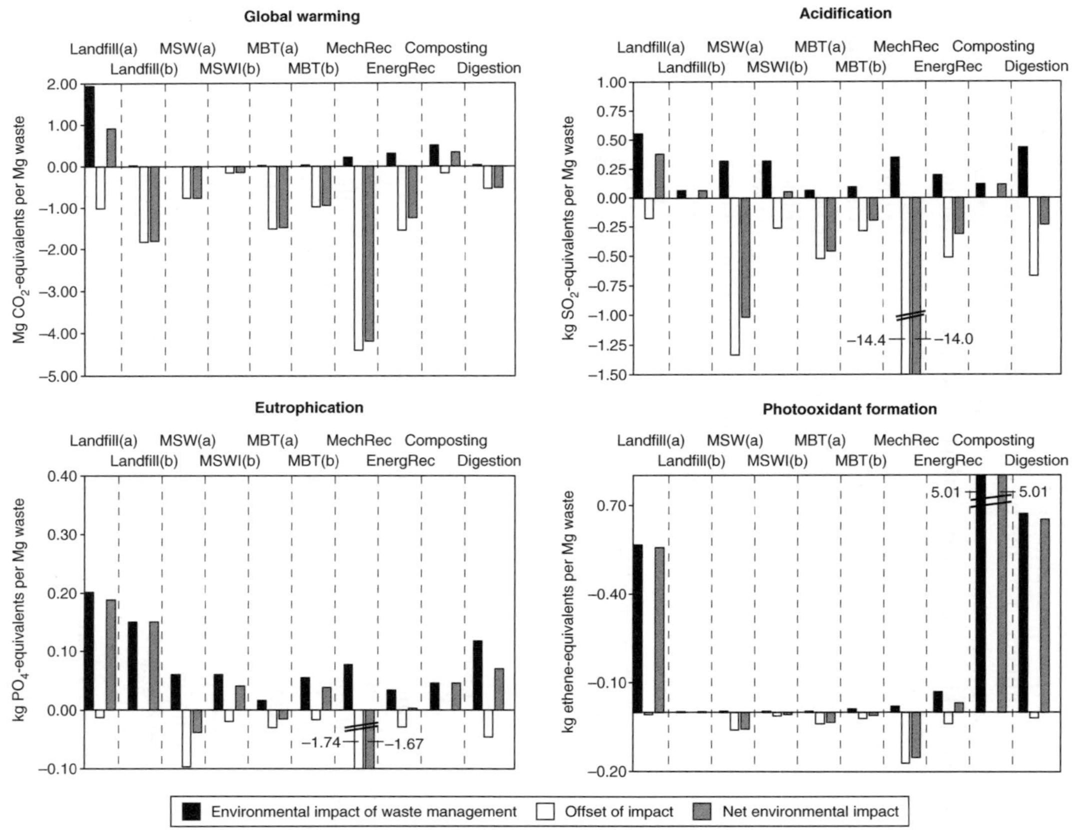

Figure 17.6 *Environmental performance of PLA in individual waste disposal scenarios. Positive values on the y-axis indicate environmental loads whereas negative values represent avoided environmental loads*

The options using thermal treatment, i.e. the scenarios MSWI, MBT and EnergRec, in general, score better than landfill due to the offset of fossil fuel use. Nevertheless, GHG emissions from thermal treatment of PET are noticeable. Despite the energy recovered in the PET scenarios, net GHG emissions occur while a considerable offset of GHG can be achieved by energy recovered during thermal treatment of PLA. This is a consequence of the fact that CO_2 from PLA combustion is regarded to be GHG neutral.

The ranking within scenarios with thermal waste treatment depends on the efficiency with which the energy content of the post-consumer packaging waste is recovered. In addition, it is influenced by the fuel type and combustion process set off by recovered energy. Consequently, due to efficient energy recovery and the assumed replacement of coal in cement kilns, the scenarios MBT(a) and EnergRec score quite well across the four environmental categories examined.

On the other hand, MSWI(a) scores particularly well concerning acidification, eutrophication and photo-oxidant formation. In this context, it is worthwhile to point out how the surrounding conditions can have a considerable influence on the results.

For the purpose of this exercise, MSWI has been modelled with a high emission reduction standard regarding air pollutants. On the other hand, the recovered heat replaces heat generation in private homes where the respective emission standards are particularly poor as compared to industrial combustion. Thus, despite the fact that MSWI(a) is less energy efficient than MBT(a), the overall net results are not so different.

The analysis of waste disposal scenarios, as discussed so far, shows quite similar patterns for PET as well as for PLA. Only landfill shows some variation as, in the PLA case, the results depends on whether degradation is likely or not to happen on a landfill site. Landfill(b), assuming practically no degradation, results in a carbon sink function and thus a net GHG offset. On the other hand, methane emissions caused by anaerobic degradation in Landfill(a) result in large GHG emissions. Now, where are composting and digestion of PLA situated within this figure?

The environmental impacts of the composting process as such are smaller than or close to those of the energy recovery scenarios and the mechanical recycling route. However, the amount of compost derived from PLA is very small, given the high degradation rate. Consequently, the environmental offset of compost application is small and the overall net environmental impact results are relatively large. This is particularly true for the potential photo-oxidant formation during composting being rather large due to the assumed release of volatile organic compounds as a product of carbon degradation.

The environmental performance of anaerobic digestion of PLA is in the range of the lower scoring thermal treatment scenarios with energy recovery. This is not surprising as the biogas generated also is converted into energy. The offset of air emissions achievable thereby is considerable (see Figure 17.6, 'Digestion', white bars). However, there are process emissions from biogas combustion and the treatment of the digestion residuals too. As already mentioned, field data for anaerobic digestion of biopolymers are lacking and results here, based on model assumptions, show that more research should be dedicated to this aspect.

The scenarios examined here indicate that PLA packaging waste, when entering into standard waste treatment pathways, performs in a similar manner as PET packaging waste, from an environmental point of view. Mechanical recycling is favourable if virgin plastics are replaced by recycled material. Environmental advantages can be obtained by thermal treatment of PLA especially if fossil CO_2 emissions are displaced by making use of recovered energy. With some limitations, this also seems to apply to anaerobic digestion.

17.4 Summary

Biopackaging made from PLA is expected to be increasingly used for packaging applications in the near future. A favourable environmental performance of PLA as compared to petrochemical counterparts such as, e.g. PET and PP will be an important factor encouraging this development. Post-consumer waste resulting from PLA packaging will have to fit into existing waste management systems and contribute to compliance with the targets set by waste policy. A waste management infrastructure optimized for the handling of biopackaging could further stimulate market acceptance.

Life cycle assessment (LCA) is the best way to examine the environmental performance of a product, covering all the steps from material production, packaging conversion and usage to post-consumer disposal. To this end, full LCAs on PLA are not publicly available. Therefore in this chapter focus has been on the assessment of the environmental performance regarding polymer production and waste disposal.

PLA has been examined, along with PET, in order to better understand the characteristics of materials from renewable and fossil-based feedstock respectively. The data provided revealed the need but also the potential for further improvement in the PLA production chain. Optimization of PLA technologies is still in progress and the reduction of fossil energy demand and related GHG emissions is likely to be achieved by enhanced fermentation technology and the implementation of biorefinery concepts. The latter would also help reduce the potential acidification and eutrophication caused by crop production.

Besides this, proper waste management strategies can affect the overall impact profile of plastics packaging in general and biopackaging in particular. Source-separated collection and subsequent mechanical recycling have substantial benefits for both PET and PLA packaging waste and are current practice in the case of PET packaging. However, only when PLA is established as a bulk packaging material and considerable amounts are present in the waste streams will the separation of PLA fractions be attractive to waste recyclers.

In the meantime, efforts should be directed at the design and development of waste management structures enabling the recovery of the energy contained in PLA and, in a way, of biopackaging in general. This scenario is of particular interest if the energy recovered can be used to displace fossil fuels as the avoided GHG emissions might largely equal the GHG caused during production of biopolymers. The potential advantages of anaerobic digestion are also founded herein. The feasibility of an efficient transformation of biopolymers into biogas still has to be proved in practice.

The role of composting is less clear. Composting can be a clever solution for individual applications like e.g. biodegradable plastic bags used for the collection of organic waste. It is also a straightforward solution for local activities as in the example of a food retailer in the USA where customers can return their biopackaging to the point-of-sale for subsequent composting at a local facility (Vink, 2003).

Compost can be a valuable soil amendment especially in regions with low soil qualities or erosion. It has, on the other hand, a relatively poor performance when it comes to potential environmental impacts as examined in the exercise presented here. This point has also been highlighted in other studies and should not be ignored.

In the end, optimized waste management depends on factors such as the local availability of the infrastructure for collection and processing, the composition of the waste stream as well as the given legal framework. For biopackaging, now entering these waste streams, it may also depend on supportive measures. As an example, the revised German Packaging Ordinance now exempts compostable biopackaging from the regulation and the related licence fees during the market introduction phase until the end of the year 2012. It thereby passes responsibility to the industry which has the task of developing an independent disposal solution.

The assessment done for the purpose of this chapter indicates that biopolymers do not *per se* contribute to more environmentally sustainable consumption. Still, they have the huge potential to do so if convenient techniques are used and the appropriate infrastructure is implemented. The results provided may be interpreted as an extended ecological footprint on a polymer mass basis. They are of preliminary character and can be complemented by full scale LCAs currently in progress covering several actual packaging applications.

References

APME (2002) Eco-profiles of the European plastics industry: polyethylene terephtalate. report for The European Centre for Plastics in the Environment, published by the Association of Plastic Manufacturers in Europe, July 2001, revised September 2002, Brussels (Accessed September 2004) Available at: www.apme.org

APME (2004) *Plastics in Europe: An Analysis of Plastics Consumption and Recovery in Europe 2002/2003*. Available at: www.apme.org

Bastioli, C. (2000) Global status of the production of biobased packaging materials, in Weber (ed.) *Proceedings of the Food Biopack Conference, held in Copenhagen, August 2000*. Available at: www.mli.kvl.dk/foodchem/special/biopack/

Crank, M., Patel, M., Marscheider-Weidemann, F., Schleich, J., Hüsing, B. and Angerer, G. (2004) *Techno-economic Feasibility of Large-scale Production of Bio-based Polymers in Europe (PRO-BIP)*, Utrecht University and Fraunhofer Institute for Systems and Innovation Research, Utrecht/Karlsruhe.

Degli-Innocenti, F. (2003) Presentation held at Packtech/EcoPac's Seminar on Eco-friendly Food and Beverage Packaging, 23–24 April 2003, Barcelona, Spain.

Detzel, A., Giegrich, J., Krüger, M., Möhler, S. and Ostermayer, A. (2004) *Ökobilanz für PET-Einwegsysteme unter Berücksichtigung der Sekundärprodukte* (*LCA of one-way PET Bottles and Recycled Products*), Institute for Energy and Environmental Research, Heidelberg/Brussels.

Ehrenberg, J. (ed.) (2002) *Current Situation and Future Prospects of EU Industry Using Renewable Raw Materials*, report prepared by Working Group 'Renewable Raw Materials', Brussels, February.

Food Production Daily (2005) Newsletter 28 January 2005. Available at: www.foodproductiondaily.com (Accessed February 2005.)

Gärtner, S. O., Müller-Sämann, K., Reinhardt, G. A. and Vetter, R. (2002) Corn to polymers: a comprehensive environmental assessment, *Proceedings of the 12th European Biomass Conference*, **2**, 1324–1326, James and James, London.

Gerngross, T. and Slater, S. (2003) Biopolymers and the environment, *Science*, **299**, 822–823.

Giegrich, J. and Fehrenbach, H. (1999) Methode zur Integration der Naturrauminanspruchnahme in Ökobilanzen, in DGfH (ed.) *Verifizierung verschiedener Methoden zur Wirkungsabschätzung des Wirkkriteriums Naturrauminanspruchnahme/Landnutzung/Landverbrauch*, Deutsche Gesellschaft für Holzforschung, Munich.

Gruber, P. and O'Brien, M. (2002) Polylactides 'Natureworks® PLA', in Doi Y and Steinbüchel, A (eds) *Biopolymers* (10 vols), vol. 4, *Polyesters III applications and commercial products*, Weinheim: Wiley-VCH, pp. 235–49.

Guinée, J. B., Gorée, M., Heijungs, R., Huppes, G., Kleijn, R., Koning, A. d., Oers, L. v., Wegener Sleeswijk, A., Suh, S., Udo de Haes, H. A., Bruijn, H. d., Duin, R. v. and Huijbregts, M. A. J. (2002) *Handbook on Life Cycle Assessment: Operational Guide to the ISO Standards*, Kluwer Academic Publishers, Dordrecht.

IBAW (2005) Highlights in bioplastics, available at *www.ibaw.org* (Accessed February 2005.)

IFEU (2005) In-house database at Institute for Energy and Environmental Research (IFEU), Heidelberg, www.ifeu.de

ISO (1997–2000) ISO series: Environmental Management – Life Cycle Assessment, ISO EN 14040 – 1997, ISO EN 14041 – 1998, ISO EN 14042 – 2000, ISO EN 14043 – 2000, ISO, Geneva.

Klauss, M. (2004) *Degradation of Biologically Degradable Packaging Items in Home or Backyard Composting Systems*, Schriftenreihe des Lehrstuhls Abfallwirtschaft und des Lehrstuhls Siedlungswasserwirtschaft No. 11, Bauhaus-Universität, Weimar, Rhombos-Verlag, Berlin.

Klöpffer, W., Renner, I., Tappeser, B., Schmidt, E., Tappeser, B., Gensch, C.-O. and Gaugitsch, H. (2001) *Methodische Weiterentwicklung der Wirkungsabschätzung in Ökobilanzen (LCA) Gentechnisch veränderter Pflanzen*, Umweltbundesamt, Vienna.

Kurth, S., Schüler, D., Renner, I., Klöpffer, W. (2004) *Entwicklung eines Modells zur Berücksichtigung der Risiken durch nicht bestimmungsgemäße Betriebszustände von Industrieanlage im Rahmen von Ökobilanzen: Endbericht zur Vorstudie*, Forschungsbericht 20148309 UBA-Texte 34/04, Berlin.

Matthews, T. (2004) Manufacturing Biopolymers: Issues and Challenges, presentation held at the 'Biodegradable Polymers and Sustainability – Insights from Life Cycle Assessment', Seminar, London, 25 May 2004. Available at: http://www.nnfcc.co.uk

Murphy, R. and Bartle, I. (2004) Biodegradable Polymers and Sustainability – Insights from Life Cycle Assessment, presentation held at the 'Biodegradable Polymers and Sustainability – Insights from Life Cycle Assessment' Seminar, London, 25 May 2004. Available at: http://www.nnfcc.co.uk

Patel, M., Bastioli, C., Marini, L. and Würdinger, E. (2003) Life-cycle assessment of bio-based polymers and natural fibres, in *Biopolymers*, Vol. 10, Wiley-VCH, pp. 409–452.

Plasticker (2005) PlasTicker-News, 8 February 2005 www.plasticker.com.

Sakai, K., Taniguchi, M., Miura, S., Ohara, H., Matsumoto, T. and Shirai, Y. (2004) Making Plastics from Garbage – A Novel Process for Poly-L-Lactate Production from Municipal Food Waste, *Journal of Industrial Ecology*, **7**, 63–74.

Stevens, E. S. (2002) *Green Plastics: An Introduction to the New Science of Biodegradable Plastics*, Princeton University Press, Princeton, NJ.

Vink, E. T. H (2004) The Sustainability of NatureWorks™ PLA and Ingeo™ Fibers: The role of LCA. Presentation held at the 'Biodegradable Polymers and Sustainability – Insights from Life Cycle Assessment' Seminar, London, 25 May 2004, downloaded January 2005 at http://www.nnfcc.co.uk

Vink, E. T. H., Rabago, K. R., Glassner, D. A. and Gruber, P. R. (2003) Applications of life cycle assessment to NatureWorks™ polylactide (PLA) production, *Polymer Degradation and Stability*, **80**, 403–419.

Vink, E. T. H., Rabago, K. R., Glassner, D. A., Springs, B., O'Connor, R. P., Kolstad, J. and Gruber, P. R. (2004) The Sustainability of NatureWorks™ Polylactide Polymers and Ingeo™ Polylactide Fibers: An Update of the Future, *Macromolecular Bioscience*, **4**, 551–564.

Würdinger, E., Roth, U., Wegener, A., Borken, J., Detzel, A., Fehrenbach, H., Giegrich, J., Möhler, S., Patyk, A., Reinhardt, G. A., Vogt, R., Mühlberger, D. and Wante, J. (2002) *Kunststoffe aus nachwachsenden Rohstoffen: Vergleichende Ökobilanz für Loose-fill-Packmittel aus Stärke bzw. aus Polystyrol* (final report, DBU-Az. 04763), Bayrisches Institut für Angewandte Umweltforschung und -technik, Augsburg (BIFA), Institut für Energie- und Umweltforschung, Heidelberg (IFEU), Flo-Pak GmbH.

18

Assessment of Biotechnology-Based Chemicals

Peter Saling and Andreas Kicherer

18.1 Introduction

Today chemical industry faces many challenges and opportunities in terms of sustainable development. Analytically assessing technology options for their environmental, economic and social aspects can be difficult. Renewable resource-based technologies are an example of a high-profile topic in the sustainability arena. Among a multitude of complex and sometimes conflicting issues are topics such as reducing fossil fuel dependency, supporting the agricultural economy, employment effects in the petroleum industry, greenhouse gas emissions, soil erosion, and durable goods produced from potential food sources, etc.

In accordance with sound scientific principles, BASF, the world's leading chemical company, has taken a practical approach towards renewable resource-based products. Using our eco-efficiency analysis tool, we have demonstrated that case-by-case analysis is necessary in order to fully assess the benefits of different technologies including biotechnology-based chemicals.

BASF is the world's leading chemical company. In 2003, BASF had sales of approximately \$42 billion and over 87,000 employees worldwide. One of the company's guiding principles is ensuring sustainable development (BASF, 2003a). In BASF this means pursuing economic success, environmental protection and social responsibility. Thereby future generations will benefit from the way business is conducted in the present.

Eco-efficiency is one of BASF's tools to ensure sustainable development. It addresses environmental and economic impacts of products and processes. The tool was developed in partnership with an external consultant in 1996, and since then has been used internally and further developed (BASF, 2004).

Renewables-Based Technology Edited by J. Dewulf and H. Van Langenhove

This chapter provides an overview of eco-efficiency analysis and presents case studies involving renewable resource-based products that use biotechnology-based production pathways.

18.2 Explanation: What is Eco-Efficiency Analysis?

18.2.1 ISO Standards for LCA

Following the ISO Standards, ISO 14040 for life cycle assessment, four main LCA phases can be set out (Marsmann, 2005):

- goal and scope definition;
- inventory analysis;
- impact assessment;
- interpretation.

Furthermore, a review by interested parties, independent experts and parties who will be affected by conclusions is considered (ISO 1997a, Chapter 7.3.3). This process should end in a fair, complete and accurate reporting to third parties.

LCA is understood as a process to respond to the goal and scope of the study, to learn about a product system's interrelations between the different phases, to take into account the legitimate interests of third parties, and to help the intended audience to comprehend the results, limitations, complexity and trade-offs of a study.

LCA is one of several environmental management techniques and may not be the most appropriate technique to use in all situations (ISO 14040). There is no single procedure that can provide the right answer to all questions. LCA studies can support environmental policy decisions but further aspects have to be included (e.g. other environmental aspects, economy, social affairs) (Marsmann, 2005).

LCA attempts to predict the overall environmental burdens associated with providing a specific product or service to society on a cradle to grave basis. Compared to other environmental management tools, LCA has two unique benefits (EUROPEN, 1999; Hindle *et al.*, 1993):

- LCA attempts to consider the whole life cycle of the product or service. This helps prevent 'problem shifting' in which an apparent improvement in one part of a life cycle can merely lead to further problems at another time or place.
- LCA tries to calculate the burdens related to the function provided by the product or service. This allows a value:impact assessment to be considered (De Smet *et al.*, 1996).

Balanced against these benefits, there are also limitations to the usefulness of LCA, not least of which is that LCA does not address the economic or social factors mentioned previously.

These areas are outside the scope of an environmental LCA, but must be taken into account in any policy or decision-making process, especially in the context of sustainable development. Other limitations include:

- LCA comparisons rarely produce clear 'winners and losers'.
- LCA is but one tool in the 'environmental management toolbox'. Although it takes a life cycle approach, LCA does not address all areas of environmental management. It

cannot, for example, assess site-specific human and environmental safety (this requires a risk assessment). Consequently a framework of different environmental management tools is needed to support decision-making; LCA alone is not sufficient (White *et al.*, 1995; SETAC, 1996). This is often a point of confusion between LCA scientists and others, with the latter considering that LCA is, can be, or should be the single, holistic tool for making environmental decisions (SPOLD, 1995).
- An LCA study relates to one specific product system at one defined point in time. A study of current systems would not indicate what would be the better option in five years' time.
- The recent addition by ISO of guidelines for impact assessment and interpretation to those for goal definition and the inventory stages of the process (ISO, 1997), has firmly established LCA methodology. However, it will take time before comprehensive databases are established. Some EU states have begun to formulate their own rules for performing LCAs based on the ISO 14040 Standards (ISO, 1998; Federation of German Industries BDI, 1998).

18.2.2 Link between ISO and the BASF Method

There are limitations to the usefulness of LCA as described before. A typical LCA does not address economic or social factors. But these areas must be taken into account in any policy or decision-making process, especially in the context of sustainable development. All three dimensions need to be assessed in a product or process evaluation due to sustainability. Therefore, BASF has developed a tool for eco-efficiency analysis. This instrument provides early recognition and systematic detection of economic and environmental opportunities and risks in existing and future business activities. In BASF, the tool is part of the decision-making process for new investments.

The eco-efficiency results are presented as aggregated information on costs and environmental impact and show the strengths and weaknesses of a particular product or process. This method uses the main ideas and regulations of the ISO Standards for the basic LCA. The ecological calculations of the single results in each category follow the ISO Standards 14040 in the main. The quantitative weighting step to find the ecological fingerprint and the portfolio are not covered in the ISO Standards. The eco-efficiency analysis has more features than are mentioned by the ISO Standards.

The methodology has been approved by the German TÜV ('Technischer Überwachungsverein, Technical Inspection Association'), a leading technical service company active in the industry, product and transport sectors. Its range of services includes consultancy, inspections, tests and expert opinions as well as certification and training. This methodology was used by the Öko-Institut (Institute for Applied Ecology) in Freiburg Germany in different APME studies (APME now: Plastics Europe, formerly the Association of Plastics Manufacturers in Europe). TNO in the Netherlands are using the BASF standard method with a different weighting system. The Wuppertal Institute accepts the method: 'Basically, the large number of indicators used in the eco-efficiency analysis of BASF make relatively reliable statements possible'. The method was initially developed by BASF and Roland Berger Consulting, Munich, in 1996.

18.2.3 Eco-Efficiency Methodology at BASF

The main outline of BASF's eco-efficiency analysis method is provided next, while a more detailed discussion is available elsewhere (Saling *et al.*, 2002). Every eco-efficiency analysis passes through several key stages. This ensures consistent quality and the comparability of different studies. Environmental impacts are determined by life cycle assessment (LCA) and economic data are calculated using the usual business or, in some instances, national economical models.

The basic preconditions in eco-efficiency analysis are:

- Products or processes studied have to meet the same defined customer benefit.
- The entire life cycle is considered.
- Both an environmental and an economic assessment are carried out.

The eco-efficiency analysis is worked out by following specific and defined ways of calculations:

- calculation of total cost from the customer viewpoint;
- preparation of a specific life cycle analysis for all investigated products or processes following the main principles of ISO 14040;
- determination of impacts on the health, safety and risks to people, assessing use of area over the whole life cycle;
- calculation of relevance and calculation factors for specific weighting;
- weighting of life cycle analysis factors with societal factors;
- determination of relative importance of ecology versus economy;
- creation of an eco-efficiency portfolio;
- analyses of weaknesses, scenarios, sensitivities, and business options;
- optionally: inclusion of social aspects.

Basic Preconditions

The specific customer benefit always lies at the centre of eco-efficiency analysis. In the majority of cases, customers with particular needs and requirements are able to choose between a number of alternative products and processes. In the context of this choice, eco-efficiency analysis compares the economic and environmental data of each solution over the entire life cycle or within the compartments in which the systems differ in life cycle.

Calculation of Total Cost from the Customer Viewpoint

The costs dimension as a part of sustainability, is given equal weight to the environmental dimension in the BASF eco-efficiency method. Therefore total costs are likewise summarized over the life cycle. The costs in question are the real costs that occur and the subsequent costs, which may occur in the future (due to tax policy changes, for example). Costs having ecological aspects, for example, water treatment plant costs, are likewise included in the overall calculation. The ways to calculate costs vary from study to study. When chemical products of manufacture are being compared, the sale price paid by the customer is used. When different production methods are compared, the relevant costs include the purchase and installation of capital equipment, depreciation, and operating costs.

Preparation of a Specific Life Cycle Analysis for All Investigated Products or Processes

- *Primary energy consumption*. The energy consumption category of impact includes all energies used to fulfil the customer benefit. Fossil energy resources are included before production as is renewable energy before harvest or use. This captures conversion losses from electricity and steam generation. The energies from biomass feedstocks are included, however, not included is the sun energy that is needed to produce the biomass. The energy consumption category in the BASF eco-efficiency analysis method covers i.e. Green Engineering Principles 3, 4, and 12 of Anastas and Zimmerman (2003), and as a result encourages the selection of energy-efficient products and processes.
- *Raw materials consumption*. The raw materials consumption category considers all materials that are used over the entire life cycle of the product under study. The consumption of the materials in a mass unit (kilograms) are weighted due to each one's reserves according to the statistical calculations of the US Geological Survey (1997) and other sources (Römpp *Chemie Lexikon*, 1998; Hargreaves *et al.*, 1993; World Resources (1996); German Institute for Economic Research, 1998). These sources provide data on how long a particular raw material will remain in production assuming today's economic methods of extraction and assuming that consumption remains constant. Renewable materials have an advantage because of a zero 'Resource factor'. In cases when renewable raw materials are not sustainably managed (e.g. rainforest logging), the appropriate resource factor is applied.
- *Emissions*. Emission values are initially calculated separately as air, water and soil emissions (waste). The calculation includes not only values from, for example, electricity and steam production and transports but also values due to direct emission from the process.

  Global Warming Potential (GWP), Photochemical Ozone Creation Potential (POCP), Ozone Depletion Potential (ODP) and Acidification Potential (AP) are the categories of air emissions. The GWP is linked to emissions of carbon dioxide, methane, halogenated hydrocarbons and di-nitrogen oxide. Different impact factors are linked to the specific emissions, resulting in carbon dioxide equivalents. In the same manner the other categories are linked to specific emissions and are aggregated to defined equivalence factors in each air emission category.

  For emissions to water, there is at present no comparable standardized, scientifically documented method for calculating the impact potentials available as for the emissions to air. For the inventory of emissions to water, we therefore use the method of critical volumes or critical limits for discharges into surface waters (BUWAL, 1991). Each pollutant emitted into water contaminates a sufficient volume until the statutory limit for this substance is reached (critical load). The limits used for the respective emission to water are the limits listed in the schedule of the wastewater regulations (Abwasserverordnung, 1997).

  The results of the inventory on solid wastes are combined to form four waste categories: special wastes, wastes resembling domestic refuse, building rubble/gangue material, and overburden. Absent other criteria, impact potentials for solid wastes are formed on the basis of the average costs for the disposal of the wastes. Wastes with dangerous contents are assessed with higher factors than non-hazardous waste.

- *The toxicity potential.* To calculate the toxicity potential, each product to be calculated is balanced from the cradle to the grave. To score the toxicity of substances, a consideration of all possible effects is needed. The most widely used system in Europe is the classification of different toxic effects and the assignment of R-phrases of the European Directive 67/546/EEC. R-phrases have been used in different LCA applications, for example, in determining eco-toxicity burdens (Brackmann, 1997; Walz *et al.*, 1996). Not only the substance is considered, but also all the raw materials and reactants needed in its manufacture. Exposure to the substance is included in two ways: due to its manufacture and during its use (Landsiedel and Saling, 2002).
- *The risk potential.* The risk potential in the eco-efficiency analysis is established using assessments in the sense of an expert judgement. The focus is always on the question of the severity of the damage that an operation can cause, multiplied by the probability of it happening. In the risk potential, the damage considered is that which can be attributed, for example, to physical or chemical reactions. Examples would be explosion or fire hazards and transportation accidents. The criteria of the risk potential are variable and may be different in each study, because they are adapted to the circumstances and special features of the particular alternatives.
- *Use of area.* Area is not consumed like a raw material but, depending on the type, scope and intensity of the use, areas are changed so radically that they are impaired or even destroyed in their ability to perform their natural functions. Apart from the direct loss of fertile soil, there are a series of consequential impacts, for example, loss of living space for flora and fauna, etc. The area requirement includes production sites, transportation, and treatment/disposal. The area is included and it is assessed through a weighting of the various area categories (Bastian and Schreiber, 1994; ENET, 1996).

Normalization and Environmental Fingerprint

After normalization and weighting have been carried out for the emissions, the appropriate computed values are collected in a specific plot, the environmental fingerprint, as shown in (Figure 18.1). This diagram shows the environmental advantages and disadvantages of the considered alternatives in a relative comparison with each other. The alternative that lies furthest out and has the value of 1 is the least favorable alternative in the category in question. The further in an alternative lies, the more favorable it is.

The pivotal point of an eco-efficiency analysis is a specific customer benefit. Examples of the questions to be asked, therefore, are, 'What is the most eco-efficient method for packaging dairy products?' or 'How can an end user most eco-efficiently whitewash a wall?'

The overall cost calculation and the calculation of the environmental fingerprint constitute independent calculations of the economic and environmental considerations of a complete system, possibly with different alternatives. If it is assumed that ecology and economy are equally important in a sustainability study (as is done in the eco-efficiency analysis method of BASF), a system that is less advantageous economically can compensate for this disadvantage by a better ecological assessment, and vice versa. Alternatives

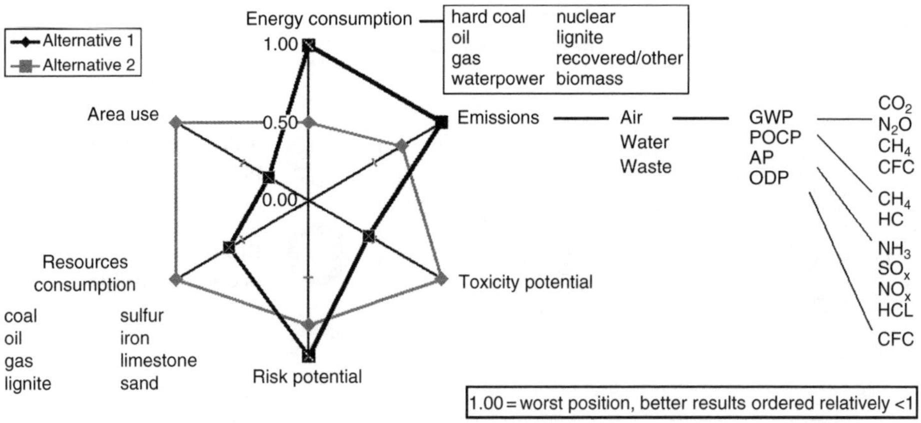

Figure 18.1 *The environmental fingerprint*

whose products are identical when assessed economically and environmentally are considered to be equally eco-efficient.

The Eco-Efficiency Portfolio

In order to be able to illustrate eco-efficiency, BASF has developed the eco-efficiency portfolio. The most eco-efficient products lie in the upper right-hand quadrant of the portfolio, which means they have the least overall environmental impact and the greatest economic benefit. The first example of such an eco-efficiency portfolio is shown in Figure 18.2 for the indigo example.

Therefore, relevance factors will be calculated and included. They indicate how important the individual environmental category is for a particular eco-efficiency analysis. Those factors are 'scientific weighting factors' because they are not influenced by a definition but only are calculated. The greater the contribution of an impact to the total category of the same impact (in Germany, for example), the higher the scientific weighting (relevance) factor. National data for other countries can also be included, for example, for the United States, Europe, Morocco, Japan, etc. Equation 18.1 defines the relevance factors for environmental categories:

$$\frac{\text{Environmental impact of an option}}{\text{Total environmental impact in Germany}} = \text{relevance}_{\text{environmental category}} \qquad (18.1)$$

The scientific weighting factors are linked to societal weighting factors which vary for different regions of the world, for each category of environmental impact. This linkage results in a final weighting scheme that combines scientific and social weighting factors.

Similar to the way that the environmental relevance factors are calculated, the total costs of a system can be related to the total sales of the manufacturing industry in the field under study. This, as in the case of the calculation of relevance factors for total environmental

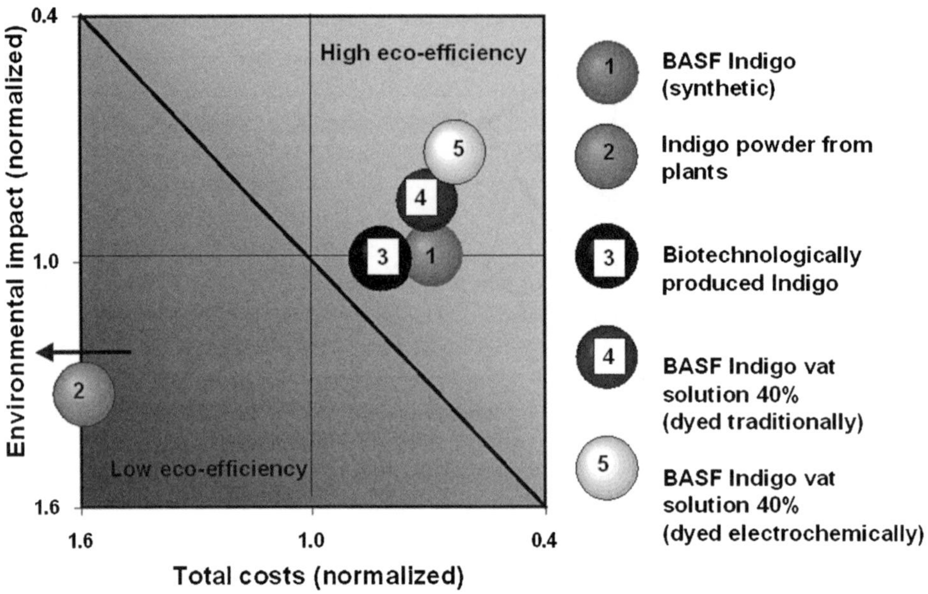

Figure 18.2 *Eco-efficiency portfolio for alternative dyeing systems*

impact, will give a relevance factor that reflects total costs. This factor reflects to what extent the alternatives studied contribute, for example, to the gross domestic product of a country. In absolute terms, the value is very small, but can be used for comparative purposes. The environmental factors and the cost relation factors are combined into an overall factor that shows whether in this study the economic or the environmental impacts are more important. Readers interested in learning more of the relevance calculations are referred to the 'methods' paper (Saling *et al.*, 2002). Figure 18.2 shows a portfolio of the Case Study 18.4.1.

Scenario Analysis

The value of the eco-efficiency analysis tool, apart from its description of the current state, lies in the recognition of dominant influences and in the illustration of 'What if...?' scenarios. Questions such as 'What is the minimum yield the new process has to have in order for it to be similar in eco-efficiency to the old process?' or 'Which site can manufacture a product most eco-efficiently?' are typical task statements for scenarios in an eco-efficiency analysis. From experience, the largest influence on the result by far is possessed by the input data and the system boundaries.

Starting with the base case portfolio, it is possible to show several different scenarios to illustrate what happens by changing a lot of input factors. This can be used to evaluate the robustness of a study, or to answer different questions dealing with future options or future developments.

Furthermore, an eco-efficiency manager tool can be added. With this manager it is possible to follow different development strategies for a product or process over the whole period of development. This could be used very effectively to determine the most important impact factors to improve the different systems due to the sustainability. The system can be evaluated by an eco-efficiency study that allows the user an online optimization by calculating a good amount of scenarios.

18.3 Evaluation of Decision-making Processes with Eco-Efficiency Analysis

Since 1996 BASF has carried out eco-efficiency studies in numerous key fields such as paints and dyes, plastics, life science, oil and gas, and chemicals. In doing so, eco-efficiency analysis has been employed in four major fields of application; (1) strategic decisions; (2) research and product development; (3) communication with policy-makers; and (4) marketing.

In strategic decisions it is possible for the application investigated to distinguish products with a promising future from products with a less promising future. Even in investment decisions eco-efficiency analysis provides valuable perspectives. For example, a potential BASF investment in a new fibre-board technology (Strawboard) in the United States did not take place due to an eco-efficiency analysis showing that the new strawboard product was only marginally better for the environment yet more costly than a conventional chipboard product with a formaldehyde-containing resin. Meanwhile, another large chemical manufacturer decided to invest heavily in that new technology. However, some years later, many of these new plants had to be closed (27) due to a slump in the construction market and in price, which put the strawboard product at a cost disadvantage.

While eco-efficiency is not the only technique that can be used in decision-making, benefits derived from its use are based on its key features; that it is a systematic methodology for incorporating a broad range of environmental impacts and costs into decisions regarding processes and products. The method is capable of handling a large number of environmental impact categories over the entire product life cycle, rather than making decisions based on just a single criterion (formaldehyde toxicity, for example).

The second field of application relates to research and product development. Promising products can be identified at an early stage, thus facilitating decision-making about the prime thrust of the development. At BASF, major R & D projects are accompanied by eco-efficiency analyses during the development phases – mini-plant, pilot plant, and basic design of a production facility – and the projects are evaluated at each milestone. The business unit funding the R & D is also involved.

The third field of application is the drawing up of position papers for discussions with opinion-makers and policy-makers. Eco-efficiency analysis makes it possible to present the complex, holistic interconnections in industrial production and product use in a graphic and readily communicable form. In this context it is possible to conduct quantitative discussions, with politicians, for instance, about the effects of planned legislation.

Eco-efficiency analysis is even used in marketing, the fourth main application area. Since the entire life cycle of a product is considered, the effects for customers are integrated into the analysis. As a result, the total vision inherent in products can be communicated to customers.

In addition to these internal uses within BASF, the eco-efficiency analysis facilitates communication and collaborative projects beyond the reach of the company's core business. Training in the use of the eco-efficiency analysis has been provided to many groups, including non-governmental organizations and the United Nations Industrial Development Organization (UNIDO) Cleaner Production Center programme.

The eco-efficiency analysis is now and will continue to be one important assessment method for R & D, production and marketing for BASF. Nearly all business units and all global regions of operation have performed studies.

Eco-efficiency analysis improves the competitiveness of BASF's products. In a recent study, it was shown that eco-efficient products perform much better in the market that non-eco-efficient products. Eco-efficiency analysis, as one important strategy and success factor in sustainable development, will continue to be a very strong operational tool at BASF (Shonnard *et al.*, 2003).

18.4 Case Studies

The eco-efficiency analysis creates different case studies to position BASF, its customers, authorities and NGOs for movement along the continuum from the current proactive phase to a sustainability strategy. Eco-efficiency is to be used as a strategic product and process development tool, which incorporates sustainability issues into planning and marketing. This chapter explains some designated examples for eco-efficiency case studies.

18.4.1 Indigo Processes

The first example chosen is the eco-efficiency analysis of indigo. Indigo is the dye that is used exclusively for dyeing blue denim. After dyeing, blue denim is further processed into jeans. BASF was the first and, until October 2000, the largest producer of synthetic indigo worldwide. Since October 2000, BASF indigo has been marketed by DyStar Textilfarben GmbH and Co. KG.

The first step in the eco-efficiency analysis is to define not only the customer benefit (functional unit) but also the possible alternatives. As many as possible of the alternatives represented in the marketplace should be included, bearing in mind that small market shares can be disregarded, depending on the problem posed.

The customer benefit was defined as follows in the indigo analysis: dyeing of blue denim to manufacture 1000 pairs of jeans. The use of indigo powder requires a relatively large amount of sodium hydrosulfite to convert the water-insoluble powder into a water-soluble form during dyeing. Even the use of indigo solutions requires the reducing agent (sodium hydrosulfite), however, in the DyStar process the reducing equivalents are provided electrochemically. Indigo from plants or produced in a biochemical process are alternatives to the traditional synthetic route of indigo production. Input data for each alternative for energy consumption, raw material utilization, emissions, land use, and safety risk were provided by the different producers of indigo and by DyStar.

The eco-efficiency portfolio for indigo production and dyeing alternatives is shown in (Figure 18.2). The positions of the alternatives refer to their relative order to each other after different normalization steps. The position 1.00 on costs axis and 1.00 on the environmental axis are linked to the overall average of the calculated alternative systems. Lower values than 1.00 refer to lower costs and environmental burdens, higher values than 1.00 refer to the opposite direction of those two categories.

Alternative 5 (synthetic indigo through the electrochemical dyeing process) is the most eco-efficient, exhibiting lower costs and environmental impacts than the other alternatives. While the production costs for Alternative 5 are similar to 1, 3, and 4 and much less than 2, the dyeing costs of 5 are dramatically lower than any of the other alternatives. Alternative 5 exhibits a marked reduction in toxicity potential compared to the others, and superior though similar impacts for emissions, energy consumption, material consumption, and risk potential. The bio-based indigo production and dyeing systems, Alternatives 2 and 3, are generally less eco-efficient compared to the synthetic indigo systems. Energy requirements for Alternatives 2 and 3 are higher, particularly in the production of the dye material, as are global warming potential, weighted material consumption, and water emissions. Toxicity potential is markedly lower for Alternatives 2 and 3 for the dye production part of the assessment, but the use of hypochlorite during the denim dyeing process part increases toxicity potential above the electrochemical process (Alternative 5). Through this case study it is shown that a biochemical-based production process for indigo production and denim dyeing is not the most eco-efficient. Although the eco-efficiency of the biochemical and plant indigo alternatives can be improved by coupling these production steps with electrochemical denim dyeing, the improvement is not enough to elevate their eco-efficiencies above the synthetic route (Alternative 5). Because bio-based processes are commonly thought to be better for the environment, this case study demonstrates the usefulness of the eco-efficiency analysis in identifying superior process alternatives.

The claim that the novel electrochemical dyeing process provides a 10% reduction in faulty batches was investigated in a scenario analysis. It is clear which eco-efficiency advantages come about as a result for this system. For clarity, the indigo powder from alternative plants was excluded. As a result, the change in one ball position will lead to changes in the positions of all alternatives involved. It was shown that in this case the electrochemical alternative was much more favourable than in the base case calculations.

The following variants and sensitivities can be calculated, modified and visualized, for example:

- determination of target corridors for research;
- breakeven point calculations;
- improvement potentials;
- strengths/weaknesses analysis;
- societal factor variations;
- testing of robustness of results;
- capital expenditures, product costs, process costs.

18.4.2 Vitamin B_2 Case Study

Vitamin B_2 is produced by BASF's Agricultural Products and Nutrition division for use as a vitamin for human and animal nutrition. As a component of animal feed, it is vital to ensure the animals' health and fitness; vitamin B_2 deficiency leads to slower growth and poor feed conversion (BASF, 2003b).

Eco-efficiency demonstrated which vitamin B_2 production process is the most eco-efficient. Three 'bio-technological' processes and one 'chemical' process were evaluated for the production of 100 kilograms of vitamin B_2 for use in animal feed pre-mix. All the processes include renewable resources such as plant oil or glucose as a raw material (Figure 18.3). The bio-technological processes use fermentation, while the chemical process starts with a bio-technological precursor and afterwards uses traditional chemistry to produce vitamin B_2.

As Figure 18.4 shows, Biotech process 1 was the most eco-efficient. It had the least overall environmental impact, and was one of the lowest cost alternatives. Biotech process 3 had noticeably higher environmental impact and higher costs. In this case the chemical process alternative had the highest cost and greater environmental impact than Biotech process 1, resulting in the lowest eco-efficiency. BASF produces vitamin B_2 via a one-step fermentation from vegetable oil with the help of the fungus Ashbya gossypii. Fermentation involves the transformation of substances with the aid of micro-organisms. BASF pioneered the move from chemical to biotechnological vitamin B_2 production on industrial scale.

The eco-efficiency analysis in this case study was able to outline and describe new goals of research activities. It was able to highlight the most important factors that influence the system and decrease its sustainability. From this point of view it is possible to define new strategies and milestones. As a consequence thereof, new sustainable processes will result after finishing the successful research activities.

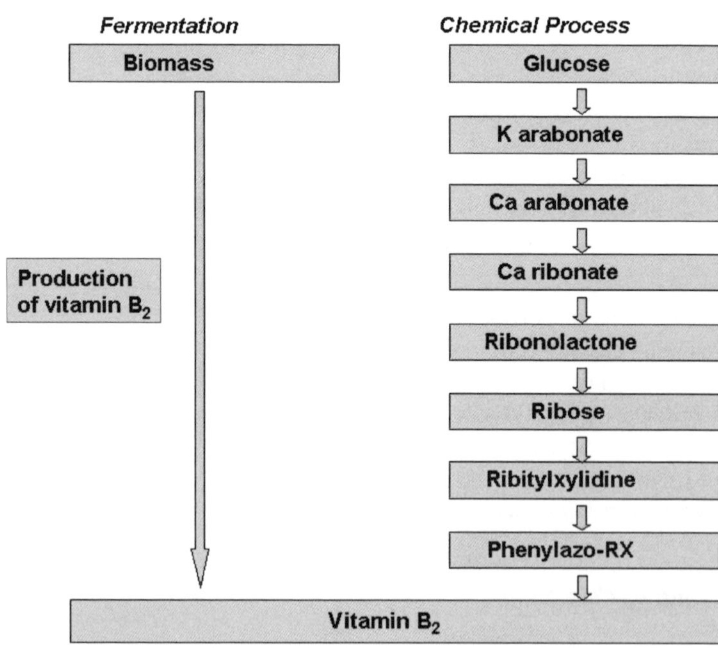

Figure 18.3 Chemical and biotechnological production pathways for vitamin B_2

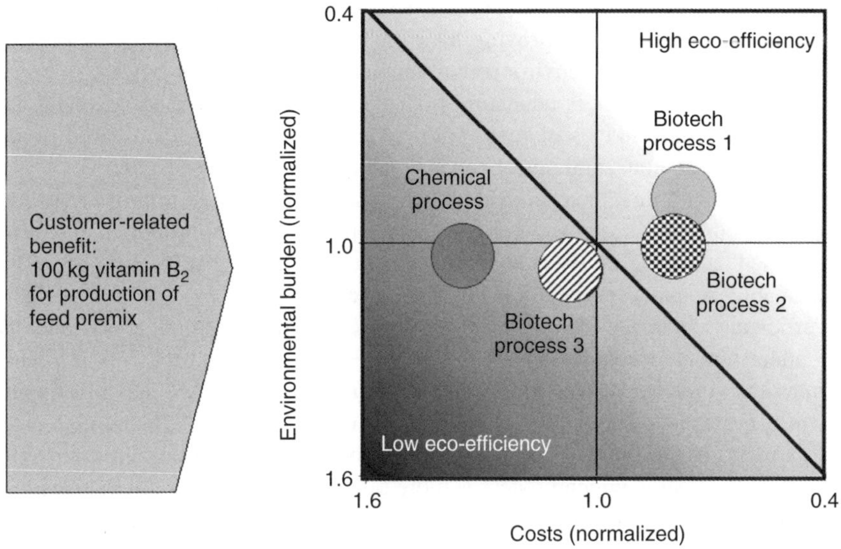

Figure 18.4 *Portfolio of vitamin B_2 production for the feed segment*

BASF recently brought a new world-scale vitamin B_2 production facility on-line in Korea. This facility, with an annual capacity of up to 3000 metric tons, uses one-step fermentation from vegetable oil. It is an excellent example of industrial-scale production using the most eco-efficient technology currently available. In this example the most eco-efficient technology used biotechnology and plant-based raw materials.

This final result can be taken in the decision-making processes and in the evaluation of sustainable processes in the area of 'white biotechnology'. In different studies, strengths and weaknesses of biotechnological processes can be compared. New processes in 'white biotechnology' and 'green biotechnology' can be assessed and compared to the actual chemical processes. The eco-efficiency study is able to support strategic decision-making processes with a different life cycle-based view of the different technologies. Without any prejudices and preferences for a certain technology, the most sustainable process can be selected and realized at BASF.

18.5 Summary

In summary, the eco-efficiency portfolio concisely represents the relative overall and economic impact of various alternatives. The ecological fingerprint provides additional details on impacts in specific environmental categories. These data can be further divided into detailed data such as air, water and solid waste emissions. Thereby eco-efficiency enables the user to understand all the effects, both 'macroscopic' and 'microscopic'.

Products derived from renewable resources are most likely to be competitive in the marketplace if they demonstrate comparable or better product quality and price versus the synthetically produced alternatives. There is significant opportunity for growth as these technologies mature, further improving their environmental fingerprint, and consumers become more sensitive to the environmental impacts of the products they use in everyday life. However, a key factor is educating both the public and industry as to the actual advantages and disadvantages of renewable resource versus petroleum-based products. The answer is not obvious, and must be evaluated on a case-by-case basis.

In accordance with this, the chemical industry should continue to improve existing and develop new petroleum-based technologies, as well as pursue opportunities in renewable resource-based technologies. By doing this, industry will ensure that the most eco-efficient and sustainable products succeed in the marketplace.

Eco-efficiency analysis can be used in a large number of applications and yields readily understandable conclusions in the case of multifactorial problems in relatively short times and at relatively low cost. Eco-efficiency analysis by BASF has already proved its worth in more than 250 studies involving not only internal BASF but also external project partners. In the future, eco-efficiency will become more important in the context of sustainability to show which process is more favourable than other alternatives.

The analysis allows a life cycle-based view of chemical products and processes that combines assessment of life-cycle environmental impacts with economic performance in equal measure to achieve a greater level of sustainability. The relevance of the eco-efficiency analysis on internal strategic decisions is very high in BASF and most analyses are presented to upper management. Nearly all business units and all global regions of operation have performed studies.

Besides promoting good corporate citizenship, eco-efficiency analysis also improves the competitiveness of the chemical industry products. In a recent study, it was shown that eco-efficient products perform much better in the market than non-eco-efficient products. Eco-efficiency analysis, as one important strategy and success factor in sustainable development, will continue to be a very strong operational tool now and in the future at BASF and other industries.

References

Abwasserverordnung (1997) *Regulation on Requirements for the Discharge of Wastewater into Surface Waters*, 27 March 1997.

Anastas, P. T. and Zimmerman, J. B. (2003) Design through the 12 Principles of Green Engineering. *Environmental Science Technology*, **37**, 94A–101A.

BASF (2003a) *BASF Corporate Report 2003*, BASF, p. 10.

BASF (2003b) BASF Group Press release, 'A big step forward in the extension of BASF's vitamins business', 10 November 2003, P-03–495.

BASF (2004) www.basf.de/en; www.oekoeffizienzanalyse.de.

Bastian, O. and Schreiber, K.-F. (1994) *Analyse und Ökologischen Bewertung der Landschaft*, Gustav Fisher Verlag, Jena.

Brackmann, H. (1997) Abteilung von Wassergefährdungsklassen (WGK) aus den Einstufungen des Gefahrstoffrechts: Ein Beitrag zur Harmonisierung der Stoffbewertung, UTA 4/97, 295.

BUWAL (Bundesamt für Umwelt, Wald und Landschaft) (1991) *Schriftenreihe Umwelt* Nr. 132, BUWAL, Bern.

De Smet, B., White, P. R. and Owens, J. W. (1996) Integrating lifecycle assessment within an overall framework for environmental management, in Curran, M. A. (ed.) *Lifecycle Assessment*, Mc-Graw Hill, New York.

ENET (1996) *Gruppe Energie-Stoffe-Umwelt: Ökoinventare von Energiesystemen*, Bundesamt für Energiewirtschaft (Switzerland), Bern.
EUROPEN (1999) *Use of Life Cycle Assessment (LCA) as a Policy Tool in the Field of Sustainable Packaging Waste Management*, EUROPEN.
Federation of German Industries (BDI) (1998) *Innovation in the Market Economy: The Political Role of Life Cycle Assessments (LCAs)*, German Institute for Economic Research, Berlin.
Hargreaves, D., Eden-Green, M. and Devaney, J. (1993) *World Index of Resources and Population*, Dartmouth Publishing, Aldershot.
Hindle, P., White, P. R and Minion, K. (1993) Achieving real environmental improvements using value: impact assessment, *Long Range Planning*, **26**, 36–48.
ISO (International Standard Organization) (1997a) *ISO 14040 Environmental Management – Life Cycle Assessment – Principles and Framework*, ISO, Geneva.
ISO (1997b) *ISO 14041 Environmental Management – Life Cycle Assessment – Goal and Scope*, ISO, Geneva.
ISO (1998) *Definition and Inventory Analysis (1998). ISO Draft International Standard 14042 Environmental Management – Life Cycle Assessment – Life Cycle Impact Assessment* ISO (1999) *ISO 14043 Environmental Management – Life Cycle Assessment – Life Cycle Interpretation*, ISO, Geneva.
Landsiedel, R. and Saling, P. (2002) Assessment of Toxicological Risks for Life Cycle Assessment and Eco-efficiency Analysis, *International Journal of LCA*, **7**, 261–268.
Marsmann, M. (2005) Bayer AG, www.europen.be/issues/lca/LCA_workshop/Presentations/Marsmann,%20Manfred.ppt.
McCoy, M. (2001) The business fallout of 'ecoefficiency', *Chemical Engineering News*, **79**, 36–36.
Römpp *Chemie Lexikon* (1998) Thieme, Stuttgart; Institute for World Economy, Keil.
Saling, P., Kicherer, A., Dittrich-Krämer, B., Wittlinger, R., Zombik, W., Schmidt, I., Schrott, W. and Schmidt, S. (2002) Eco-efficiency Analysis by BASF: The Method, *International Journal of LCA*, **7**, 203–218.
SETAC (1996) Working Group on Conceptually Related Programmes, Draft report, Taormina, Sicily, May 1996.
Shonnard, D. R., Kicherer, A. and Saling, P. (2003) Industrial Applications Using BASF Eco-Efficiency Analysis: Perspectives on Green Engineering Principles, *Environmental Science and Technology*, **37**, 5340–5348.
SPOLD (The Society for Promotion of Lifecycle Development) (1995) *Synthesis Report on the Social Value of LCA*, SPOLD Workshop, Brussels, June 20–21, 1995. SPOLD, Brussels.
US Geological Survey (1997) *Mineral Commodity Summaries*, US Department of Agriculture, Washington, DC.
Walz, R., Herrchen, M., Keller, D. and Stahl, B. (1996) Impact category ecotoxicity and the valuation procedure. Ecotoxicological impact assessment and the valuation step within LCA – Pragmatic approaches, *International Journal of LCA*, **4**, 193–198.
White, P. R, De Smet, B., Owens, J. W. and Hindle, P. (1995). Environmental management in an international consumer goods company, *Resources, Conservation and Recycling*, **14**, 171–184.
World Resources (1996) *Guide to the Global Environment*, World Resources, Oxford.

19

Assessment of Bio-Based Pharmaceuticals

The Cephalexin Case

Alle Bruggink and Peter Nossin

19.1 Introduction

Pharmaceuticals and fine chemicals were among the first industrial products to feel the impact of biocatalysis in their manufacturing processes. After more than an age of stoichiometric chemistry these industry segments started to feel the need for better environmental control. Catalysis was the obvious answer to this demand, but a simple translation of the vast experience in catalysis in the petrochemical industry was by no means an easy solution. In particular, the elevated temperatures, quite common and often conditional to successful chemocatalysis, are not compatible with most of the multifunctional molecules in pharmaceuticals and fine chemicals. The use of enzymes, mostly employing ambient reaction conditions, paved the way to bio-based processes for many products starting in the 1980s. The semi-synthetic antibiotics, i.e. the penicillins and cephalosporins, were the leading group of products to prove the efficiency, both economically and environmentally, of biocatalysis. Several other products have followed to feel the benefits of biocatalysis and among the present level of successful examples, are acryl amide, glucose/fructose, lactic acid, D-phenylalanine and propanediol, in the industrial segment of bulk chemicals (Liese *et al.*, 2000). In the sector for fine chemicals and pharmaceuticals, biocatalytic processes are available nowadays for most generic drugs, mostly as second-generation processes in order to meet economic and environmental demands. Although, from an academic viewpoint, biocatalysts are available for the manipulation of all synthetically meaningful chemical bonds, the use of biocatalysis is stagnating in the synthesis of new drugs (Schoemaker *et al.*, 2003). The lack of a wide and timely

Renewables-Based Technology Edited by J. Dewulf and H. Van Langenhove

availability of robust biocatalyst formulations with a proven range of applications (like chemocatalysts in the petrochemical sector) and the ever-increasing pressure for shorter 'time to market' for new drugs are causing this undesirable situation. At the same time, chemocatalysis is increasingly able to operate at more ambient reaction conditions but by no means fast enough to take over the dominant role of biocatalysis in the present and future synthesis of drugs and fine chemicals. In particular, the limited selectivity of chem-catalysts when confronted with the multi-functionality of most pharmaceutical molecules is hampering their breakthrough. Further development of biocatalysts, including robust formulations and simple reactor devices, will lead the way towards more sustainable processes for pharmaceuticals. Eventually, biocatalysis and biosynthesis will allow the pharmaceutical industry to escape from the present dilemma in which the complexity of the molecules is increasing faster than the catalytic methods available for sustainable processes, forcing the industry to continue with multi-step stoichiometric chemistry (Bruggink *et al.*, 2003).

Next to biocatalysis, full biosynthesis (fermentation) is making its way in the production of drugs and fine chemicals. However, biosynthesis of non-natural products as a more sustainable alternative to existing chemical processes is still very much in its infancy. Biopharmaceuticals, which are only available through fermentation or extraction from purely natural sources, are not discussed in this chapter. β-Lactam antibiotics, however, are in fact biopharmaceuticals '*avant la lettre*', as the basis of all these products is in the natural penicillins and cephalosporins. Industrial and therapeutic development in the past 35 years, however, has evolved only around the semi-synthetic product variants in which stoichiometric, synthetic organic chemistry has played a pivotal role. Much of the advancements of organic chemistry in the past decades were inspired by the molecular challenges of important drugs such as antibiotics and cardiovasculars. The replacement of this type of modern chemistry by (bio-)catalytic methods is the answer to the desire for more efficient and environmentally compatible processes. As for biocatalysis, the β-lactam antibiotics have provided also the first examples of chemical processes being replaced by biosynthesis. Whereas biocatalysis is mainly restricted to employing enzymes, catalyzing a single molecular conversion, biosynthesis employs living cells capable of multi-step conversions starting with simple renewable feedstocks (i.e. fermentations based on sugar). Further research in genomics and metabolic pathway engineering will have its impact on this type of biosynthesis of drugs and pharmaceutical building blocks. It is to be expected, however, that the first industrial results will be seen in improved and increased availability of enzymes and biocatalysts followed by improvements of existing fermentation processes. Full biosynthesis of non-natural products will have to wait a decade or so before seeing industrial reality.

19.2 Assessment Methods During Process Development and Technology Transfers

19.2.1 History and Growth of the Need for Adequate Assessment Methods at DSM

On 4 April 2001 the DSM Board set up a Steering Group on Sustainability at corporate level with the aim to actively pursue the principles of sustainability in its internal activities and procedures and its external communications. A well-working corporate policy on sustainability, annual reports based on the sustainability principles of people, planet and profit and a number one position in the Dow Jones Index Sustainability Group for the chemical sector are the gratifying results. Such achievements can only be reached with a history of awareness of social and environmental aspects next to economic parameters. Over one hundred years of transformations at DSM, from coal and fertilizers through petrochemicals to performance materials and life sciences-based products, could only be successful when integrated views on economic and social parameters were available. Full awareness of environmental responsibilities emerged some 30 years ago, resulting in product stewardship practices and a quantified approach as equally valid as economic factors in product and process developments. The present DSM policy or sustainability is governed by the balancing of its three main elements:

- essential contributions to its social surroundings (People);
- efficient usage of all resources (Planet);
- profitable economic growth (Profit).

The corporate mission to go with this policy is formulated as: 'Supply of products and services enabling a sustainable performance of our customers and end-users, thereby applying sustainable processes and procedures in manufacturing our products with a clear and transparent communication about our behavior and performances.' It is DSM's vision to serve this mission by developing a competitive edge through the integration of biology, chemistry and physics. This corresponds to a major shift of focus in DSM activities towards life sciences-based products.

Translated into technological development, this constitutes a desire to improve the main parameters in our processes by at least a factor of 10; the factor 10 being the result of a growing world population by a factor 1.5 to 2 and a more evenly spread welfare by a factor 4 to 5, compensated by a 6 to 10 times more efficient use of all resources. Thus, by incorporating modern molecular biology in our commercial and technological portfolio, we expect to improve important product and process positions to the order of a factor of 10, thereby stimulating a move towards a more sustainable society and environment. Given this clear and quantified goal, the need for a quantitative and verifiable measurement of technological plans and achievements was apparent.

19.2.2 Sustainability Assessment in the Early Phases of Development

As is common in a business-to-business industry, process development and process improvements are the most important R&D activities. In the early phases several process profile analyses are used to assess the value of various process options. In its most simple form a matrix is constructed, showing the key parameters for a number of process alternatives vs its score in qualitative terms. In fact, this approach is applicable to most R&D projects and proposals. The quality of many research proposals, both industrial and academic, would improve considerably if these simple matrices were used more often. The majority of research projects are aiming at improving a particular existing situation. Thus, in a matrix similar to Figure 19.1, it should be no problem to compare the existing situation with the research proposal, on one hand, and with a competing alternative, on the other.

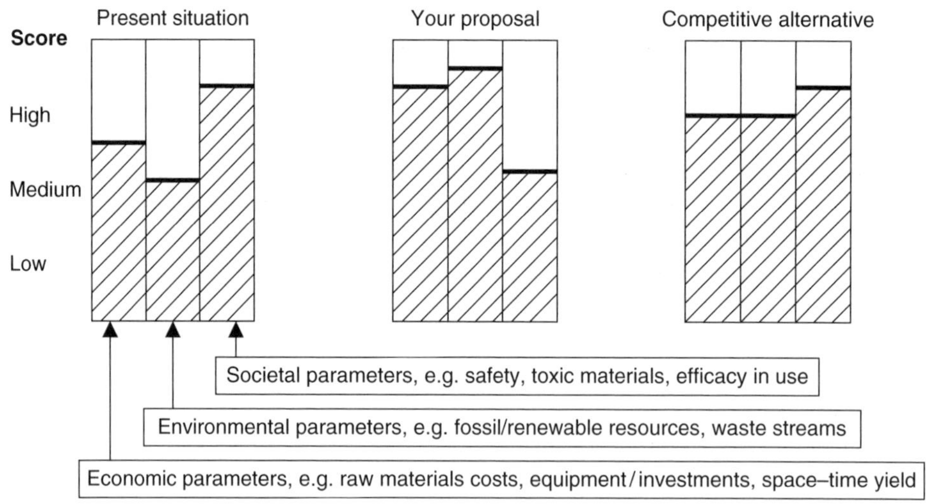

Figure 19.1 *Sustainability check in early R&D phases; comparing existing situation with the proposed research and a competing alternative; equal development stages should be assumed or projected*

This would also ease the work of many referees in showing the added value of the research proposal in a more qualified manner. It would also allow for more scientifically sound introductions in most scientific publications about the economic and/or societal relevance of the published work. If so desired, the matrix of Figure 19.1 can also be drawn with the existing situation as reference and showing the potential advantages of alternatives as scores on an arbitrary scale.

In the next step the parameters can be subdivided, specified and given a relative weight, leading to a semi-quantitative arrangement of process routes. In a more advanced stage the parameters and weighing can be replaced by expected or real cost elements, leading to a cost curve as exemplified in Figure 19.2. These profile analyses have been improved and applied successfully for a period of over 25 years at DSM in:

- choosing the best process routes for further development in the earliest R&D stages;
- reducing the number of alternatives to be explored or continued in the costly practical stages of development;
- assessing the optimal process variant for a given location;
- comparing competitor processes for purposes of technology assessment (see Figure 19.2a);
- comparing competitor processes at actual manufacturing conditions (see Figure 19.2b).

The main advantages of these approaches are the clear focus in all activities together with quantified and verifiable parameters in non-technical language. The power of these analyses lies also in the consistency of its application over a number of years, in particular, in competition assessment. Very often, companies change their methods too easily, blurring the value of historic comparisons. Personal bias can play a role in assessments in the early development phases. Proper parameter definition and group exercises can reduce these aspects.

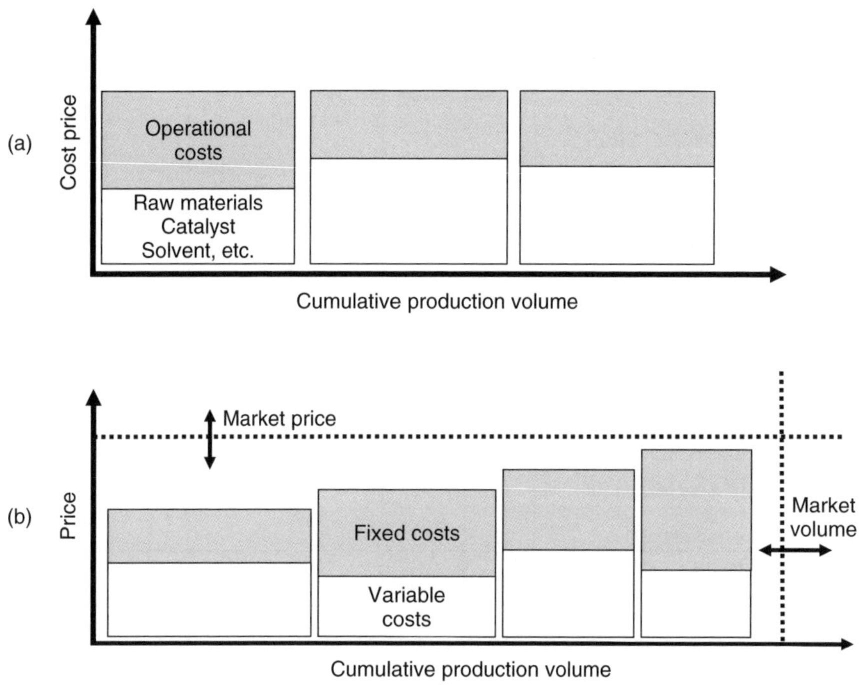

Figure 19.2 *(a) Technology assessment: three industrial processes under identical socio-economic conditions showing very similar economics, i.e. no differentiation for technology reasons (D-p-hydroxyphenylglycine); (b) Competition analysis showing four producers at different locations employing one of the processes shown in Figure 19.3a (D-p-hydroxyphenylglycine)*

19.2.3 Sustainability Assessment in the Engineering and Commercial Stages

In principle, the methods used in the final stages of development and the early stages of commercialization are the same. Only the level of quantification is increased, requiring a lot of concrete data, and the analyses are extended as much as possible to the full value chain. Key suppliers, key customers and end markets (consumers or patients) should be included. DSM has developed a set of quantifiable parameters for the three main elements of sustainability: People, Planet and Profit. The tool can be used as a self-assessment tool and allows a simple presentation of a preferred solution vs a competing or existing solution. It can also be used in portfolio analyses. Figure 19.3 shows the details of our sustainability metrics.

Figure 19.3a shows Cephalexin as an example of possible parameters on the People component in sustainability. Parameters should be chosen carefully with maximum relevance to the type of business. A maximum of 10 to 15 parameters are recommended, the majority of which should be broadly applicable within the given market sector. A limited number of parameters specific to the project can be chosen. Agreement on which parameters will be used in the scoring should be made beforehand. A maximum score of five points can be reached per parameter. The scoring should be based on

(a)

Score **Project Performance Indicators**	0 <0%	1 0–5%	2 5–20%	3 20–30%	4 30–50%	5 >50%
Mandatory PIs						
1. Market need (**Niche=1; Basic=3; Unmet=5**)			**2**			
Decrease in toxicity potential						
2. of upstream raw materials						**5**
3. of downstream materials or products		**1**				
Decrease in risk potential						
4. of maximum pressure		**1**				
5. of maximum temperature		**1**				
6. of no. of hazardous compounds to handle		**1**				
7. of no. of hazardous compounds to transport		**1**				
8. of non-contained GMOs		**1**				
Facultative PIs						
9. Decrease in toxicity potential of additives			**2**			
10. Reduction of disposal risk						
Business Specific PIs						
11.						
12.						
People score **(based on 8 mandatory + 1 facultative PIs)**						**Σ=15** **max = 45**

(b)

Score **Project Performance Indicators**	0 <0%	1 0–5%	2 5–20%	3 20–30%	4 30–50%	5 >50%
Mandatory PIs						
Reduction in raw material consumption		**1**				
1. Oil		**1**				
2. Natural gas		**1**				
3. Metals						
Reduction in energy consumption		**1**				
4. Steam		**1**				
5. Electricity						
Reduction in emissions to air		**1**				
6. CO_2 equivalents		**1**				
7. Hydrocarbons		**1**				
8. Halogenated organics		**1**				
9. Acid compounds						
Reduction in emissions to water		**1**				
10. Oxygen demand (BOD/COD)		**1**				
11. Kjeldahl-N						
Reduction in emission to soil		**1**				
12. Special solid waste						
Business Specific PIs						
13.						
Planet score **(based on 12 mandatory PIs)**						**Σ = 12** **max = 60**

Figure 19.3 *(a) Sustainability metrics of People (b) Sustainability metrics of Planet (c) Sustainability metrics of Profit*

(c)

Score Project Performance Indicators	0 <0%	1 0–5%	2 5–20%	3 20–30%	4 30–50%	5 >50%
Mandatory PIs						
Reduction in investment capital						
1. At (your) company level				3		
2. At customer level		1				
Reduction in process costs						
3. At company level		1				
4. At customer level		1				
Reduction in variable costs						
5. At company level		1				
6. At customer level		1				
Margin increase						
7. At company level		1				
8. At customer level		1				
Facultative PIs						
9. Reduction of time-to-market						
10. Increase of company market sales			2			
Business Specific PIs						
11.						
12.						
Profit score (based on 8 mandatory + 1 facultative PIs)						

Figure 19.3 *(Continued)*

concrete, measurable performance indicators. Thus, most parameters are defined as a potential increase or decrease for an important performance indicator. The scores given in Figure 19.3a are representative of the Cephalexin case discussed below. Please note that the scores given in Figure 19.3 are only an example and are of no relevance to the Cephalexin case discussed below.

Depending on the type of business, a completely different set of parameters can be envisaged, i.e. in transferring a process to a developing country, education chances, employability, rural development, child labour, etc. can be used. If desired, transport and storage issues or personal safety issues could be given extra attention. The parameters shown in Figure 19.3b for the Planet component of sustainability are rather straightforward. Again, parameters should be chosen with a maximum of relevance to the processes or products involved and allow for a maximum discriminative power. The scoring is preferably in terms of relative reductions (or increases) on well-known indicators. The scores given in Figure 19.3b are representative of the Cephalexin case discussed below. Also on the Profit component (Figure 19.3c) the metrics are rather self-explanatory. Moreover, most companies will have sufficient experience in assessing the economic value of their business and possible or competing alternatives. Full attention to the advantages for the customer should be noted.

The total score can be reached in different ways. Individual scores on People, Planet and Profit can simply be added or a particular weighting, agreed in advance, can be used. The results can be presented in simple tables or graphs understandable to a wide range of interested decision-makers. In most cases the process of filling out the various tables is more important and produces more awareness raising than the end result. Missing data, non-comparable development stages, insufficient discrimination of parameter scores give rise to many fruitful discussions and additional research. All lead to increased insights and awareness for a much greater part of the full value chain and the life cycle of the products involved. Although the tool should not be used to calculate absolute scores on sustainability (which are of very limited use anyway), it can be applied to a business portfolio or a particular market sector. Special attention should then be given to an agreed set of parameters, the weightings and comparable development stages (i.e. a level playing field for all options).

During the development process the tool can be applied at several stages. Confirmation of earlier choices and expectations are always important issues. Also new ideas or upcoming competitors and their patent publications can be assessed in this way. In the latter cases unequal development stages often form a big hurdle. A group discussion will be needed to assess and agree on the development potential of a new idea or a competitor threat.

19.3 Assessment of Bio-Based Routes to Cephalexin

Cephalexin, with an annual volume of over 4000 tons, is the largest semi-synthetic cephalosporin on the market and an important representative of the group of β-lactam antibiotics, the world's largest single cluster of molecular closely related anti-infectives (Bruggink, 2001). Continuous market growth, the present global volume of lactam antibiotics amounts to over 50 000 tons/year, and ongoing product improvements and renewals for over a period of 40 years have made the manufacturing of these products into a global business, feeling and following all the major trends in economy and technology. Lapsed patent protections and stiff competition from the fast-growing markets in India and China have forced the producers into a continuous battle of process improvements and renewals together with increasing volumes and backward and/or forward integrations to reach economy of scale. Lack of serious competition from sizeable new anti-infectives allow the β-lactams a continued healthy life cycle with probably another 20 to 25 years, helped by increasing attention to more conscious applications in its various therapies to combat resistance. This stimulates very much continuous process innovations despite the generic and mature character of most commercial β-lactams. The overall result is the worldwide availability of these life-saving drugs for very attractive prices as low as $6/kg for Pen.G, the basic building block for most producers (see also Figure 19.4) and $50–60/kg for bulk Cephalexin.

19.3.1 Traditional Routes to Semi-Synthetic Antibiotics

The onset of the therapeutic use of β-lactam antibiotics through the natural penicillins Pen.G and Pen.V, all prepared by fermentation, could have been the basis of a very early example of bio-based industrial productions, both in its feedstock and the further conversion processes. The quickly emerging need for non-natural semi-synthetic antibiotics and the lack of success in early attempts with enzymes in synthesis schemes favoured the application of synthetic organic chemistry, the latter

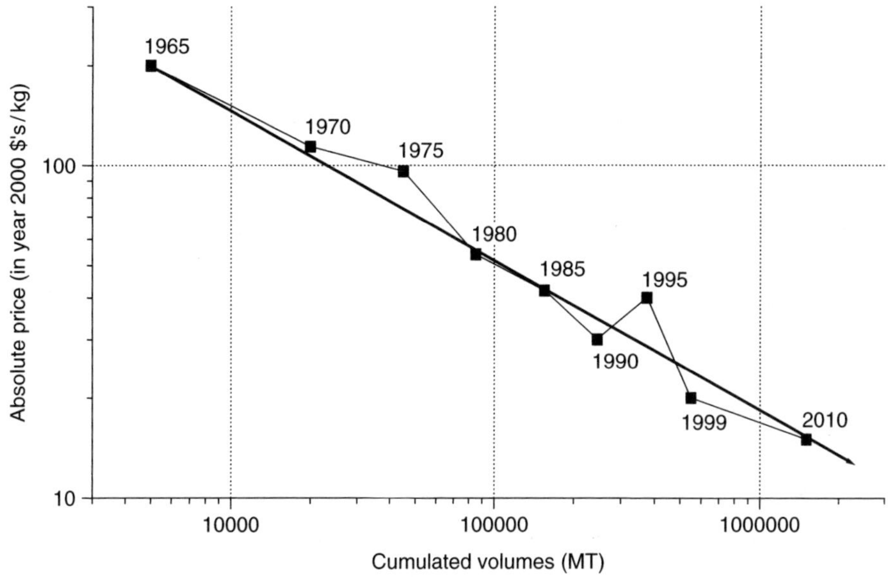

Figure 19.4 *Experience curve for Pen.G prices 1965–2010*

being spurred by the memorable breakthroughs of organic synthesis in the period 1960–80. In fact, only the crucial β-lactam nucleus remained devoid of synthetic approaches on an industrial scale, whereas all other parts of the molecules were manipulated through (modern) organic synthetic methods. The reaction scheme for Cephalexin as originally practised by DSM and several others is shown in Figure 19.5. The fully integrated scheme starting with non-renewable resources such as oil for the synthetic parts and renewable feedstock, i.e. sugar, for the lactam nucleus is shown. The scheme is the result of a 25-year period of route improvements and integrations and acquisitions of several manufacturing sites. Whereas the early process variants showed overall yields below 50% (based on Pen.G) and produced well over 50 kg of waste per kg of end product, the process nowadays reaches yields of around 80% though still producing ca. 15 kg of waste per kg of Cephalexin. Most improvements were achieved through continuous process optimizations and an increasing number of recycling of solvents and reagents.

Whereas up to six different companies were involved in the early periods of Cephalexin manufacturing, present productions are limited to one to three or four companies serving the complete product chain. Fully integrated plants, however, have not been built yet. Mostly four to six different plants are operated to cover the full chain.

19.3.2 Bio-Based Routes to Cephalexin

As mentioned before, the semi-synthetic antibiotics were among the first industrial products to experience the benefits of biocatalysis. The historic background in fermentation in

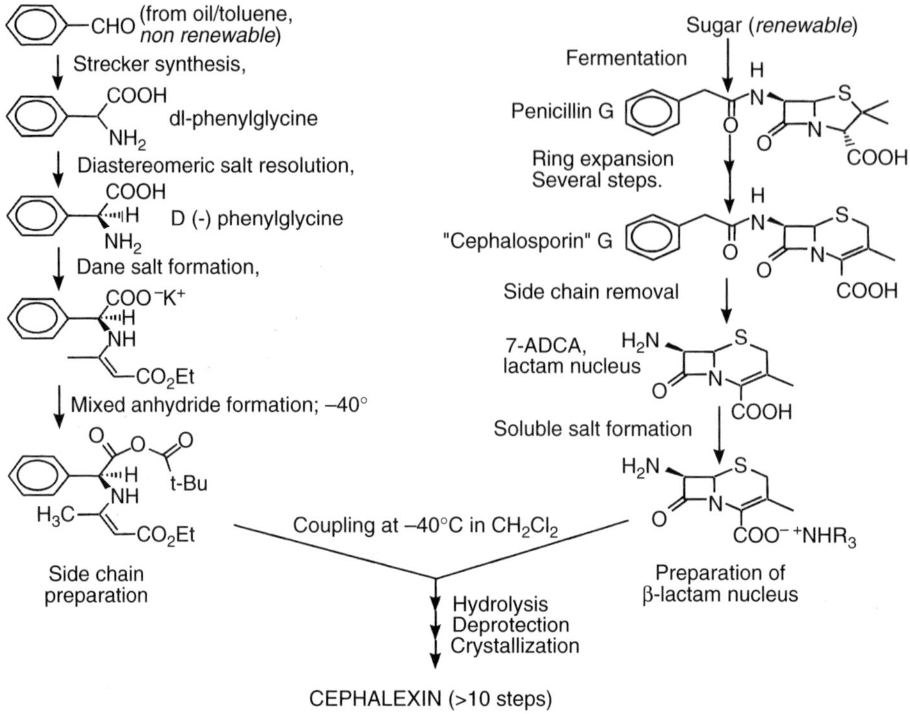

Figure 19.5 *Integrated, semi-synthetic route for Cephalexin (DSM); note stoichiometric chemistry in all steps except sugar-based Penicillin G fermentation*

many companies together with the drivers presented in the previous section laid the basis for this early involvement. The relatively simple hydrolytic steps in which the natural side chains were removed from the β-lactam nucleus were the first to be done enzymatically, employing well-known amide acylases such as derived from *E.Coli.* As early as 1985 almost all 6-APA, the β-lactam nucleus for the semi-synthetic penicillins Ampicillin and Amoxicillin, was produced from Pen.G with biocatalysis. A decade was needed to bring the reverse reaction, coupling of a β-lactam nucleus such as 6-APA or 7-ADCA with a non-natural side chain, i.e. phenylglycine, to industrial practice. Cephalexin was not by accident the first representative. Product prices, market volumes and the commercial expectations at that time, 1990–2000, all just came together in these process innovations. Other cephalosporins were too small to cover the development costs whereas the large penicillins like Amoxicillin and Ampicillin were too large to take the entrepreneurial risk. A broad collaboration with the Dutch universities, well stimulated by the Dutch government, eventually provided a strong technological base for successful enzymatic coupling of 7-ADCA with D-phenylglycine amide at a few hundreds ton scale at the DSM site in Barcelona in 1997. Presently, industrially reliable enzyme catalyzed coupling procedures are available for all β-lactam nuclei with all commercially relevant synthetic side chains. Biocatalysis also made its way into the synthesis of the required side chains. Figure 19.6 summarizes the present (bio)synthesis route for Cephalexin, including the biosynthesis of

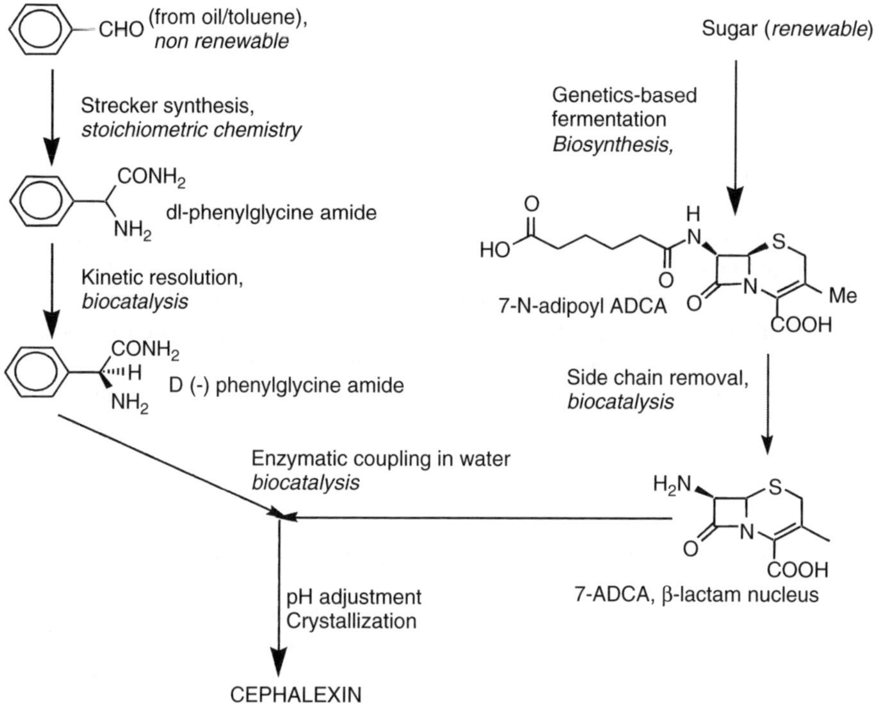

Figure 19.6 *Bio-based process scheme for Cephalexin*
Source: DSM.

7-ADCA to be discussed below. After a few years of practice the biocatalytic process met all expectations in terms of economic and environmental advantages (discussed in more detail in Section 19.3.3). Moreover, product quality improved with respect to the level and number of impurities, crystal form and reproducibility.

A more deep-seated process change was underway in the meantime: the biosynthesis of the β-lactam nucleus 7-ADCA. The very elegant multi-step process developed by Gist Brocades in the 1970s lost competitive power because of its completely stoichiometric character demanding costly and laborious recycling of solvents and reagents or unacceptable environmental consequences. Genetic engineering of *P. Chrysogenum*, the well-known penicillin-producing mould, allowed ring expansion of the five-membered thiazoline ring in the penicillin skeleton to the six-membered ring of the desired cephalosporin moiety inside the cells of the micro-organism. The process performs best when the traditional side chain precursor, phenylacetic acid, is replaced by adipic acid. The latter can readily be removed by enzymatic hydrolysis employing a glutaryl-related enzyme. Product purification and isolation proved to be quite different from the experiences in the chemical process, eventually resulting in similar advantages in economy, ecology and quality as found with the enzymatically prepared Cephalexin. Industrial production of 7-ADCA through this bio-route was started successfully in 2000 at the DSM Gist site in Delft and is now running at full scale.

The remarkably short period of ca. 10 years needed to replace the traditional chemical routes by bio-based processes poses the question of the next process changes. Present state of the art in metabolic pathway engineering allows for bio-based synthesis of all intermediates, nuclei and side chains alike, of all large-scale commercialized lactam antibiotics. Even full biosyntheses of end products like Ampicillin or Amoxicillin are no longer utopistic. Whether these processes will see industrial reality is quite a different question. Product life cycles and market potentials must be large enough to earn back the substantial development costs, which is not an easy task given the already very low prices of the bulk end products. In fact, only full biosynthesis of complete end products meeting the efficiencies of the well-proven biosynthesis of natural Pen.G would promise sufficient economic and environmental advantages to allow for all development costs and investments. A more likely development could be a further integration of process steps on a single production site. Today, the process shown in Figure 19.6 is operated at three different sites: Delft for the 7-ADCA nucleus, Geleen (NL) for the side chain and Barcelona for the enzymatic coupling.

19.3.3 Sustainability Metrics Applied to the Cephalexin Processes

Alongside the development of technology and sustainability assessments as presented above, the assessments of the green process routes to Cephalexin were evaluated. Throughout the development of the enzymatic process, the projected advantages in costs for chemicals and environmental factors were confirmed. Major breakthroughs during the overall development period of 6–8 years were achieved in productivity and stability of the required biocatalysts, allowing space–time yields (kg/l/hr) equal or better than the traditional, fully optimized stoichiometric processes. In more general terms, the advantages of catalysis over stoichiometric processes were fully confirmed. Approval for the required investments, millions of dollars for the enzymatic coupling and tens of million dollars for the direct biosynthesis of 7-ADCA, were in fact the most difficult hurdle. Given the stiff international competition, mainly stemming from India and China, short payback times for all investments were crucial. The bio-routes were implemented at plant scale in the period 1996–2001 and assessed for sustainability at their stable and robust performance in 2003. A comparison was made with the traditional routes at their top performance between 1990 and 1996. Data were taken from running plants in Europe and projected to a single site in The Netherlands, thus excluding transport costs. Assessments were done from 'cradle to gate', i.e. starting from the most basic sources for energy and materials and up to the drum of bulk active ingredient Cephalexin leaving the factory on its way to the customer for formulation and further distribution. To prevent any DSM bias, all assessments were done with the well-known independent Öko Institute for LCA (life cycle analysis) in Germany.

The spider's web diagram in Figure 19.7 shows a clear picture of the profound advantages of biocatalysis over traditional stoichiometric chemistry. Improvements of over a factor 2 are achieved at every parameter, all related to the production of 1 kg of Cephalexin. Parameters are briefly discussed below. All parameters are normalized according to existing LCA standards (Oeko, 2004).

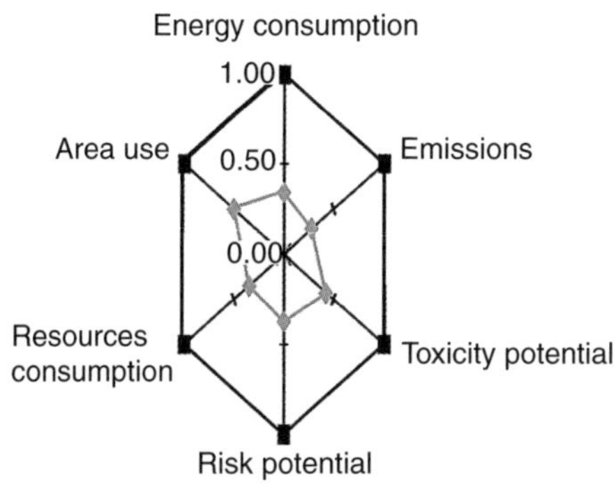

Figure 19.7 Sustainability assessment of Cephalexin processes: bio-based route vs stoichiometric chemistry as reference

Resource Consumptions

The inputs of all non-renewable resources such as coal, oil and minerals are measured. They are weighted according to their expected lifetime and added. Oil has an expected lifetime of 40 years and therefore is weighted four times higher as coal with a lifetime of ca. 160 years. Renewables are considered to have an unlimited lifetime. The bio-route shows a 66% reduction on resources. Main advantages are: less oil, mainly because phenylacetic acid as a large oil-consuming agent, is omitted and less natural gas as source for generation of silylating agents (which is no longer required). The bio-based process consumes, surprisingly, 5% less sugar, reflecting the high process efficiency. The overall use of water is reduced by 90%! As most bio-based processes employ water as solvent, this is at first sight an unexpected outcome.

Energy Consumption

All renewable and non-renewable energy carriers are added according to their heating value, amounting to a reduction of 66% in the bio-route. Electricity consumption is reduced by 54% and steam consumption by 68%. Consequently less coal and natural gas at source are used as depletable source because of less process heat and on-site electricity generation. The reduced energy consumption has the largest impact on the total sustainability improvement: 17% out of a total of 64%. Resources and energy together represent half of the total effect.

Emissions

This aggregated parameter constitutes waste (90% reduction in landfill waste), emissions to water (reduced by 52%) and emissions to air. The latter are quantified in terms of contribution to global warming potential (reduced by 83%), photochemical ozone creation potential (reduced by 50%) and acidification potential (reduced by 55% in the bio-route). For example, halogenated solvents are completely eliminated in the bio-based process.

Toxicity potential

Elimination of numerous toxic solvents and reagents results in a 58% reduction, the third largest impact on the total score.

Risk potential

The risk potential is reduced by 63% for virtually the same reasons as for toxicity.

Area demand

The area demand is reduced by 50%, mainly due to lower electricity demand for the bio-route. Less area demand is the consequence of decreased transport and generation of both electricity and energy carriers, such as coal, natural gas and/or biomass. The higher process efficiency results in a 5% lower sugar demand and consequently a 5% lower area demand for the production of agricultural feedstocks.

Due to the limitations of the LCA methodology and the restriction of our analyses from basic materials to bulk active ingredient, the People component in this sustainability assessment is slightly under-estimated. Apart from reductions in risk and toxicity potential, the bio-based Cephalexin also has some distinct advantages in its further downstream processing and applications. The absence of residual solvents in the formulated product, together with a much simpler impurity profile is an advantage to all patients taking these types of antibiotics. In particular, the well-known bitter taste of most penicillins and cephalosporins is no longer present in the products resulting from the green routes. Improvements in crystal properties, crystal size reproducibility and particle size distribution allow more reliability and robustness in formulation processes. Of greater advantage are the absence of the bad smell of traditionally prepared products and the much longer shelf life. Virtually all the advantages and improvements as described above for Cephalexin have also been realized for the related penicillins and cephalosporins. DSM is harvesting these achievements through their recent launch of a range of brand name products: DSMPureACTIVES™. Thus, Cephalexin is marketed as Purilex™ together with Purimox® for Amoxicilln and Puridrox™ for Cefadroxil.

19.4 Summary

Renewable resource-based biotechnologies are having a threefold impact on pharmaceuticals:

1. Molecular biology, including genomics, proteomics, metabolic pathway engineering etc., is are a rich source for new cell-based medicines, i.e. biopharmaceuticals. Although renewable based, the real motive for this development is new drugs rather than rational resource employment.
2. The same and similar techniques are used to improve existing or develop new fermentation processes (biosyntheses) for enzymes and existing (mainly nature-based) products.
3. To a large extent these enzymes are used to replace stoichiometric, non-catalytic chemistry.

The latter two, i.e. biosynthesis and biocatalysis, have a direct link to renewable resources and sustainability. In particular, the fine chemical industry, providing the pharmaceutical industry with its active ingredients, has experienced the effects. This chapter shows how biocatalysis, and to a lesser extent biosynthesis, have changed the character of the manufacturing processes for the world largest semi-synthetic antibiotics. Cephalexin is shown as a learning example. In fact, these penicillins and cephalosporins have led the way to full acceptance and development of biocatalysis as the strongest tool for environmentally compatible processes for a great number of other fine chemicals for pharmaceuticals (although competition-driven economic issues were the initial, underlying motive). For penicillins and cephalosporins the results were impressive: improvements of a factor 2 or more have been achieved in all sustainability parameters, such as energy consumption, emissions, resource consumptions, land area use and toxicity potential. Ozone-depleting agents and halogenated solvents and reagents were completely eliminated from all manufacturing processes. Improved product quality, in particular, product smell and taste, was found to be an additional bonus. As development times for these biocatalytic processes are still relatively long compared to pressurizing time-to-market demands in the research for new drugs, full exploitation has not been achieved in modern medicinal chemistry.

Alongside the development of these biocatalytic processes for the antibiotics, a semi-quantitative methodology has been worked out for the assessment of sustainability potential. Based on the three principles of sustainability, People, Planet and Profit, a rational set of parameters for all three groups has been developed. All parameters are defined in terms of improvements (decrease, increase, etc.) relative to an existing situation or existing process. All principles of LCA are used to arrive at meaningful and broadly applicable definition of parameters and their metrics.

References

Bruggink, A. (2001) Industrial synthesis of semi-synthetic antibiotics, in Bruggink, A. (ed.) *Synthesis of β-lactam Antibiotics*, Kluwer, Dordrecht, pp. 12–56.

Bruggink, A., Schoevaart, R. and Kieboom, T. (2003) Concepts of nature in organic synthesis: cascade catalysis and multi-step conversions in concert, *Organic Process Research and Development*, **7**, 622–640.

Liese, A., Seelbach, K. and Wandrey, C. (2000) *Industrial Biotransformations*, Wiley-VCH, Weinheim.

Oeko (2004) Oeko for LCA data; see http://www.oeko.de/service/gemis/en/index.htm.

Schoemaker, H. E., Mink, D. and Wubbolts, M.G. (2003) Dispelling the myths – biocatalysis in industrial synthesis, *Science*, **299**, 1694–1697.

Part IV
Conclusions

20
Conclusions

Jo Dewulf and Herman Van Langenhove

20.1 Introduction

Although the word did not exist as such, renewables have been the main sources for energy and material supply in societies for many centuries, and this roughly until 1850. With the start of the era of power machines based on coal and other fossil fuels (first coal, later oil), the predominance of renewables decreased. The explosive growth of mainly the organic chemical industry after World War II, driven by the production of fossil fuel-based polymers, further reduced the role of renewables. However, in the last decades of the twentieth century, some drawbacks of the fossil resource-based society became obvious. Although non-renewables undoubtedly have contributed tremendously to the welfare, it became obvious that gaseous, liquid and solid emissions into the environment had detrimental effects (Chapter 1). Environmental concerns expressed by pressure groups and supported by the public forced the authorities to undertake environmental regulations on emissions. This resulted mainly in the 1980s and 1990s in environmental technology (mainly end-of-pipe approach), enabling gas, liquid and solid waste treatment. At the end of the twentieth century, the sustainability issue became widespread in the society. It became accepted wisdom that technology not only should perform well in terms of preventing emissions, but also that it had to be rethought in a broader perspective. Based on the concept of sustainability, it is clear that, with a growing world population, an increasing standard of living and with long-term detrimental effects caused by fossil resources, particularly global warming, renewable resources are ready to make a comeback after 150 years. According to scenarios developed by academics and industry, the share of renewables of about 50% is expected in 2050.

Renewables-Based Technology Edited by J. Dewulf and H. Van Langenhove

Chapter 2 demonstrated that renewables indeed have the potential to play a significant role as resources for energy and materials. In our free market economy, the condition sine qua non for the introduction of more renewable resource-based technology will be cost effectiveness. The growing industrialization of especially Asian countries is raising the demand for energy and materials, whereas it is expected that fossil resources will not be able to cope fully with this growing demand. This situation may lead to an acceleration in research and development of renewables-based technology.

Attention is drawn to the fact that 'renewables' as such is a quite vague term, covering biomass, wind, solar and hydropower. Basically, the engine of renewable resource production is the sun, inducing biochemical processes (biomass growth) and physical processes in the atmosphere resulting in wind and precipitation. Both biochemical and physical processes lead to a large diversity of 'renewables'. Chapter 2 illustrates that this wide range of resources can be used both for energy and material purposes. It is shown that the renewable resources can be divided into two main groups. First, there are the pure energy resources delivering predominantly electricity: solar radiation, wind and hydropower. Second, there is biomass, which can be a resource for both energy and materials. Biomass is a much more complex renewable resource, with traditional applications of food and materials such as wood and cotton. However, in the past decade, two major new applications of biomass have been developed. First, new materials have been developed from biomass. A typical example is biodegradable polymers such as corn-based polylactic acid plastics. Through fermentation of corn, followed by polymerization, renewables-based plastics are manufactured and commercialized by NatureWorks LLC (formerly Cargill Inc.). Finally, regulations on solid waste disposal and growing energy prices have given an incentive to valorize the energy content of biomass material in solid waste streams.

20.2 The Available Sustainability Metrics

Measuring sustainability of technology is complex. One reason for this has to do with sustainability itself: it is a broad holistic issue with environmental, economic and social dimensions. A second reason for the complexity is the large number of levels at which the sustainability concept can be applied, and thus the definition of the technology to be assessed. Do we want to assess a product, a service, a production process, a production facility, a company or an industrial sector? Chapter 3 deals with these critical questions in a clear manner. As a third focus for our attention we have to be aware that 'technology' is not an isolated system delivering only one product. Modern technology is usually embedded in an industrial ecosystem delivering a set of products, ending up in so-called joint production. This allocation issue is well discussed in Chapter 4: several methods, which allow attributing resources and emissions to a single product or service, are described.

20.2.1 The Themes in Sustainability Metrics

Looking at the available metrics, it turns out that indicators are often limited to one or two dimensions of sustainability, predominantly environment and economics. Economic metrics are available in the market economy and are typically integrated in company and national reports. With respect to environmental sustainability, the predominant metric that was developed in the 1980s and 1990s is environmental life cycle analysis (Chapter 6). In this era, concern about the environmental effects of technologies was focused on emissions. As an illustration, typically seven out of eight themes of the life cycle assessment method are emission-related: global warming, ozone depletion, acidification, eutrophication, photochemical oxidant formation, human toxicity and ecotoxicity; one single theme was resource depletion. Today, in an era where sustainability principles dominate and where waste reduction technologies are common practice, growing emphasis on resource depletion is expected. In this sense, environmental metrics need to be rethought. Finally, integration of social indicators in renewables-based technology assessment is much more limited and will need further exploration.

20.2.2 Definition of Functional Unit and System Boundaries

With respect to the question 'What technology has to be assessed?', two major definitions have to be considered. First, there is the definition of the functional unit. If one aims, for example, at comparing a fossil resource-based technology with a renewable resource-based technology, one needs to define a functional unit allowing comparison. In the past, products have been considered frequently as the functional unit. However, the sustainability idea, which, according to the UN definition of it in 1987, must be 'fulfilling the needs', makes a shift from product to service: products are only a vehicle to deliver the service one uses to fulfil the needs of the population. This fits with the one-liner 'doing more with less' and the dematerialization concept.

Next to an appropriate definition of the functional unit, sustainability metrics also need to be based on a good definition of system boundaries of the technology to be assessed. System boundaries chosen in a too restrictive way may lead to erroneous conclusions. This is nicely illustrated in Figure 3.2 with hydrogen as energy resource. Hydrogen is nowadays perceived as a clean energy source, resulting only in water emissions when it is combusted. However, hydrogen as such is neither a renewable nor a non-renewable resource. It is only an energy carrier that can be manufactured both from renewable and non-renewable resources. This means that assessment of hydrogen-based technogies is well served by a cradle to grave approach. Only in this way, does it become obvious what type of resources (renewable or not) are involved in the overall production chain.

20.2.3 The Basic Metrics

Chapters 5–10 show six different metrics that can be put into practice as sustainability metrics. Chapter 5 focuses on metrics especially designed to assess renewables-based energy: biofuels. The presented net energy balancing method may be the preferred method in systems where the input of renewables versus non-renewables is so obvious. Chapter 6 presents the typical environmental life cycle analysis, where a major emphasis is on emissions into the environment, rather than resource extraction from the environment. Chapter 7, on the other hand, presents a thermodynamics-based method – exergy analysis – where emphasis is put on the nature (renewable versus non-renewable) and technical potential of natural resources. Chapter 8, explaining the substance and material flow analysis method, makes clear how natural resources flow into our industrial enterprises and how they flow partially back into the environment as waste material. Chapter 9 on the ecological footprint and Chapter 10 on the sustainable process index both present a single indicator quantifying the impact that technology and mankind have on the environment, by taking into account both the nature of used resources and the generated emissions.

20.2.4 Case Studies of Assessment Metrics

Chapters 11–19 show applications of different sustainability metrics for different sectors. Since biomass is a predominant renewable resource and since its production requires substantial land area, effects on land use are discussed in the first chapter of this part of the book. Chapter 12 focuses on a rather traditional renewable resource-based sector: forestry. An assessment diagram of renewables and non-renewables-based doorframes is presented in Figure 12.1.

The assessment of renewable resources for energy production purposes is dealt with in Chapters 13–15. Chapter 13 focuses on the emissions that are generated when biomass is employed for energy purposes. The possible contribution of biofuels to greenhouse gas emission reduction is the topic of Chapter 14. By making use of the ORWARE model, effects on greenhouse gas emissions and both renewable and non-renewable energy intake are quantified for the organic waste treatment industry in Chapter 15.

Chapter 16–19 show renewables-based material production assessment, as they are performed in three companies and a research institute. Chapter 16 shows the assessment of surfactant manufacturing, a sector where both renewables and non-renewables are competitive as a resource. The overall assessment of emerging bioplastics and fossil resource-based plastics, considering their overall life cycle, is well illustrated in Chapter 17. Application of (micro)biologically based processes in the chemical industry instead of chemical processes in, for example, vitamin production is discussed in Chapter 18. It demonstrates the eco-efficiency approach: balancing environmental and economic benefits. Finally, different production routes for pharmaceuticals are compared in Chapter 19, paying attention to the triple bottom line of sustainability through presenting performance indicators for people, profit and planet.

20.3 Where Are We Going in Assessing Renewables-Based Technology?

This book has shown that sustainability assessment of renewables-based technology is still at an early stage. It has been shown in particular that there are a large number of metrics available, being quite diverse in nature, presented by academics, public and private financed research institutes, and industry. The question may be raised if, in the long run, we are moving towards one single generic assessment tool. This may be very doubtful given the diversity of renewables-based technology and the questions that have to be answered. According to Levett (1998), we should take a modest 'fitness-for-purpose' approach in developing indicators, i.e. using different indicator sets for different purposes, rather than straining to produce a single definitive set of sustainable development indicators.

Anyhow, there is an obvious need to have appropriate sustainability metrics for renewables-based technology. Renewable energy, for example, is a key element in energy policy worldwide nowadays. Renewable energy is important in contributing to sustainability and security; it helps to meet emissions targets, it stimulates research and development of new technologies; it provides a set of diverse energy and material sources; and it contributes to rural development by putting forward new opportunities for agriculture. Last but not least, it makes national economies more independent of oil-producing countries. In order to quantify the sustainability of new renewables-based technology that claim to be effective here, metrics are of key importance.

A nice example of the need is the 2003 EU Directive on 'biofuels and other renewable fuels', stating that 2% of the fuels for transportation should be biofuels at the end of 2005, and 5.75% at the end of 2010. This directive presumes a black-and-white situation: either fuels are renewable or non-renewable. A careful analysis shows that different types of biofuels rely on renewables to a different extent. In other words, a refined approach relying on adequate sustainability metrics, could be very welcome here (Chapter 5, Chapter 7, Figure 7.5).

The example shows that our industrial society is an industrial network. One conclusion that can be drawn is that we may need sustainability metrics for both 'renewables-based' and 'non-renewables based' technology, including prominent information on the degree of renewables being used. This will help us better understand to what extent and how we rely on natural resources. Bearing the industrial ecology theory in mind, a second conclusion may be that we should put effort into quantifying to what extent technology relies on recovered waste materials instead of both virgin renewable and non-renewable resources.

Reference

Levett, R. (1998) Sustainability indicators – integrating quality of life and environmental protection, *Journal of the Royal Statistical Society A*, **161**, 291–302.

Index

Note: Page numbers given in italics refer to figures and numbers in bold refer to tables

Renewables-Based Technology Edited by J. Dewulf and H. Van Langenhove